BELTSVILLE SYMPOSIA in AGRICULTURAL RESEARCH

A Series of Annual Symposia Sponsored by
THE BELTSVILLE AGRICULTURAL RESEARCH CENTER
Northeastern Region, Agricultural Research Service
United States Department of Agriculture

[2] Biosystematics in Agriculture

PREVIOUS SYMPOSIA IN THIS SERIES:

[1] Virology in Agriculture

May 10-12, 1976

Published, 1977

Beltsville Symposia in Agricultural Research

[2] Biosystematics in Agriculture

Invited papers presented at a symposium held
May 8-11, 1977, at the Beltsville Agricultural
Research Center (BARC), Beltsville, Maryland 20705

Organized by THE BARC SYMPOSIUM II COMMITTEE
Lloyd Knutson, Chairman

Sponsored by
THE BELTSVILLE AGRICULTURAL RESEARCH CENTER
Northeastern Region, Agricultural Research Service
United States Department of Agriculture

ALLANHELD, OSMUN & CO.
MONTCLAIR

A HALSTED PRESS BOOK

JOHN WILEY & SONS
NEW YORK CHICHESTER BRISBANE TORONTO

ALLANHELD, OSMUN & CO. PUBLISHERS. INC.
19 Brunswick Road, Montclair, N.J. 07042

Published in the United States of America in 1978
by Allanheld, Osmun & Co.
Distribution: Halsted Press,
a division of John Wiley & Sons, Inc., Publishers,
605 Third Avenue, New York, New York 10016

Library of Congress Cataloging in Publication Data

Main entry under title:

Biosystematics in agriculture.

(Beltsville symposia in agricultural research; 2)
"Invited papers presented at a symposium held May 8-11,
1977 at the Beltsville Agricultural Research Center
(BARC), Beltsville, Md."
Includes index.
1. Biology—Classification—Congresses. 2. Agriculture—Congresses. I. Romberger, John Albert,
1925- II. United States. Agricultural Research
Center, Beltsville, Md. I. Series.
S494.5.B56B56 574'.01'2 77-84408
ISBN 0470-26416-0

Printed in the United States of America

Editors and Consultants

GENERAL EDITOR

John A. Romberger
Forest Service
Northeastern Forest Experiment Station
Forest Physiology Laboratory

ASSOCIATE EDITORS

Richard H. Foote
Agricultural Research Service
Insect Identification and Beneficial Insect Introduction Institute
Systematic Entomology Laboratory

Lloyd Knutson
Agricultural Research Service
Insect Identification and Beneficial Insect Introduction Institute

Paul L. Lentz
Agricultural Research Service
Plant Protection Institute
Mycology Laboratory

SUBJECT MATTER CONSULTANTS

Lloyd Knutson
Suzanne W. T. Batra
John M. Burns
James A. Duke
Richard H. Foote
A. Morgan Golden
Ronald W. Hodges
Leo E. LaChance
J. Ralph Lichtenfels
Paul M. Marsh
Douglass R. Miller
William R. Nickle
Flora G. Pollack
Maynard J. Ramsey
F. Christian Thompson
Francis A. Uecker
Donald R. Whitehead

Contributors and Their Affiliations

Lekh R. Batra
U.S. Department of Agriculture
Agricultural Research Service
Plant Protection Institute
Mycology Laboratory, BARC
Beltsville, Maryland 20705

Guy L. Bush
University of Texas
Department of Zoology
Austin, Texas 78712

James A. Duke
U.S. Department of Agriculture
Agricultural Research Service
Plant Genetics and Germplasm Institute
Plant Taxonomy Laboratory, BARC Beltsville, Maryland 20705

Richard H. Foote
U.S. Department of Agriculture
Agricultural Research Service
Insect Identification and Beneficial Insect Introduction Institute
Systematic Entomology Laboratory, BARC Beltsville, Maryland 20705

T. Gary Gautier
Smithsonian Institution
National Museum of Natural History
Automatic Data Processing Program
Washington, D.C. 20560

A. Morgan Golden
U.S. Department of Agriculture
Agricultural Research Service
Plant Protection Institute
Nematology Laboratory, BARC
Beltsville, Maryland 20705

Charles R. Gunn
U.S. Department of Agriculture
Agricultural Research Service
Plant Genetics and Germplasm Institute
Plant Taxonomy Laboratory, BARC
Beltsville, Maryland 20705

J. Heslop-Harrison
Royal Society Research Professor
University of Wales
Welsh Plant Breeding Station
Aberystwyth, United Kingdom

David Kavanaugh
California Academy of Sciences
Golden Gate Park
San Francisco, California 94118

Bryce Kendrick
University of Waterloo
Biology Department
Waterloo, Ontario N2L 3G1, Canada

G. Barrie Kitto
University of Texas
Department of Zoology
Austin, Texas 78712

Leo R. LaSota
University of Maryland
Horticulture Department
College Park, Maryland 20742

Norman D. Levine
University of Illinois
College of Veterinary Medicine and Agricultural Experiment Station
Urbana, Illinois 61801

J. Ralph Lichtenfels
U.S. Department of Agriculture
Agricultural Research Service
Animal Parasitology Institute
Parasite Classification and Distribution Unit, BARC
Beltsville, Maryland 20705

E. S. Luttrell
University of Georgia
Plant Pathology Department
Athens, Georgia 30602

Armand R. Maggenti
University of California
Department of Nematology
Davis, California 95616

Eugene G. Munroe
Agriculture Canada
Biosystematics Research Institute
Research Branch
Central Experimental Farm
Ottawa, Ontario K1A OC6, Canada

William R. Nickle
U.S. Department of Agriculture
Agricultural Research Service
Plant Protection Institute
Nematology Laboratory, BARC
Beltsville, Maryland 20705

David Rosen
The Hebrew University
Faculty of Agriculture
Rehovot, Israel

Beryl B. Simpson
Smithsonian Institution
Department of Botany
National Museum of Natural History
Washington, D.C. 20560

Edward E. Terrell
U.S. Department of Agriculture
Agricultural Research Service
Plant Genetics and Germplasm Institute
Plant Taxonomy Laboratory, BARC Beltsville, Maryland 20705

Lynn H. Throckmorton
University of Chicago
Department of Biology
Chicago, Illinois 60637

Donald R. Whitehead
U.S. Department of Agriculture
Agricultural Research Service
Insect Identification and Beneficial Insect Introduction Institute
Systematic Entomology Laboratory, BARC Beltsville, Maryland 20705

Foreword

The value of all scientific literature derives from the premise that it will be possible for the authors and others to relate reported findings to earlier discoveries, to other ongoing research, or to the various applications of science. In agriculture and the biological sciences, this obvious tenent necessitates accurate identifications of the organisms under investigation. Failure to obtain reliable identifications is akin to reporting data on an unknown proprietary product, the composition of which could conceivably change from batch to batch depending on the cost and availability of ingredients. Although this is a reasonable parallel, there is a feeling among some managers and scientists that the research results and service capabilities of taxonomists are of minor importance in modern science.

In truth, it is impossible for me to consider some otherwise comprehensive studies as serious research because of their failure to consider the possibility of contamination among organisms studied, to base the research on reliable taxonomic information, or, where needed, to preserve voucher specimens. Perhaps my observations are atypical, but I have encountered enough examples to convince me that a disregard for sound taxonomy has sometimes resulted in unfortunate losses in time and resources.

The symposium on which this volume is based was not intended to resolve all questions that relate to the broad and complex field of biosystematics. It was our hope, however, that the symposium would help to underscore the importance of taxonomy, to give increased visibility to an essential area of study, and to illustrate the critical role of biosystematics to long-term advances in agriculture. Thus, in a very real sense, we have held to the original objective of these symposia which is to stress the interdependence of the various disciplines of science in agricultural research and development.

A. A. Hanson
Director
Beltsville Agricultural Research Center

Symposium Organization

Beltsville Agricultural Research Center (BARC)

A. A. HANSON, DIRECTOR

BARC SCIENCE SEMINAR COMMITTEE

J. R. Lichtenfels, Chairman
G. R. Beecher, Cochairman
J. D. Anderson
N. J. Chatterton
T. O. Diener
R. A. Kilpatrick
L. Knutson
R. D. Lumsden
D. P. Morgan
L. D. Owens
J. R. Plimmer
R. L. Powell
J. A. Romberger
R. L. Smiley

BARC SYMPOSIUM II SUBCOMMITTEE

L. Knutson, Chairman

Task Groups

PROGRAM

L. Knutson, Chairman
L. R. Batra, Mycology
J. A. Duke, Plant Taxonomy
W. Friedman, Nematology
A. M. Golden, Nematology
R. W. Hodges, Systematic Entomology
J. R. Lichtenfels, Animal Parasitology
S. Nakahara, Systematic Entomology
J. A. Romberger, Forest Physiology

PUBLICITY

J. A. Duke, Chairman
L. R. Batra
A. M. Golden
J. R. Lichtenfels
D. P. Morgan

POSTER SESSION

R. W. Hodges

LOGO

R. B. Ewing

PUBLICATION

J. A. Romberger, Chairman
R. H. Foote
P. L. Lentz

ARRANGEMENTS

R. L. Smiley, Chairman
D. P. Morgan
W. R. Nickel

FINANCE

J. D. Anderson, Chairman
J. R. Lichtenfels

EXHIBITS

T. J. Spilman, Chairman
A. E. Fusonie
R. E. White

Preface

The field of biosystematics—from the classical taxonomic studies to the quantitative, experimental studies of populations—is approaching a new stage. In the days of Linneaus, Darwin, and the explorers, it was "big science"—later mostly a handmaiden supporting other research. Now biosystematics is being recognized as both the essential first step and the final integrator for the broad spectrum of biological, agricultural, and medical sciences. From the quarantine inspector with an intercepted, unknown, and possibly new pest species to the researcher studying colonies of organisms that look similar but behave differently, there is the same need for biosystematic data. Identification is the key to the scientific literature, and the predictive powers of classifications are available only after correct identifications have been made.

The present accelerating evolution of theory and methods in biosystematics, increasingly interdisciplinary in approach, is occurring in an environment more stressed than during previous periods of change. Great opportunities and urgent demands for solutions to critical problems are everywhere. The current "new systematics" is certainly characterized by a new degree of relevancy to "real world" problems.

Mankind's basic physical needs are mostly provided by about 100,000 species of plants and animals. In another sense, thousands of other species compete with us for the same resources. Environmentally safe control of our competitors depends heavily on precise knowledge of the diversity of life. Refined systematics information can lead to the discovery of additional useful species and to new uses for presently known species. For example, of the estimated 300,000 species of parasitic wasps, only a few hundred have been tried as biological control agents of insect pests. There is here a clearly underutilized resource. Biosystematics in agriculture is especially oriented toward solving problems of this nature, while maintaining high standards of quality in research.

With such considerations, we, in the five systematics laboratories and other concerned laboratories at Beltsville, invited other biologists to join us for a three-day meeting on biosystematics. The discussions were informative and frequently challenging. We hope that this Symposium was a good experience for all participants and that this printed record will convey much of the factual content of the meeting and also some of its less tangible character and benefit.

Lloyd Knutson, Chairman
BARC Symposium II
Biosystematics in Agriculture

Contents

one

INTRODUCTION

1] Once Again, What Is a Species?

by ARTHUR CRONQUIST*

ABSTRACT

The species is an ancient folk concept, which biologists have adopted and shaped to their own use. In the last several decades most zoologists have come to rely on reproductive isolation as the sole specific criterion. As a result, they have been forced to recognize the existence of sibling species, which are reproductively isolated but not otherwise obviously different. The necessity to admit sibling species is an embarrassment to zoological taxonomy, but the embarrassment has not yet become so severe as to force a theoretical re-evaluation. Efforts to use reproductive isolation as the sole specific criterion in plants lead us into such a morass that the vast majority of botanical taxonomists have had to put them aside and seek other criteria. There is a large measure of agreement among working plant taxonomists about specific concepts, but these concepts have not yet been satisfactorily verbalized. I suggest that the criteria we actually use are best codified as follows: *Species are the smallest groups that are consistently and persistently distinct, and distinguishable by ordinary means.* Disputes about specific delimitation in plants mostly come down to questions of how consistent is consistent, how persistent is persistent, and how ordinary are ordinary means. There must be some room for flexibility in using these criteria to cope with problems in particular groups, but they remain the essential criteria. The existence of consistent, persistent differences implies reproductive isolation. In practice such isolation is usually merely inferred, by both botanists and zoologists, rather than being rigorously demonstrated. It remains to be seen whether the number of sibling species will eventually become so large as to force zoologists to accept the more pragmatic specific criteria that botanists have found more useful.

INTRODUCTION

After all that has been written about the nature of species during the past two hundred years, and especially during the past fifty years, the necessity of another rehash may seem doubtful. Following on a period of extreme

*The New York Botanical Garden, Bronx, New York 10458.

differences of opinion, especially during the early decades of the twentieth century, many evolutionary biologists have come to a specific concept that sounds very much like the traditional naturalist's concept in modern dress. Many zoologists, in particular, seem to believe that all the important questions have been resolved by the development of this "biological species" concept. Yet the biological species concept, so-called, works rather badly for plants. Botanists have had to develop their own way of reconciling the traditional naturalist's concept of species with genetic knowledge. Thus another look may be in order.

Most of what I have to say in this paper has been said more eloquently and in more detail by various authors in a number of other places. A formidable bibliography could be compiled. My recapitulation of some well-known facts is necessary to provide the setting in which the modern botanical concept of species can be properly understood, and to help demonstrate that the "biological species" concept has a more limited usefulness than many biologists have supposed.

This paper is concerned with the recognition and definition of species as they now exist, within a time-frame of a few hundred or at most a few thousand years. It is not concerned with the definition of what are called "successional" or "evolutionary" species, which are continuous through evolutionary time, with a span of many thousands or even millions of years. For example, many anthropologists consider that *Homo erectus* and *Homo sapiens* form a pair of successional species. In this view, evolutionary changes accumulated in *H. erectus* over a period of several million years, very gradually transforming it into *H. sapiens.* In such a successional line there are no breaks, and the conceptually necessary organization of the phylad into separate species is purely arbitrary. At any given time, on the other hand, there are gaps in the pattern of diversity among organisms, and these gaps are of critical importance in the delimitation of species.

THE SPECIES AS A FOLK CONCEPT

The species is an ancient folk concept, which biologists have adopted and shaped to their own use. It is the common experience of naturalists, in whatever part of the world, that the individual plants and animals they see can be mentally grouped into a number of taxa, in each of which the individuals are basically alike. In societies that are still close to nature, each of these taxa that has any importance to the society has a name.

The idea that species consist of reproductively compatible individuals has a firm foundation in folk knowledge. Any farm boy knows that horses beget horses, and donkeys beget donkeys, and that if you breed a jackass to a mare you get a mule, which begets nothing. The reproductive processes in plants are not so obvious to the unlearned, but it is clear enough to all concerned that wheat grains produce wheat plants, clover seeds produce clover plants, and mustard seeds produce mustard plants. Reproductive continuity is an essential part of the folk concept of species.

The other essential part of the folk concept of species, to go along with reproductive continuity, is that you can tell them apart by looking at them. Together these two criteria result in the recognition of conceptually useful species. Each species encompasses a certain range of variability, but within any local area the distinctions between species are usually clear. Folk taxonomy of course does not address itself to things that are not perceived as being of any importance or intrinsic interest, but for the things that do attract attention it serves its purposes well enough. It operates in blissful ignorance of the possibility that there may be reproductive barriers between things that are otherwise essentially alike.

Folk nomenclature is sometimes essentially monomial. For example, *cerris, ilex, robur,* and *suber* were the classical Latin names for four species of oak. Often the nomenclature is binomial instead of monomial, and there may even be several hierarchical ranks, as has been pointed out by Berlin *et al.* (*2*).

The binomial system of Linnaeus is an attempt to formalize a way of naming things that was well known to people at least in northern European countries. The English colonists who came to America in the seventeenth and eighteenth centuries were well acquainted with binomial nomenclature. They had no difficulty in recognizing the genera *Acer* and *Quercus,* for example, under the names maple and oak, and they proceeded to distinguish such species as the sugar maple, silver maple, and red maple, and the red oak, white oak, black oak, etc.

THE IRREGULAR PROGRESSION OF IDEAS LEADING TO THE "BIOLOGICAL SPECIES" CONCEPT

It is no secret among biologists that there are frequently practical problems in applying the folk concept of species to the often bewildering diversity among individuals in nature. In attempting to meet such problems, biologists have sought a more precise formulation of the concept. One can read in the works of various nineteenth-century biologists definitions that sound very modern, emphasizing either reproductive isolation or discontinuity of variation, or both, as the essential specific criteria. These were produced at a time when heredity was only vaguely understood, but they are nonetheless very respectable efforts at definition.

One might suppose that the burst of knowledge about genetics in the first two or three decades of the twentieth century would lead to an improved understanding of the nature of species. In the long run it did; but initially it fostered as much confusion as clarity. The mutation theory of De Vries (*30*), published in 1901-3, envisioned the frequent saltatory origin of species by genetic change in a single individual, or by concurrent repetition of the same change in a series of similar individuals. Biologists such as Lotsy (e.g., *21*) proposed definitions that would in effect treat each recognizable genotype as a separate species. In 1918 C. Hart Merriam (*26*) recognized 78 species of grizzly and big brown bears in North America. These are all currently considered to belong to a single species, *Ursus arctos* L., which is widespread in northern Eurasia as well as in western North America (personal communication from E. R. Hall). A school of thought fostered in the United States by N. L. Britton and in the Soviet Union by V. L. Komarov considered that anything worthy of a name should be called a species, thus raising to specific rank many taxa considered by others to be of only infraspecific importance.

I have read learned efforts to defend the naming of individual apomicts as distinct species, on the basis of absolute genetic barriers among them. This approach led to the proposal of thousands of specific names in such genera as *Taraxacum* and *Hieracium*, producing utter taxonomic confusion. Such treatments are simply not useful in understanding and communicating the pattern of diversity in nature, and not many of us take them seriously any more.

During this period of confusion there remained a considerable body of both plant and animal taxonomists who retained the traditional concepts of species, but their voices were often lost in the storm.

By 1930 it was becoming clear that any really useful concept of species would have to be grounded in the traditional folk concept. In that year Du Rietz (*9*) presented an admirable review of the fundamental units of biological taxonomy, and produced a definition of species that still has some relevance today. In Du Rietz' view, species are "the smallest natural populations permanently separated from each other by a distinct discontinuity in the series of biotypes." The existence of barriers to interbreeding between species is implicit in this definition, and explicit in his accompanying discussion. Du Rietz did not really address himself, however, to the problem of reproductive barriers not accompanied by other readily detectable differences. His definition, often cited in the decades immediately after its publication, is seldom repeated in print today.

By the late 1930's, definitions were being produced that explicitly relied on reproductive isolation as the essential specific criterion. I well remember being exposed in graduate school to the then relatively new definition by Dobzhansky (*7*) of a species as "that stage of evolutionary progress at which the once actually or potentially interbreeding array of forms becomes segregated into two or more separate arrays which are physiologically incapable of interbreeding."

It was at once obvious to me and others that Dobzhansky's formulation was too severe to be botanically useful. It would, for example, sweep great numbers of different species and even genera of orchids into a single species. The barriers to hybridization in orchids rest to a considerable extent on adaptations to specific pollinators. If you will transfer the pollen yourself you can produce hybrids between very different kinds.

By 1951 Dobzhansky (*8*) had relaxed his definition to read that species are "groups of populations the gene exchange between which is limited or prevented in nature by one, or a combination of several, reproductive isolating mechanisms. In short, a species is the most inclusive Mendelian population." That brought his formulation into essential harmony with one proposed in 1940 by Mayr (*22*) and rephrased in 1942 (*23*) to read that species are "groups of actually or potentially interbreeding natural populations which are reproductively isolated from other such groups." Here the emphasis is placed on what actually happens in nature, rather than on what can be induced experimentally.

Proponents of the specific criterion of reproductive isolation use the term "biological species" for the taxa so delimited, as if species defined on any other criteria were not biologically meaningful groups. Many of them have moved from Dobzhansky's original position—that the reproductive criterion is useful precisely because species as so defined are essentially similar to traditional taxonomic species delimited on less clear criteria—to the position that reproductively defined species are conceptually superior to species defined on other criteria. The term taxonomic species has sometimes been used in a pejorative way, to contrast with the really meaningful biological species. This attitude amounts to nothing less than attempted theft. The traditional naturalist's concept of a species has a long-established prior claim to the term, and has been an integral part of taxonomic usage and thought for more than 200 years. If the concept can be sharpened up and made more useful by emphasis on the criterion of reproductive isolation, well and good. But if so-called biological species turn out to be significantly different in many instances from taxonomic species, then a new term should be coined, one that does not invite confusion with the taxonomic species.

I am not sure who first introduced the term biological species. Mayr used it in 1942 (*23*), in such a way as to suggest that it was relatively new. By the mid-1950's there was a well-established school of thought that the real species in nature are the "biological" species (e.g., *12, 24*). There are some practical problems with the biological species concept, but for the past two or three decades the majority of zoologists (or at least the vertebrate zoologists and many entomologists) have managed to live with them. They accept the difficulties as the price for a clear and generally applicable specific concept. We should, however, take note of these difficulties as a background for considering the applicability of the biological species concept to plants.

SOME BASIC PROBLEMS WITH THE BIOLOGICAL SPECIES CONCEPT

One difficulty with the biological species concept is that reproductive isolation between species, and reproductive continuity within species, are much more often inferred than directly demonstrated. Reproductive continuity among individuals of morphologically continuous groups is assumed, until the contrary is demonstrated. The existence of gaps in the pattern of visually observable phenetic diversity is taken to indicate reproductive isolation, and the species are defined accordingly. Thus, the theoretical precision of the definition does not necessarily lead to improved precision in practice.

When reproductive isolation is demonstrated between things that are otherwise difficult or impossible to distinguish, it becomes necessary to admit the existence of what are called sibling species. Sibling species are of course an embarrassment to the system. When Dobzhansky was promoting his genetic phraseology of the specific concept (e.g., *7, 8*) he made a point of the fact, as he considered it, that species as so defined would conform in very large part to species as traditionally conceived. We do not yet know how wide of the mark that belief may be. It seems reasonable to suppose that not nearly all the sibling species have yet been discovered, so that the eventual total cannot now be predicted. A few sibling species can be tolerated, but large numbers of them would restrict the usefulness of the taxonomic system to a progressively more limited coterie of specialists. Under such conditions, other biologists would have to give up taxonomy, or develop an alternative set of specific concepts that would be more generally useful.

Another difficulty with the criterion of reproductive isolation is that it cannot be applied directly to allopatric populations. Barring the possible existence of sibling species, it is usually easy enough to determine, in one way or another, the specific or nonspecific status of things that live close enough together to have a physical opportunity to interbreed. Most species, however, do not consist of continuous, panmictic populations. Instead, they consist of numerous sets of local populations that are only tenuously connected by interbreeding. (cf. *10, 11*). The geographically peripheral local populations of a species may be separated by many, even hundreds of kilometers from the nearest other such population. The geographical range of a species may be divided into two or more segments separated by a broad gap. There are, for example, Rocky Mountain species that turn up again in northeastern Minnesota, or on the Gaspé Peninsula of Quebec, or in both places. The casual arrival of seeds of a species on a remote oceanic island may set the stage for speciation, but surely the offspring of those first seeds do not immediately constitute a new species. Likewise, a Mediterranean weed introduced into California does not thereby become a different species.

Zoologists have developed a rough and ready method for dealing with such geographical isolates. If they differ from their nearest relatives about as much as sympatric species differ from each other, they are treated as distinct species. If they differ about as much as geographic subspecies usually differ, they are treated in subspecific rank, and if they do not differ appreciably from the main bulk of the species population they are not given any separate taxonomic status.

The necessity for some such method of considering geographic isolates is obvious, but it should be equally obvious that the vaunted precision of the definition is thereby appreciably reduced. It is an ancient and necessary taxonomic principle that individual characters are only as important as they prove to be in any individual instance in marking taxa that have been recognized on the basis of all the available information (*5, 20, 29*). Extrapolation of the taxonomic value of a character from one group to another is always open to error. The assessment of the amount of difference between geographically isolated populations is subjective, and the evaluation of its taxonomic significance even more so.

In sum, the effort to refine the field-naturalist's concept of a species and express it in genetic terms has been only partly successful; but zoologists in general, and vertebrate zoologists in particular, have found the gain to be worth the cost.

SOME BOTANICAL PROBLEMS WITH THE BIOLOGICAL SPECIES CONCEPT

Let us now consider how the criterion of reproductive isolation works in defining species of plants, with emphasis on higher plants.

First, I should point out that the very existence of species as discrete units implies a barrier to free and successful interbreeding. The barrier need not be an internal, genetic barrier, but it must in any case be effective. If breeding between two groups is as easy as breeding within the groups, and if the offspring of such intergroup breeding are as successful and as fertile as members of the parent groups, then the two groups, even if initially separate, will inevitably merge into one. If two populations maintain their separate existence over any considerable period of time, then we may be sure that there is some sort of barrier to their successful interbreeding. Even if hybrids are frequently produced, and even if the hybrids produce some offspring, the hybrids and their offspring must be under some handicap, as compared to the members of the parent groups. Otherwise the two groups would become one group. Two freely interbreeding populations can coexist only for the length of time it takes them to merge.

Interspecific hybridization appears to be more frequent among many groups of plants than is customary at least among terrestrial vertebrates. In order to have a useful taxonomic system, botanists must often admit the existence of a larger number of hybrid intermediates between species than students of terrestrial vertebrates will admit (*1*). I specify terrestrial vertebrates because I understand that students of fish frequently find an amount of interspecific hybridization comparable to that in many groups of plants. Even so, the great majority of the individual plants (and fish) generally sort themselves out into recognizable groups that may be connected by a lesser number of intermediates. Perhaps such groups correspond to the adaptive peaks so often referred to in the evolutionary literature. Thus, although the amount of admissable hybridization among recognized species of sunflowers may be greater than that among sparrows, the difference is only one of degree. It does not by itself seriously limit the applicability of the biological species concept in plants.

The greater problem in using reproductive isolation as the sole specific criterion among plants comes from the frequency of barriers to interbreeding that are not accompanied by other evident differences. Not only is the determination of the reproductively isolated groups a long and difficult process (cf. *13*), but these groups often have little other practical or even theoretical significance. Differences in number of chromosomes may restrict interbreeding, without causing any other consistent, detectable differences. The literature is replete with examples, both euploid and aneuploid. Other, less readily detectable cytological differences may also introduce a barrier to interbreeding (*13*).

In some other genera, as in *Draba* (*Erophila*) and *Claytonia*, plants with different chromosome numbers may be readily interfertile (*19, 31*). Thus, even the demonstration of a difference in chromosome number does not necessarily demonstrate reproductive isolation.

Although otherwise similar plants at different ploidy levels are often virtually intersterile, polyploidy may also break down the differences between groups that are fully distinct at the diploid level. A sterile hybrid between two diploids may regain fertility by doubling the number of chromosomes. Again the literature is replete with examples. Harlan and DeWet (*17*) provide a recent review. In what have come to be called pillar complexes, polyploids may dominate the scene, forming a morphological and physiological continuum that obscures the differences among the sharply distinct relictual diploid populations.

Self-sterility is another complicating factor. Self-sterile plants are usually also sterile with a certain fraction of the population to which they belong. Thus the inability of individuals *a* and *b* to cross with each other does not necessarily prevent them both from being interfertile with individuals *c, d, e, f,* and *g*.

Members of different local populations may show a complicated and irregular pattern of interfertility and intersterility. Local population *a* may be fertile with *b, d,* and *f,* but not with *c,* and *e*. Local population *b* may be fertile with *a,* and *c,* but not with *d, e,* and *f*; whereas *c* may be fertile with *b, d,* and *e,* but not with *a* and *f*; etc. One of the most interesting studies of this sort of pattern is that reported in *Nemophila menziesii* H. & A. by Rick in 1946. His fascinating verbal presentation at the annual meetings of the American Association for the Advancement of Science is unfortunately documented in print by only a short abstract (*27*).

Self-pollination constitutes a barrier to interbreeding within many groups. Individual strains within a self-pollinated group may at first appear to be distinctive. As studies are carried forward, more and more such strains are discovered, until eventually one finds a morphological continuum. Furthermore, self-pollination is seldom 100% obligate. Usually there is an occasional successful cross-pollination. Offspring of such a cross segregate according to well-known Mendelian principles, to produce eventually a series of lines that maintain their identity until the next successful cross-pollination. Taxonomists have frequently been misled to describe these self-pollinated lines as distinct species.

Apomixis creates a still more troublesome disturbance to genetic definitions of species. It is comparable to self-pollination in that the same genotype is perpetuated in successive generations. It also resembles self-pollination in that it is very often not fully obligate; a small amount of ordinary sexual reproduction commonly accompanies apomixis. The reason that apomixis causes more taxonomic difficulty than self-pollination is that apomixis is so often accompanied by hybridization and polyploidy. In polyploid-apomictic groups the morphological distinctions between what might otherwise be well-marked species frequently become badly blurred, and an attempt to recognize each individual apomictic line as a distinct species leads only into a morass.

GILIA AS A CASE IN POINT

The efforts by Verne Grant to apply the biological species concept in the section *Arachnion* of the genus *Gilia* (Polemoniaceae) are especially instructive. I take up this matter in some detail, not to belabor Dr. Grant, but precisely because he is an able botanist who tried diligently, persistently, and ultimately unsuccessfully to apply the "biological species" concept in this group. The section *Arachnion* consists of annuals and winter-annuals, some with relatively large flowers and usually out-pollinated, others with smaller flowers and usually selfed. The taxonomic problems in the group are largely concentrated in the self-pollinated members.

Grant began his intensive studies of the self-pollinated members of the section *Arachnion* in 1949. In a series of papers extending over a period of 10 years, from 1956 to 1966, he and his associates (notably Alva Day Grant) presented a number of partial and complete treatments of the groups (e.g., *6, 13, 14, 16*). Each of these is more complicated and more difficult to use than the previous one, because of the acquisition of new information requiring the fragmentation of what had previously been considered unitary species. A series of quotations from Grant's 1964 paper (*13*) may help to show the pickle that the biological species concept had got him into by that time:

Although each species in the *Gilia inconspicua* complex has its characteristic ecological preferences, these preferences overlap, and the species commonly grow sympatrically in various combinations of two, three, or four . . . . The observed variation within a poly-species colony is a result of individual variations in the constituent species populations superimposed on the interspecific character differences. The morphological differences between biotypes of the same population are sometimes as great as the differences between sibling species. On the other hand, the genotypes of different sibling species may give rise to convergent phenotypic modifications under similar environmental conditions. The two sources of phenotypic variation within populations obscure the already slight differences between the species.

Again:

The analysis of a sympatric flock of sibling species would be difficult enough if the constituent species were genotypically and phenotypically uniform within themselves. But this happy condition is not realized in nature. Superimposed on the interspecific differentiation in the colonies are two sorts of intraspecific variation. Any given species population may contain two or more genotypically different biotypes differing in flower color, leaf shape, or some similar character. Any given class of genotypes can be represented in the population by individuals differing markedly in phenotype as a result of environmental modifications. Conversely, the genotypes of different sibling species, under similar environmental conditions, may have convergent phenotypic modifications. The two sources of intrapopulational variation produce phenotypic differences between individual plants which always obscure and often overshadow completely the slight interspecific differences.

Again:

Although the different sources of variation in a poly-species colony of small-flowered Cobwebby Gilias are known, it does not necessarily follow that they can in practice be separated by observational evidence alone . . . . The structure of the more complex colonies composed of several species is, in the final analysis, beyond the resolving power of the hand lens.

Again:

Gilia transmontana forms a single fertility group in the western and central Mojave Desert, but morphologically indistinguishable populations farther east in southwestern Utah form a separate fertility group intersterile with the former. . . . the hybrid between *Gilia sinuata* from the Mojave Desert and *G. sinuata* from the Sonoran Desert in Arizona is sterile with reduced chromosome pairing and does not get beyond the F1 generation. The two population systems are treated tentatively, pending further information, as related but non-interbreeding allopatric species, and are referred to as *G. sinuata* I and *G. sinuata* II, respectively.

About that time, Grant got tired of *Gilia*, and he has published nothing new on the group since 1966. On the basis of the evidence to date, it seems to me that the number of intersterile groups that might be discovered in the small-flowered Cobwebby Gilias is limited mainly by the amount of time and effort that botanists will devote to the study. I do not think that the possiblities have as yet been anywhere nearly exhausted.

THE BIOLOGICAL NATURE OF SPECIES

In spite of the problems he had run into, Grant in 1964 (*13*) still maintained the theoretical primacy of the biological species concept over the traditional taxonomic species concept. Again I quote:

Admittedly, a system of classifying plants into species that is convenient to use is the only system which can be employed generally and as such serves a useful purpose in the exploratory and practical phases of taxonomy. The blocking out of easily identified but artificial units of classification at the species level is not, however, a worthwhile ultimate goal of taxonomy, contrary to some taxonomists.

By 1971, in a more general work on plant speciation, Grant (*15*) had retreated to the position that biological species and taxonomic species are different though overlapping concepts, each with its own sphere of usefulness. I agree that the two concepts are overlapping but different, and that each has its own usefulness. There is still a semantic problem, however. Continued use of the two terms invites continued confusion. Taxonomists have a long-established prior claim to the use of the word species. If biological species, so-called, differ significantly from taxonomic species, then the term biological species should give way to some other term that does not use the word species. (I repeatedly affix the modifier "so-called" to the term biological species, because the term carries an air of invidious distinction, as if taxonomic species are somehow not biological if they do not conform to the biological species concept.)

As we have pointed out, the traditional concept of species has always had reproductive continuity as one of its essential components. By 1940 it was becoming apparent that there were many different bases for this reproductive continuity. In 1943 Camp and Gilly (*3*) distinguished 12 different kinds of species, based on their population structure and reproductive behavior. The names they proposed for these different sorts of species have not become widely used, and need not be revived here, but it is well to remember that not all species have the same sort of population structure.

The twentieth-century discoveries in genetics and the new tools for refined morphological and chemical analysis require some adjustment of the traditional concept of a species. Formal definitions in the past have usually relied directly or indirectly on discontinuity of variation as the essential specific criterion. Even the definitions formally based on breeding behavior had an implicit assumption that reproductively isolated populations would also be morphologically distinguishable.

So long as biologists were unaware of cryptic discontinuities in nature, it did not matter whether reproductive isolation or morphologic discontinuity was taken as the essential specific criterion. One criterion assumed the other as a corollary. Now we know that the two criteria are sometimes dissociated. The biological species concept emphasizes the criterion of reproductive isolation at the expense of the criterion of morphologic discontinuity.

During the 1940's, and continuing to some extent into the 1950's, many plant taxonomists were intrigued with the biological species concept that was being promoted by the zoologists. Sooner or later, however, they discovered, individually, that the biological species concept does not work well in plants. Too often it gives results that fly in the face of common sense, or it leads into a swamp in which it seems impossible ever to acquire enough data to come up with a definitive treatment.

THE WORKING CONSENSUS AMONG PLANT TAXONOMISTS

During the 1950's and 1960's a working consensus developed among plant taxonomists that differences which cannot be detected by an ordinary observer with ordinary equipment will not be considered of specific importance. I remember hearing Dr. Henry Thompson, one of our outstanding biosystematists, give a paper at one of the national meetings some years ago, about breeding behavior among some members of the Loasaceae. He felt it necessary to preface his remarks with what he called the "pledge of allegiance," to the effect that if you can't tell the things apart, they belong to the same species, regardless of reproductive or cryptic morphologic differences that might exist.

Current taxonomic literature amply documents the prevalence of the same idea. Let us take a single recent issue of the journal Rhodora (volume 79, January 1977) to illustrate the point. Chuang and Constance (*4*) have an article on the cytogeography of *Phacelia ranunculacea* (Nutt.) Constance, a member of the Hydrophyllaceae. They find that the species consists of two geographically separated cytotypes. The Ozarkian cytotype has n=6, whereas the Appalachian cytotype has n=14. Aside from the chromosome number, they are unable to detect even micromorphological differences between the two cytotypes. They further comment that, "The very small size of the tubular-campanulate corollas . . . and the inclusion of the stamens have unfortunately frustrated all attempts to hybridize members of the two cytotypes." They unequivocally keep both cytotypes in the same species.

In the same issue of Rhodora, Keeley and Jones (*18*) have an article on *Vernonia* in the Bahama Islands. These authors find that the Bahama Island plants, previously called *Vernonia insularis* Gleason, are morphologically very similar to the Florida plants, *V. blodgettii* Small, and that the island and mainland plants hybridize freely in the experimental garden. They therefore reduce *V. insularis* to synonymy under *V. blodgettii*. They did find some chemical differences between the island and mainland materials they examined, but they do not feel confident that the chemical differences will hold up when more plants are studied. They further comment, "In any case it is difficult to provide a rationale for distinguishing macroscopic organisms on the basis of chemical differences that require special training and a sophisticated laboratory to detect. . . . "

Although a working consensus on the nature of plant species has actually developed, at least among American botanists, I have not found an adequate formulation of that consensus. Taxonomists shy away from efforts at formal definition, or they make comments reminiscent of a statement attributed to a justice of the Supreme Court of the United States, that he had difficulty in formulating an adequate definition of pornography, but had no difficulty recognizing it when he saw it. Or they take refuge in an old in-joke, that a good species is what a good taxonomist says it is. There is just enough truth in the latter definition to make it funny, but it doesn't enlighten anybody who doesn't already have some concept of the nature of species. Indeed it fosters the notion, already abroad, that taxonomists don't really know what they are about, and reach their conclusions in the absence of any governing rationale.

A NEW EFFORT TO DEFINE SPECIES

I have tried to formulate the concept of species under which most plant taxonomists (at least in the United States) now actually work. There is nothing new in the ideas, only in the manner of expression. *Species are the smallest groups that are consistently and persistently distinct, and distinguishable by ordinary means.* That very generalized definition sets the limits within which taxonomists can reasonably disagree. We may differ as to how consistent is consistent, how persistent is persistent, and how ordinary are ordinary means, but any group that does not meet each of these tests at least to some reasonable degree is not considered to be a species.

It may be worth while to take the definition apart and see why each piece is necessary. First, species are the *smallest* groups that meet the remaining tests. If the group can be subdivided into smaller groups that meet those tests, then it is these smaller groups that should be treated as species.

Second, species must be *consistently* distinct. This means that all, or a very large proportion, of the individuals under consideration clearly belong to one group or another, and not somewhere in between. You cannot arbitrarily slice off one end of a unimodal curve of variation and call it a separate species, nor can you split such a unimodal curve in the middle or toward one side. Either there must be an absolute discontinuity of variation, or at least there must be a bimodal curve. Taxonomists may reasonably differ as to how close the low point between the peaks of the curve must approach the zero line, and different genera may call for different standards, but the test must be met at least to some degree if the group is to be called a species.

Third, species must be *persistently* distinct. There must be a reasonable assurance that all or a vast majority of the offspring of members of a given species will also belong to that species, for the foreseeable future. The so-called cinnamon bear is consistently distinguishable from the black bear, but it is not *persistently* distinct. Cinnamon bears and black bears regard themselves as the same sort of animal, and they breed together without hindrance. Thus the offspring of a cinnamon bear may be black, or vice versa. The two types, although *consistently* distinct, are not *persistently* so, and therefore they belong to the same species.

It can be seen that hybridization between groups will reduce the degree to which they meet the test of persistence, and often also the test of consistency. The more freely two groups hybridize, the more doubtful their separate specific status becomes. The precise frequency of hybridization that will invalidate a potential specific distinction varies from genus to genus, and also according to the opinion of the individual taxonomist. The place where the line is to be drawn is arbitrary, but the test is a real one.

Finally, species must be distinguishable by *ordinary means*. The phraseology here is deliberately loose. Means that are ordinary in one group may not be ordinary in another. An electron microscope and a fully equipped laboratory may be ordinary means for a bacterial taxonomist. The bacterial species that can be distinguished by such means play very different roles in nature, and our understanding of the world is much enhanced by the seemingly fine distinctions. A compound microscope with an oil immersion lens is an ordinary means for a mycological taxonomist. Mycologists find the distinctions they can draw at this level of observation to be biologically meaningful, and mostly they see no need to go further and use an electron microscope in defining species. Taxonomists considering vascular plants of ordinary size are reluctant to give specific status to things that cannot be distinguished with the aid of a hand lens, or at most with a dissecting microscope giving up to 20 or 40 diameters magnification. Even if such a microscope is needed to identify a dried specimen with certainty, the plants as they grow in nature can probably be recognized with the naked eye, after one has become familiar with the group. Vascular plant taxonomists find the distinctions they can draw at this level of observation to be biologically meaningful, and mostly they see no need to go further in search of specific characters.

To the extent that taxonomists differ as to what constitutes ordinary means, they will differ in certain instances as to the circumscription of species. Presumably everyone has an ultimate sticking point, at which he or she will cry "Hold! Enough!"; but that sticking point will vary from individual to individual. I can readily imagine that taxonomists might reasonably differ as to the level of magnification they are willing to use in examining the megaspores of *Isoetes* to determine specific differences.

It will rightly be pointed out that the specific definition I am propounding is so generalized that it leaves a great deal of room for individual interpretation and difference of opinion. In a sense, it merely sets the limits within which reasonable taxonomists may disagree. I will answer that such is the nature of the world.

Any attempt at a more precise definition will merely limit its application. Additional criteria that may help to clarify the taxonomy in one group will be useless or destructive in another. Any effort to specify the precise amount of intergradation that can be admitted between species is likewise foredoomed. The species in *Rudbeckia* are all sharply limited, if you take a broad definition of one or two of them. An insistence on the same degree of distinctness in *Senecio* would cause only confusion. It is necessary to give specific recognition to a great many North American taxa of *Senecio* that are only about as distinct as varieties in some other genera. The species one would arrive at in *Senecio,* if he insisted on really sharp interspecific distinctions, would be so large and internally diversified as to be conceptually useless. Polyploid-apomictic complexes, such as in *Crepis,* are sometimes only arbitrarily divisible into species. The definition I propose can be adapted to serve in all these cases, by varying the severity of one or another of the criteria. Ultimately, the taxonomist must produce a treatment that appeals to the mind as the best conceptual organization of the diversity that exists in nature.

We have noted that the biological species concept emphasizes the criterion of reproductive isolation, at the expense of phenetic discontinuity, when these two criteria are in conflict. The definition here presented focuses attention instead on the criterion of phenetic discontinuity, but actually it requires that both criteria be met. Phenetic discontinuity between groups cannot persist in the absence of a barrier to interbreeding. Reproductive isolation is a necessary corollary of persistent phenetic discontinuity.

INFRASPECIFIC TAXA

The nature of infraspecific taxa is germane to this paper only insofar as it is necessary to distinguish them conceptually from species. We may properly note, however, that here again the pattern of variation in plants is often somewhat different from that in animals. Vertebrate zoologists seem to be content with only one infraspecific level, which they call a subspecies. Such subspecies always have a strong geographical component; they are geographical phases of a species (*25, 28*). Many species of higher plants show a similar pattern of geographic differentiation, although there is a running argument about whether such geographical phases should be called subspecies or varieties. Plant species often have ecologically differentiated phases in the same geographic area. These may differ as significantly as the geographically assorted elements of a species, and they are usually considered to be taxonomically significant if they can be recognized by morphological markers. Often the infraspecific taxa show some combination of these two types of populational differentiation. Two such taxa may have distinctive but widely overlapping geographic ranges, and tend to occupy different habitats in the area of sympatry. In other species there may be geographically segregated infraspecific units, each of which also shows a taxonomically significant pattern of ecological variation. Thus it should not be surprising that many botanists see a need for more than one infraspecific category.

Because of the greater variety of patterns to be seen, botanists cannot usefully confine themselves to the simple definition of geographic subspecies favored by vertebrate zoologists. (I specify *vertebrate* zoologists, because I understand that taxonomists of social insects such as ants also find some problems with this simple concept.) A more generalized concept of the nature of infraspecific taxa must therefore be sought.

Infraspecific taxa are groups that are sufficiently differentiated to call for some taxonomic recognition, but do not meet the specific standards to the same degree as the species themselves. They show more intergradation, and sometimes they differ in less obvious characters. The standards and the criteria will differ from species to species. The taxonomist does his duty to the rest of the community of biologists by delimiting species that can consistently be recognized by ordinary means, and then for infraspecific taxa he uses whatever criteria and standards seem appropriate to the situation. It is only necessary that the curve of variation be bimodal or plurimodal rather than unimodal, and that there be a reasonable degree of reproductive continuity through successive generations within each of the modes.

It should not be assumed that all species contain subspecies or varieties. Many species are not usefully subdivisible.

Given the necessity for fairly loose definitions, it should not be surprising that some taxa will be classified at the specific level by the taxonomists, and at an infraspecific level by others. There is some legitimate room for differences of opinion between lumpers and splitters. I consider the black maple, *Acer nigrum* Michx. f., to be most usefully treated at the specific level, but I cannot really quarrel with those of my colleagues who consider that it is better treated as a subspecies or variety of the sugar maple, *Acer saccharum* Marsh. Darwin was right that there is every gradation in nature from the infraspecific to the specific level of evolutionary differentiation.

THE REALITY OF SPECIES

Until Darwin's time, biologists in general believed that species exist in nature and are merely recognized by taxonomists. After 1859 the situation did not seem so simple. It was recognized that at any given time there will be some groups at the threshold of specific differentiation. Biologists might reasonably differ about the taxonomic status of such groups. The developing understanding of genetics during the first two decades of the twentieth century gave a further blow to the idea that species are real entities. Many biologists came to believe that they are purely arbitrary mental concepts. The necessity for such concepts was usually accepted, but their arbitrary nature was emphasized. During the past several decades the pendulum has swung the other way.

I believe that the thesis and the antithesis here are subject to rational synthesis. Individuals exist. Each individual has a particular set of phenotypic characteristics. Successful interbreeding in nature is restricted to groups of individuals that are similar in most respects. There are gaps of all sizes in the distribution of diversity in nature. Many combinations of characters that might be imagined simply do not exist. (I never saw a photosynthetic dog.) In producing a taxonomic scheme, we try to recognize the clusters and draw lines in the gaps between them. The smallest clusters that are separated by reasonably evident gaps from other such clusters are the species. Usually, although not always, the continuity within a cluster is reinforced by interbreeding within the cluster. Interbreeding between members of different clusters, on the other hand, is restricted or impossible.

The distribution of diversity in nature, strongly influenced by breeding patterns, is such that the number of potentially satisfying mental organizations is not infinite or even large, but always very small. Very often there is only one mentally satisfying organization at the species level for a particular genus or portion of a genus. To that extent, species exist in nature and are merely recognized by taxonomists. On the other hand, sometimes there are two or even three different organizations of a small group that might be seriously entertained. One such organization might appeal to one taxonomist, and a different one might appeal to another. To that extent, the species is merely a creation of the mind, but the choices are always limited.

APPLICATION OF BOTANICAL CONCEPTS OF SPECIES TO ANIMALS

Zoologists have long tended to believe that the taxonomic and evolutionary concepts that they devise to understand what is going on in the animal world are equally applicable to the plant world. After all, organisms are organisms. Let me, for a moment, try to put the shoe on the other foot. I have pointed out that the so-called "biological species" concept, originally developed for animals, works so badly in plants that most present-day plant taxonomists don't even try to use it. Instead they have come to a much looser and more flexible concept, which I have tried in this paper to formalize. Might it not eventually become necessary for zoologists to adopt the botanical concept of species?

I suppose it depends on the number of sibling species that are discovered. Reproductive continuity, in both plant and animal species, is usually inferred rather than rigorously demonstrated. If continuing studies of animals turn up more and more sibling species to confound the field naturalist and confront common sense, there will eventually come a time when some Young Turk will say, "We have had enough of this nonsense! There is no use calling things different species if you can't tell them apart." If, on the other hand, the number of recognized sibling species remains relatively limited, then zoological taxonomists will probably see no reason to change their views. It remains to be seen.

LITERATURE CITED

1. Anderson, E. 1949. *Introgressive hybridization.* Wiley, New York. 109 pp.

2. Berlin, B., D. E. Breedlove, and P. H. Raven. 1973. *General principles of classification and nomenclature in folk biology.* Amer. Anthropologist 75: 214–242.

3. Camp, W. H., and C. L. Gilly. 1943. *The structure and origin of species*. Brittonia 4: 324–385.

4. Chuang, T. S., and L. Constance. 1977. *Cytogeography of Phacelia ranunculacea* (Hydrophyllaceae). Rhodora 79: 115–122.

5. Cronquist, A. 1968. *The evolution and classification of flowering plants*. Houghton Mifflin, Boston. 396 pp.

6. Day, A. 1965. *The evolution of a pair of sibling allotetraploid species of cobwebby Gilias (Polemoniaceae)*. Aliso 6: 25–75.

7. Dobzhansky, Th. 1937. *Genetics and the origin of species*. Columbia University Press, New York. 364 pp.

8. Dobzhansky, Th. 1951. *Genetics and the origin of species*, 3rd ed. Columbia University Press, New York. 364 pp.

9. Du Rietz, G. E. 1930. *The fundamental units of biological taxonomy*. Svensk Bot. Tidskr. 24: 333–428.

10. Ehrlich, P. R., and P. H. Raven. 1969. *Differentiation of populations*. Science 165: 1228–1232.

11. Gleason, H. A., and A. Cronquist. 1964. *The natural geography of plants*. Columbia University Press, New York. 420 pp.

12. Grant, V. 1957. *The plant species in theory and practice*. Pages 39–80 *in* E. Mayr, ed., *The species problem*. Publ. 50, Amer. Assoc. Adv. Sci., Washington.

13. Grant, V. 1964. *The biological composition of a taxonomic species in Gilia*. Pages 281–328 *in* E. W. Caspari and J. M. Thoday, eds. *Advances in Genetics*. 12. Academic Press, New York and London.

14. Grant, V. 1966. *The selective origin of incompatibility barriers in the plant genus Gilia*. Amer. Naturalist 100: 99–118.

15. Grant, V. 1971. *Plant speciation*. Columbia University Press, New York. 435 pp.

16. Grant, V. and A. 1956. *Genetic and taxonomic studies in Gilia. VIII. The cobwebby Gilias*. Aliso 3: 203–287.

17. Harlan, J. R., and J. M. J. DeWet. 1975. *On Ö Winge and a prayer: the origins of polyploidy*. Bot. Review 41: 361–390.

18. Keeley, S. C., and S. B. Jones, Jr. 1977. *Vernonia (Compositae) in the Bahamas—reexamined*. Rhodora 79: 147–159.

19. Lewis, W. H., and R. L. Oliver. 1971. *Meiotic chromosomal variation in Claytonia virginica*. J. Hered. 62: 379–380.

20. Linnaeus, C. 1751. *Philosophia botanica*. Kiesewetter, Stockholmiae. 362 pp.

21. Lotsy, J. P. 1914. *Prof. E. Lehmann über Art, reine Linie und isogene Einheit*. Biol. Centralbl. 34: 614–618.

22. Mayr, E. 1940. *Speciation phenomena in birds*. Amer. Naturalist 74: 249–278.

23. Mayr, E. 1942. *Systematics and the origin of species from the viewpoint of a zoologist*. Columbia Univ. Press, New York. 334 pp.

24. Mayr, E. 1957. *Species concepts and definitions*. Pages 1–22 *in* E. Mayr, ed., *The species problem*. Publ. 50, Amer. Assoc. Adv. Sci., Washington.

25. Mayr, E. 1969. *Principles of systematic zoology*. McGraw-Hill, New York. 428 pp.

26. Merriam, C. H. 1918. *Review of the grizzly and big brown bears of North America (genus Ursus) with description of a new genus, Vetularctos*. N. Amer. Fauna 41. 136 pp.

27. Rick, C. M. 1946. *Studies of natural populations in the Nemophila Menziesii-atomaria complex*. Amer. J. Bot. 33: 825.

28. Simpson, G. G. 1961. *Principles of animal taxonomy*. Columbia Univ. Press, New York. 247 pp.

29. Stebbins, G. L. 1975. *Flowering plants. Evolution above the species level*. Belknap Press of Harvard Univ. Press, Cambridge, Mass. 399 pp.

30. Vries, H. de. 1901–03. *Die Mutationstheorie: Versuche und Beobachtungen über die Entstehung von Arten im Pflanzenreich*. Veit. Leipzig. 2 vols. (648 pp., 752 pp.)

31. Winge, Ö. 1940. *Taxonomic and evolutionary studies in Erophila based on cytogenetic investigations*. Compt. Rend. Trav. Lab. Carlsb. (Sér. Physiol.) 23: 41–74.

two

BIOSYSTEMATICS AND AGRICULTURAL RESEARCH

2] The Importance of Cryptic Species and Specific Identifications as Related to Biological Control

by DAVID ROSEN*

ABSTRACT

Reliable systematics is the basis for any meaningful research in biology. This is especially true in applied biological control, inasmuch as most successful natural enemies are rather highly host-specific. "Cryptic," closely-related species of pests may harbor different natural enemies, and their correct identification may determine the success or failure of importation projects aimed at bringing them under biological control. On the other hand, closely-related or sibling species of natural enemies may differ markedly in their host preferences, as well as in other biological attributes that may determine their success as biological control agents. Recognition of such cryptic species or strains may add effective weapons to the arsenal of biological control. Inadequate systematics of either pests or natural enemies was the cause of prolonged delays in the ultimate success of several important biological control projects. Correct specific identification is essential also in the mass-production and recovery phases of biological control. Unfortunately, systematic knowledge of major natural-enemy groups is still very inadequate. Efforts should be made to encourage and support systematic research of such groups, with special emphasis on the biosystematic approach.

INTRODUCTION

Correct identification and classification of organisms are essential for any intelligent interpretation of biological and ecological information. Harold Compere (*8*) expressed this very succinctly when he pointed out that, "Theories are no more sound than the systematics on which they are based." Biologists should, therefore, be thoroughly acquainted with the

*The Hebrew University, Faculty of Agriculture, Rehovot, Israel.

basic concepts, principles, and methodology of systematics. In the words of the late F. S. Bodenheimer (*5*):

> It is essential that every zoologist—and certainly every ecologist—acquire good knowledge of some special animal group, no matter how small. Any inquiry into a branch of biology can be reliably conducted only on the basis of accurate taxonomy. . . . Basic taxonomic information on any group gives the biologist the *Dos moi po sto* of Archimedes, the point from which he may manipulate the lever for any biological problem. A zoologist is definitely not a man who does not know any animal.

If in general biology the importance of systematics may sometimes be viewed as academic, in economic entomology it may often have an immediate practical value. Upon encountering an unknown pest, our first question usually is: "What is it?" Correct identification provides the key to any available information about the species, its distribution, biology, habits, and possible means of control, which would otherwise have to be independently investigated at a considerable expense of time and effort.

No other aspect of applied entomology is more intimately dependent on sound systematics—more precisely, biosystematics—than is biological control. In its classical form, this method of control usually involves the introduction of exotic parasites or predators against a pest of foreign origin. When live natural enemies are being sought or are actually transferred from one region to another in order to bring about the biological control of an arthropod pest, the correct identification of both pest and natural enemy species and the recognition of cryptic species or infraspecific entities may, of course, be of utmost importance (*33*). Indeed, the great importance of reliable biosystematics has been illustrated time and again by the success or failure of projects aimed at the biological control of many serious pests. This has been mainly due to the remarkable specificity of natural enemies.

IMPORTANCE OF SPECIFIC IDENTIFICATIONS

Specificity of natural enemies. Although some parasites or predators are truly polyphagous, a survey of successful biological control projects shows that most effective natural enemies are either virtually monophagous or only narrowly oligophagous. Their host or prey spectrum is effectively limited by an elaborate process of host selection.

In parasites, this selection process is seen as consisting of four distinct steps: (1) habitat selection, (2) host finding, (3) host acceptance or "psychological" selection, and (4) host suitability or "physiological" selection (*17, 18, 35*). Of these, habitat selection is always the first, sometimes the most important eliminative step. Many Ichneumonidae, for instance, are known to be habitat-specific, attacking a wide range of hosts that occur in their preferred microhabitat (*47*). Aphidiidae, too, are basically habitat-dependent, most of the species being characterized by their occurrence in rather defined habitats (*44*). The Chalcidoidea, on the other hand, usually also exhibit marked host-specificity (*35*).

Predators are generally regarded as more polyphagous than parasites. However, they, too, may be considerably more specific than has been commonly realized. Many species of Coccinellidae are rather strictly habitat specific, whereas others are very particular in prey selection. In many cases, certain species of prey, although acceptable to a coccinellid, may be unsuitable for its development or reproduction (*21*). In fact, disregard for the specificity of Coccinellidae was probably one of the main reasons for the failure of an extensive biological control project directed against the Bermuda cedar scales, *Carulaspis minima* (Targioni Tozzetti) and *Lepidosaphes newsteadi* (Sulc) (Homoptera: Diaspididae). Scores of coccinellid species were introduced into Bermuda for that purpose, but most of them failed to become established (*36, 45, 46*).

The suitability of hosts or prey to natural enemies may be affected by the host plant. Thus, certain endoparasites are incapable of completing their development in the California red scale, *Aonidiella aurantii* (Maskell) (Homoptera: Diaspididae), when the latter is growing on *Cycas*, whereas the same host on citrus is quite suitable (*15, 19, 43*). Similarly, the coccinellid *Rodolia cardinalis* (Mulsant) rejects its common prey, the cottony-cushion scale, *Icerya purchasi* Maskell (Homoptera: Margarodidae) on *Spartium* or *Genista* plants (*21*).

Identification of pest species. Correct identification of the pest is of supreme importance in bioloical control. When a pest is misidentified, considerable time and effort may be wasted on the introduction of natural enemies of the wrong species, which, if at all specific, may fail to become established in the new habitat. In Canada, for instance, the lodgepole pine needle miner, *Evagora starki* (Freeman) (Lepidoptera: Gelechiidae), was confused with *E. milleri* (Busck) and other needle miners, resulting in attempts to introduce several unsuitable parasites (*48*). Although certain biological control projects have been successfully concluded in spite of inadequate taxonomy of the pest species, more often this was the main cause of prolonged delays in ultimate success.

Locating the native habitat of an exotic pest is one of the main problems of biological control. Misidentification of an African pest as a related Oriental species may misdirect the search for natural enemies to the wrong continent, again resulting in useless waste of time, resources, and effort. Misclassification may have the same effect. The sugar-beet leafhopper, *Circulifer tenellus* (Baker) (Homoptera: Cicadellidae), was at first considered to be a member of the genus *Eutettix* and of South American origin, but early surveys in South America were unsuccessful. Only after it had been concluded that *tenellus* actually belonged to a group of species of the genus *Circulifer* occurring around the Mediterranean was the search directed to that region and several natural enemies were indeed found (*39, 41*).

Identification of natural enemies. Correct identification of natural enemies is as essential to the success of biological control projects as is the correct identification of the pest. Obviously, natural enemies cannot be utilized for biological control unless they are discovered and recognized as such. To cite one well-known example, it was the discovery and description by Silvestri of certain whitefly parasites in tropical Asia that provided the initial stimulus for a successful biological control project carried out against the citrus blackfly, *Aleurocanthus woglumi* Ashby (Homoptera: Aleyrodidae), in the West Indies and Central America (*6, 9*). In a recent revision of the genus *Aphytis* (Hymenoptera: Aphelinidae) (*14, 34, 37*), some 35 species have been recognized and described as new. They are all primary parasites and are now available for biological control of armored scale insect pests.

Misidentification of an exotic natural enemy as one already present in the new habitat may delay indefinitely its utilization in biological control projects. On the other hand, misidentification of indigenous and exotic natural enemies as distinct, when they are actually identical, may lead to fruitless importations and unnecessary confusion.

A potentially undesirable organism may be misidentified as a useful natural enemy. This actually happened in one of the earliest intentional transfers of natural enemies from one continent to another. *Tyroglyphus phylloxerae* Riley, a tyroglyphid mite, was described in the United States as a predator of the notorious grape phylloxera, *Viteus vitifoliae* (Fitch) (Homoptera: Phylloxeridae), and in 1873 was introduced as such into France, where it reportedly became established (*22*). However, some time later it was determined to be a synonym of *Rhyzoglyphus echinopus* Fumouze and Robin, a well-known pest of various bulbs and already present in Europe (*28*). Moreover, this species has been implicated as one of the causative agents of a deadly root disease of grapevines known as phthiriosis (*4*). So, at best, this was a useless transfer of a species already present in Europe, but it could have been a serious blunder.

The opposite may also occur. *Tytthus mundulus* (Breddin) (Hemiptera: Miridae), an effective egg predator of the sugar-cane leafhopper, *Perkinsiella saccharicida* Kirkaldy (Homoptera: Delphacidae), was so poorly known as to be considered a plant-feeding insect. This caused a 15-year delay in the eventually outstanding biological control of the leafhopper as a serious pest in Hawaii (*13*).

Primary and secondary parasites may be closely related systematically (e.g., the genera *Diversinervus* and *Cheiloneurus* in the Encyrtidae, or *Aphytis* and *Marietta* in the Aphelinidae). Species of the large genus *Tetrastichus* (Hymenoptera: Eulophidae) may be either primary or secondary, sometimes even on the same host species, as is the case with *T. ceroplastae* (Girault) and *T. ceroplastophilus* Domenichini on *Ceroplastes* spp around the Mediterranean (*2, 32*). Misindentification (or misclassification) of a primary parasite as hyperparasitic may again delay its utilization in biological control, whereas the opposite may result in the inadvertent introduction and establishment of an injurious hyperparasitic species. Often, even cursory examination of imported material by an experienced taxonomist may suffice to segregate known hyperparasites. Close collaboration with a taxonomist is therefore indispensable during the importation phase of any biological control program. A taxonomist would certainly not have recommended the introduction of an *Azotus* (a well-known genus of hyperparasitic Aphelinidae) into Bermuda for use against the Bermuda cedar scales (*36*).

The role of the taxonomist is by no means limited to the early phases of a biological control project. Correct identification is equally essential in any subsequent evaluation of the establishment, dispersal and efficacy of introduced natural enemies, as well as in any attempts at their mass production. Mass cultures of natural enemies may be readily contaminated by undesirable species. Unless continuous collaboration with a taxonomist is maintained, such contamination may not be noticed for long periods. This apparently occurred in the biological control project directed against the San Jose scale, *Quadraspidiotus perniciosus* (Comstock) (Homoptera: Diaspididae), in Europe. In Germany, mass cultures of an imported parasite, *Prospaltella perniciosi* Tower (Hymenoptera: Aphelinidae), were apparently invaded and taken over by a related, ineffective species, *P. fasciata* Malenotti. About 4 million wasps of the latter species were released in the Heidelberg region during 1956–58, before its identity was revealed by taxonomists, but it never became established (*33, 36*).

RECOGNITION OF CRYPTIC FORMS

Ever since the advent of the so-called "new systematics" (*23*), most practicing taxonomists have been more-or-less aware of the importance of biosystematics and of the need to recognize biological differences between closely-related entities. This need is all the more apparent in applied biological control. As pointed out by DeBach (*12*), "From the standpoint of practical biological control, we are vitally interested in whether natural enemies differ from one another, regardless of our ability to tell them apart morphologically. All grades of specific or subspecific genetically-based difference may be important."

Cryptic species. As more observational and experimental information about intraspecific variation is accrued, numerous cryptic species of both pests and natural enemies are revealed, and so-called sibling (i.e., morphologically indistinguishable) species are differentiated on morphological grounds. On the other hand, cases of host-determined, geographical or seasonal polymorphism or polychroism are elucidated. Biological differences often provide the first indication to the possible distinctness of closely-related entities. In Europe, for instance, *Elachertus geniculatus* (Hartig) (Hymenoptera: Eulophidae), parasitic upon lepidopterous larvae on conifers, was considered synonymous with *E. nigritulus* (Zetterstedt), attacking larvae on herbs and deciduous trees. Because of the discordant host-plant records, reared material was examined and morphological differences between the two species were eventually discovered (*1*).

Closely related, "cryptic" or sibling pest species may harbor different natural enemies. In fact, parasitic Hymenoptera sometimes prove to be more discriminative than are expert systematists, and parasitological data may serve as reliable taxonomic criteria for the recognition of their hosts (*35*). An interesting case in point is furnished by a complex of two morphologically indistinguishable but biologically distinct scale insects, both referable to *Chionaspis pinifoliae* (Fitch) (Homoptera: Diaspididae), infesting different pine species in California. They are attacked by entirely different parasites belonging to different hymenopterous families (*27*). Similarly, closely-related species of *Aonidiella* (Homoptera: Diaspididae) are attacked by different parasite species, and the same is apparently true for closely-related species of *Saissetia* (Homoptera: Coccidae) (*35*).

Closely allied pest species may differ in their suitability as hosts to certain parasites. The Egyptian alfalfa weevil, *Hypera brunneipennis* (Boheman) (Coleoptera: Curculionidae), is very difficult to separate morphologically from the alfalfa weevil *H. postica* (Gyllenhal), but its larvae exhibit a much higher capacity to encapsulate the eggs of the ichneumonid parasite *Bathyplectes curculionis* (Thomson) (*50, 51*).

On the other hand, closely-related or sibling species of parasites may differ markedly in their host preferences, as well as in various other biological attributes that may determine their success as biological control agents. Recognition of such cryptic species may amount to the addition of new weapons to the arsenal of biological control. Clausen (*6*) quotes the following example. *Exorista larvarum* (L.) (Diptera: Tachinidae), a European parasite of the gypsy moth, *Lymantria dispar* L. (Lepidoptera: Lymantriidae), and various other Lepidoptera, is practically indistinguishable from the American species, *E. mella* (Walker), which only seldom attacks the gypsy moth, and which also differs in its site of pupation. After *E. larvarum* was recognized as distinct on the basis of these biological differences, it was successfully introduced and established in the United States.

Another case in point is that of *Trioxys utilis* Muesebeck (Hymenoptera: Aphidiidae), introduced into California against the spotted alfalfa aphid, *Therioaphis maculata* (Buckton), and *Trioxys pallidus* (Haliday), introduced against the walnut aphid, *Chromaphis juglandicola* (Kaltenbach). The two parasites are near-siblings and were actually synonymized for a while, but *T. utilis* attacks only aphids of the subfamily Therioaphidinae, mostly on short, herbaceous perennials, whereas *T. pallidus* attacks only aphids of the subfamily Callaphidinae in arboreal habitats *(20)*.

Biosystematic studies of the genus *Aphytis* have furnished some excellent examples of the importance of cryptic species in biological control. In fact, each of the originally-described, more common species in that genus is now known to include a group of sibling or near-sibling species that may exhibit different biological capabilities. Consider, for instance, a complex of four near-identical species: *A. holoxanthus* DeBach is the most effective natural enemy of the Florida red scale, *Chrysomphalus aonidum* (L.); *A. yasumatsui* Azim is parasitic upon the bifasciculate scale, *Chrysomphalus bifasciculatus* Ferris; *A. melinus* DeBach is the most effective known parasite of the California red scale, *Aonidiella aurantii* (Maskell), whereas *A. fisheri* DeBach, which is indistinguishable from *melinus* except for pupal pigmentation, also attacks the California red scale but has failed to become established in California. Until only a few years ago, all these and several other species were masquerading under the name *Aphytis chrysomphali* (Mercet) *(10, 13, 37)*.

Sibling species can be recognized as distinct only through biosystematic research. With biparental forms, this can normally be achieved by reciprocal crossing tests, reproductive isolation being the ultimate criterion for the determination of their systematic status *(12, 30)*. In some instances, further study may reveal some cryptic morphological differences, as was the case with several closely-related species of *Muscidifurax* (Hymenoptera: Pteromalidae) introduced into California *(25)*. In others, the species remain indistinguishable.

Uniparental ("ethological") species present a much more perplexing systematic problem, especially if they also happen to be siblings. Should a uniparental "form" be regarded as a distinct species in relation to a morphologically indistinguishable biparental "form," on the premise that they are reproductively isolated? Recent studies by Rössler and DeBach *(38)* have demonstrated that even this is not always true. They found that a uniparental form of *Aphytis mytilaspidis* (Le Baron), differing in host preference from a sibling biparental form, is nevertheless capable of mating with it and producing viable hybrids. Whatever their systematic status, such entities should be recognized for possible utilization in biological control projects.

Infraspecific categories. Hybridization tests sometimes reveal that certain sibling forms are only partially reproductively isolated. Such entities are termed "semi-species," and are different from subspecies or strains, which by definition are reproductively compatible. Semi-species may exhibit different host preferences, as well as various reproductive relations. Studies with forms of *Aphytis* illustrate how complicated this can get. Rao and DeBach *(12, 30)* investigated, among others, a complex of three allopatric,

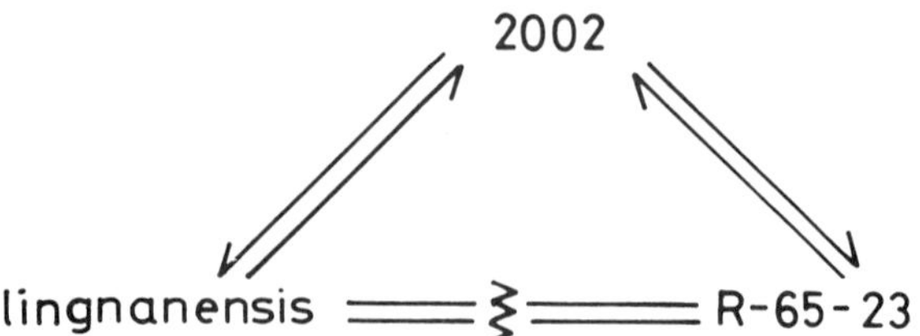

Figure 2.1. Reproductive relations between three allopatric forms in the *Aphytis lingnanensis* complex (see text). Adapted from Rao and DeBach (*30*).

sibling forms: *A. lingnanensis* Compere (ex *Aonidiella aurantii* from southern China), *A.* "2002" (ex the coconut scale, *Aspidiotus destructor* Signoret, from Puerto Rico), and *A.* "R-65-23" (ex the snow scale, *Unaspis citri* (Comstock), from Florida). Careful study has shown that, whereas *lingnanensis* and "R-65-23" are only partially isolated from "2002," they are completely isolated from one another. In other words, *lingnanensis* and "R-65-23" are good species in relation to one another, but both are semispecies in relation to "2002."

H. S. Smith (*42*), in an excellent review of insect races published in 1941, predicted that, with further research, many supposed races may eventually turn out to be distinct species. This was undoubtedly correct, as is illustrated by the following two examples. *Trichopoda pennipes* (Fabricius), a tachinid parasite of Hemiptera, is known to exist in three host-determined "races" in the eastern, southern, and western United States. Crosses between the eastern and western races failed, indicating that they may indeed represent sibling species (*16, 41*). Similarly, races of *Hypera postica* (Gyllenhal) from the eastern and western United States, differing in their susceptibility to the parasite *Bathyplectes curculionis* (Thomson), also exhibit partial reproductive isolation (*3, 29*).

However, genuine races or strains do exist, and are the equivalent of distinct species as far as practical biological control is concerned. *Bathyplectes curculionis*, for instance, apparently exists in two races in northern and southern California, differing in their ability to avoid encapsulation by different *Hypera* spp. (*40*). When the larch sawfly, *Pristiphora erichsonii* (Hartig), developed resistance to the ichneumonid parasite *Mesoleius tenthredinis* Morley in parts of Canada, a Bavarian race of the parasite was introduced that could overcome that resistance to a much greater degree, and could transmit that ability to the progeny of crosses with the locally-established race (*49*). Two races of the encyrtid parasite *Comperiella bifasciata* Howard, introduced into California, represent yet another example. The Chinese race may prefer *Aonidiella aurantii* (Maskell) as a host but develops well also in *A. citrina* (Coquillett), whereas the Japanese race strongly prefers *A. citrina* and virtually never completes development in *A.*

aurantii. These races are morphologically indistinguishable and cross readily in the laboratory, but maintain their distinctness in the field (*13*).

In conclusion of this section, the following quotation from DeBach (*12*) seems apropos:

> Biological control workers should never consider, *a priori*, that two similarly appearing populations in different geographical areas represent the same species... It is better to err on the safe side, than it is to fail to import a natural enemy because it is considered to be already present.

SOME CASE HISTORIES

A few selected case histories of biological control will serve further to illustrate some of the salient points discussed in the preceding paragraphs.

The Califonia red scale. In no other project was potential success so dependent on correct specific identifications as it was in the biological control of the California red scale, *Aonidiella aurantii* (Maskell). Attempts to introduce natural enemies of that important pest into California failed repeatedly for half a century, mainly due to inadequate systematic knowledge of both the pest itself and some of its natural enemies.

From 1877 to 1937, *Aonidiella aurantii* was repeatedly confused with some of its congeners, especially *A. citrina* and *A. taxus* (Leonardi). Certain endoparasites of these species do not attack *A. aurantii*, and their introduction from the Orient into California was therefore unsuccessful. The effects of various host plants on the susceptibility of the scale to parasites (see above) were unknown at the time, and exposure of infested *Cycas* plants in the Orient led to further failures and to the erroneous conclusion that no effective parasites of the California red scale existed in the Far East. The misclassification of the scale as a species of *Chrysomphalus*, a genus once believed to be of South American origin, then misdirected the search for natural enemies toward that continent (*7, 15, 36*).

As early as 1906, a species of *Aphytis*—in all probability *A. lingnanensis*—was recognized by George Compere as an effective parasite of the California red scale in China. Unfortunately, that species was misidentified as one already present in California, and no attempt was made to introduce it. Species of *Aphytis* were therefore subsequently ignored for several decades, and when one of them—apparently *A. melinus*—appeared in cultures of the scale exposed in the Orient, it was painstakingly destroyed as an undesirable contaminant. *A. lingnanensis* and *A. melinus* are presently recognized as the two most effective natural enemies of the California red scale. They were not discovered until 1947 and 1956, respectively (*7, 15, 36*).

The Florida red scale. During the last 20 years or so, *Chrysomphalus aonidum* has been successfully brought under complete biological control in Israel and elsewhere by *Aphytis holoxanthus*, an excellent parasite described by DeBach in 1960 (*11, 15, 36*). Attempts to obtain natural enemies against this serious pest were initiated in Florida as early as 1879, but when *A. holoxanthus* was reared from it by Albert Koebele in Hong Kong near the turn of the century, it was misidentified as another species and ignored. *Aphytis holoxanthus* was again reared from the Florida red scale in Java in the 1920's, but was again misidentified (*13, 15*). Needless to say, this effective natural enemy might have been available for biological control for decades, had its distinct identity been recognized much earlier.

The coffee mealybug. A serious pest of coffee in Kenya, *Planococcus kenyae* (Le Pelley) was first misidentified as the citrus mealybug, *P. citri* (Risso), then as the cacao mealybug, *P. lilacinus* (Cockerell), and numerous unsuccessful attempts were made to control it by the introduction of predators and parasites of those species from four different continents. One whole year was spent on a fruitless search through the Pacific area. Only after the pest was correctly identified as an undescribed species, known only from East Africa, was a search for its parasites conducted in neighboring Uganda and Tanganyika. Several effective parasites were then discovered and introduced into Kenya, and complete biological control was achieved (*26, 33*).

The citriculus mealybug. One of the most outstanding successes in the history of biological control was achieved in spite of inadequate systematics. *Pseudococcus citriculus* Green invaded Israel in the late 1930's and soon became a serious pest of citrus. It was misidentified as *P. comstocki* (Kuwana), and a parasite of that species, *Clausenia purpurea* Ishii (Hymenoptera: Encyrtidae), was therefore introduced against it from Japan. Spectacular biological control was attained, and only after the successful conclusion of the project was the true identity of the pest revealed (*31*). Obviously, this should be regarded as atypical.

The olive scale. The importance of cryptic species is best illustrated by the successful biological control of *Parlatoria oleae* (Colvée) in California. Extensive surveys throughout the range of distribution of the pest yielded four so-called "biological strains" of *Aphytis*, all morphologically indistinguishable from *A. maculicornis* (Masi) but possessing distinct biological attributes. One of these "strains," introduced from Iran, soon proved to be superior to all the others. This "Persian *Aphytis*"—recently described as a distinct species, *A. paramaculicornis* DeBach and Rosen (*14*)—was a major factor in the biological control of *P. oleae* (*15, 36*).

CONCLUSIONS

Obviously, reliable systematic knowledge of all organisms involved in a biological control project is essential for success. Unfortunately, such knowledge is generally inadequate, especially with regard to natural enemies. Major groups are in need of revision, and experts are few. In the parasitic Hymenoptera, for instance, it has been estimated that between 70 and 90% of all extant species are still undiscovered, and that no biological information whatsoever is available for 97% of all species (*24, 47*). Moreover, modern-day taxonomists must cope with a burdensome legacy of past work that was often conducted with poor optical equipment and with no regard for intraspecific variation or biosystematics. The great majority of described species in many important groups cannot be reliably identified on the basis of existing descriptions (*9*). Great efforts should be made to encourage and support thorough revisional work, with special emphasis on modern techniques such as scanning electron microscopy and on various biosystematic approaches.

Collaboration with a taxonomist should be a matter of routine in all biological control projects. However, biological control workers should be aware of the great difficulties with which the taxonomist is continually confronted, especially when called upon to identify dead specimens of exotic species, as is so often the case. Collaboration should definitely be bilateral. Biologists should always furnish the taxonomists with as much biological information as is available to them. Ideally, the biological control expert should personally do at least part of the biosystematic work on the organisms with which he or she is most concerned.

LITERATURE CITED

1. Askew, R. R. 1965. *Host relations in the Chalcidoidea (Hymenoptera) and their taxonomic significance.* Proc. Twelfth Internat. Congr. Entomol. (London, 1964): 94.

2. Bénassy, C., and E. Biliotti. 1963. *Ceroplastes rusci L. (Homoptera, Coccoidea, Lecaninae) exemple intéressant pour l'étude de la dynaminque des populations.* Entomophaga 8: 213–217.

3. Blickenstaff, C. C. 1965. *Partial intersterility of Eastern and Western U.S. strains of the alfalfa weevil.* Ann. Entomol. Soc. Amer. 58: 523–526.

4. Bodenheimer, F. S. 1951. *Citrus entomology in the Middle East.* W. Junk, The Hague. 663 pp.

5. Bodenheimer, F. S. 1959. *A biologist in Israel.* Biological Studies, Publ. Jerusalem. 492 pp.

6. Clausen, C. P. 1942. *The relation of taxonomy to biological control.* J. Econ. Entomol. 35: 744–748.

7. Compere, H. 1961. *The red scale and its insect enemies.* Hilgardia 31: 173–278.

8. Compere, H. 1969. *Changing trends and objectives in biological control.* Proc. First Internat. Citrus Symp. (Riverside, 1968) 2: 755–764.

9. Compere, H. 1969. *The role of systematics in biological control: a backward look.* Israel J. Entomol. 4: 5–10.

10. DeBach, P. 1959. *New species and strains of Aphytis (Hymenoptera, Eulophidae) parasitic on the California red scale, Aonidiella aurantii (Mask.), in the Orient.* Ann. Entomol. Soc. Amer. 52: 354–362.

11. DeBach, P. 1960. *The importance of taxonomy to biological control as illustrated by the cryptic history of Aphytis holoxanthus n. sp. (Hymenoptera: Aphelinidae), a parasite of Chrysomphalus aonidum, and Aphytis coheni n. sp., a parasite of Aonidiella aurantii.* Ann. Entomol. Soc. Amer. 53: 701–705.

12. DeBach, P. 1969. *Uniparental, sibling and semi-species in relation to taxonomy and biological control.* Israel J. Entomol. 4: 11-28.

13. DeBach, P. 1974. *Biological control by natural enemies.* Cambridge Univ. Press. 323 pp.

14. DeBach, P., and D. Rosen. 1976. *Twenty new species of Aphytis (Hymenoptera: Aphelinidae) with notes and new combinations.* Ann. Entomol. Soc. Amer. 69: 541-545.

15. DeBach, P., D. Rosen, and C. E. Kennett. 1971. *Biological control of coccids by introduced natural enemies.* Pages 165-194 *in* C. B. Huffaker, ed. *Biological control.* Plenum Press, New York and London.

16. Dietrick, E. J., and R. van den Bosch. 1957. *Insectary propagation of the squash bug and its parasite Trichopoda pennipes Fabr.* J. Econ. Entomol. 50: 627- 629.

17. Doutt, R. L. 1959. *Biology of parasitic Hymenoptera.* Ann. Rev. Entomol. 4: 161-182.

18. Doutt, R. L. 1964. *Biological characteristics of entomophagous adults.* Pages 145-167 *in* P. DeBach, ed. *Biological control of insect pests and weeds.* Chapman and Hall, London.

19. Flanders, S. E. 1942. *Abortive development in parasitic Hymenoptera induced by the host-plant of the insect host.* J. Econ. Entomol. 35: 834-835.

20. Hall, J. C., E. I. Schlinger, and R. van den Bosch. 1962. *Evidence for the separation of the "sibling species" Trioxys utilis and Trioxys pallidus (Hymenoptera: Braconidae, Aphidiinae).* Ann. Entomol. Soc. Amer. 55: 566-568.

21. Hodek, I. 1973. *Biology of Coccinellidae.* W. Junk, The Hague. 260 pp.

22. Howard, L. O. 1930. *A history of applied entomology.* Smithsonian Misc. Coll. 84: 1-564.

23. Huxley, J. S., ed. 1940. *The new systematics.* Clarendon Press, Oxford. 583 pp.

24. Kerrich, G. J. 1960. *The state of our knowledge of the systematics of the Hymenoptera Parasitica, with particular reference to the British fauna.* Trans. Soc. Brit. Entomol 14: 1-18.

25. Kogan, M., and E. F. Legner. 1970. *A biosystematic revision of the genus Muscidifurax (Hymenoptera: Pteromalidae) with descriptions of four new species.* Can. Entomol. 102: 1268-1290.

26. Le Pelley, R. H. 1943. *The biological control of a mealy bug on coffee and other crops in Kenya.* Empire J. Exp. Agric. 11: 78-88.

27. Luck, R. F., and D. L. Dahlsten. 1974. *Bionomics of the pine needle scale, Chionaspis pinifoliae, and its natural enemies at South Lake Tahoe, Calif.* Ann. Entomol. Soc. Amer. 67: 309-316.

28. Michael, A. D. 1903. *British Tyroglyphidae,* Vol. II. The Ray Soc. London. 183 pp.

29. Puttler, B. 1967. *Interrelationship of Hypera postica (Coleoptera: Curculionidae) and Bathyplectes curculionis (Hymenoptera: Ichneumonidae) in the Eastern United States with particular reference to encapsulation of the parasite eggs by the weevil larvae.* Ann. Entomol. Soc. Amer. 60: 1031-1038.

30. Rao, S. V., and P. DeBach. 1969. *Experimental studies on hybridization and sexual isolation between some Aphytis species (Hymenoptera: Aphelinidae). I. Experimental hybridization and an interpretation of evolutionary relationships among the species.* Hilgardia 39: 515-553.

31. Rivnay, E. 1968. *Biological control of pests in Israel (a review 1905-1965).* Israel J. Entomol. 3: 1-156.

32. Rosen, D. 1967. *The hymenopterous parasites of soft scales on citrus in Israel.* Beitr. Entomol. 17: 251-279.

33. Rosen, D., and P. DeBach. 1973. *Systematics, morphology and biological control.* Entomophaga 18: 215-222.

34. Rosen, D., and P. DeBach. 1976. *Biosystematic studies on the species of Aphytis.* Mushi 49: 1-17.

35. Rosen, D., and P. DeBach. 1977. *Use of scale-insect parasites in Coccoidea systematics.* Va. Polytech. Inst. and State Univ. Res. Div. Bull. 127: 5-21.

36. Rosen, D., and P. DeBach. 1978. *Diaspididae.* Pages 78-128 *in* C. P. Clausen, ed. *Introduced parasites and predators of arthropod pests and weeds: a world review.* U.S. Dept. Agric., Agric. Handbook 480.

37. Rosen, D., and P. DeBach. 1978. *Species of Aphytis of the world (Hymenoptera: Aphelinidae).* Israel Univ. Press, Jerusalem (in press).

38. Rössler, Y., and P. DeBach. 1972. *The biosystematic relations between a thelytokous and an arrhenotokous form of Aphytis mytilaspidis (Le Baron) (Hymenoptera: Aphelinidae) I. The reproductive relations.* Entomophaga 17: 391-423.

39. Sabrosky, C. W. 1955. *The interrelations of biological control and taxonomy.* J. Econ. Entomol. 48: 710-714.

40. Salt, G., and R. van den Bosch. 1967. *The defense reactions of three species of Hypera (Coleoptera, Curculionidae) to an ichneumon wasp.* J. Invertebr. Pathol. 9: 164-177.

41. Schlinger, E. I., and R. L. Doutt. 1964. *Systematics in relation to biological control.* Pages 247-280 *in* P. DeBach, ed. *Biological control of insect pests and weeds.* Chapman and Hall, London.

42. Smith, H. S. 1941. *Racial segregation in insect populations and its significance in applied entomology.* J. Econ. Entomol. 34: 1-13.

43. Smith, J. M. 1957. *Effects of the food plant of California red scale, Aonidiella aurantii (Mask.) on reproduction of its hymenopterous parasites.* Can. Entomol. 89: 219-230.

44. Starý, P. 1970. *Biology of aphid parasites (Hymenoptera: Aphidiidae) with respect to integrated control.* W. Junk, The Hague. 643 pp.

45. Thompson, W. R. 1951. *The specificity of host relations in predacious insects.* Can. Entomol. 83: 262-269.

46. Thompson, W. R. 1954. *Biological control work on cedar scales in Bermuda.* Rept. Sixth Commonw. Entomol. Conf. London, pp. 89-93.

47. Townes, H. 1971. *Ichneumonidae as biological control agents.* Proc. Tall Timbers Conf. Ecol. Anim. Contr. by Habitat Mgmt. 3: 235-248.

48. Turnbull, A. L., and D. A. Chant. 1961. *The practice and theory of biological control of insects in Canada.* Can. J. Zool. 39: 697-753.

49. Turnock, W. J., and J. A. Muldrew. 1971. *Pristiphora erichsonii (Hartig), larch sawfly.* Pages 175-194 *in Biological control programmes against insects and weeds in Canada 1959-1968.* Tech. Commun. Commonw. Inst. Biol. Control 4.

50. van den Bosch, R. 1964. *Encapsulation of the eggs of Bathyplectes curculionis (Thomson) (Hymenoptera: Ichneumonidae) in larvae of Hypera brunneipennis (Boheman) and Hypera postica (Gyllenhal) (Coleoptera: Curculionidae).* J. Insect Pathol. 6: 343-367.

51. van den Bosch, R., and E. J. Dietrick. 1959. *The interrelationships of Hypera brunneipennis (Coleoptera: Curculionidae) and Bathyplectes curculionis (Hymenoptera: Ichneumonidae) in southern California.* Ann. Entomol. Soc. Amer. 52: 609- 616.

3] Taxonomy of Nematodes That Parasitize Insects, and Their Use as Biological Control Agents

by WILLIAM R. NICKLE*

ABSTRACT

Given adequate moisture and suitable temperature, insect-parasitic nematodes debilitate, sterilize, or kill significant numbers of many kinds of pest insects such as mosquitoes, black flies, chironomids, grasshoppers, moths, and Japanese beetles. The search for these self-perpetuating biological control agents and other environmentally safe pest suppression techniques is currently underway as we seek alternative methods of pest control. Preparations of one species of nematode are already on the market for mosquito suppression and the door is open for others to follow. The systematics of these parasites is still in the "alpha" stage, but is in an accelerated growth period. The classification system makes possible the recognition of potentially important insect-parasitic nematodes. The parasites' morphology, infective stages, life cycles, pathology, and host ranges are useful in determining whether each newly developed parasitic relationship is worthy of further study. The Mermithidae, Sphaerulariidae, and Neoaplectanidae are three families of nematodes that parasitize insects. The systematics and biological control potential of some of the important representatives of these families are discussed in detail. Lists of currently accepted genera and their hosts are included. Members of the Mermithidae presently seem to have the most potential as biological control agents.

INTRODUCTION

Nematodes are roundworms. They are very abundant and often constitute as much as 90% of the invertebrate multicellular fauna in agricultural soils. Billions of nematodes are present in the top 15 cm of an acre of agricultural land. Most of these nematodes are worm-like and under a millimeter in

*Nematology Laboratory, Plant Protection Institute, Beltsville Agricultural Research Center, U. S. Department of Agriculture, Beltsville, Maryland 20705.

length. They have evolved to feed on many different types of organisms such as: bacteria, fungi, higher plants, animals, and even humans. We know and have recorded probably less than 10% of the nematodes that exist. Much of the land and water area of the world has not even been surveyed for these interesting animals which, as far as species number is concerned, occupy third place in the animal kingdom, behind arthropods and molluscs. It is not too surprising, therefore, to find nematodes killing, sterilizing, or otherwise debilitating insects. In habitats where adequate moisture is available, we find some insect-parasitic nematodes significantly suppressing populations of pest insects, such as black flies, mosquitoes, chironomids, grasshoppers, gypsy moths, Japanese beetles, and corn rootworms.

The taxonomy of plant, soil, and insect-parasitic nematodes is less advanced than entomological taxonomy. There have never been in this country, for example, more than ten full-time professionals engaged in nematode taxonomy. We have no subspecies or sibling species in the insect-parasitic nematodes and few in plant-parasitic nematodes such as were discussed by the preceding author, though we are quite aware of species complexes, races, and biotypes, some of which will probably eventually find their place as siblings and subspecies after more information becomes available. Contemporary nematode systematists are mostly engaged in "alpha" taxonomy, a refinement of some existing "alpha" taxonomy, identifications, and quarantine work.

The purpose of this review is to make available a brief insight into the status and importance of the biosystematics of insect-parasitic nematodes as they affect the selection on these biological control agents of pest insects. These self-perpetuating biological control agents, along with other environmentally safe pest suppression techniques, offer an alternative to the use of hazardous pesticides.

REVIEW OF THE LITERATURE

More papers have been published on insect-parasitic nematodes in the last 25 years than in all prior history. Taylor (*22*) summarized the fossil record on insect-parasitic nematodes from beetles and chironomids in Rhine lignite (Eocene) and Baltic amber.

Gould's (*10*) classic paper on a mermithid killing the ant *Lasius flavus* (F.) is one of the first published papers. He states that:

> Amongst other incidents that tend to lessen and destroy ant-flies, it is observable that [an] abundance of them are demolished by a white and long kind of worm, which is often met within their bodies. You may frequently take three from the insides of the large, but seldom more than one from the small ant-fly. These worms lie in a spiral form, and some may be extended half an inch.

Another early paper on mermithid nematodes was written by Goeze (*9*). He stated: "In June of 1781 after a hard rain storm, they (the worms) were widely distributed in the garden in the newly dug flower beds and wound or coiled by the hundreds like the filaments around the border of boxwoods." These worms were probably *Mermis nigrescens* Dujardin, 1842, from grasshoppers.

Hagmeier (*11*) wrote the largest and most complete revision of the mermithids known up to that time (1912). The quality of his drawings and descriptions was excellent and the revision serves as a cornerstone and major starting point for current systematic work.

The early American work on nematode systematics was done in the U.S. Department of Agriculture primarily by Christie (*6*), and Steiner (*21*). The life cycle studies by Christie were well done and the Steiner Mermithid Collection, USDA Nematode Collection, Beltsville, Maryland, originated from this period of research.

Bovien (*4, 5*) recognized the intricate life cycles of some insect-parasitic nematodes from Denmark. Five types of adult nematodes were shown to be involved in his new genus *Heterotylenchus*.

The early Soviet work was done by Polozhentsev (*16, 17*) on mermithids of May and June beetles and gypsy moths. More recently, Artyukhovsky (*1, 2, 3*) and Rubtsov (*18, 19*) have made outstanding contributions in their field, with Artyukhovsky working with terrestrial mermithids and Rubtsov with aquatic mermithids. Rubtsov has described more taxa of insect-parasitic nematodes than anyone else.

Fuchs (*7, 8*), Wachek (*23*), and Rühm (*20*) working in Germany contributed some excellent works on sphaerulariid parasites of bark beetles and other insects. Rühm's research on the life cycles of the bark beetle parasites was useful in the development of this area of research in the United States.

The Canadians have been active in the development of biological control agents, an effort in which Welch (*24, 25*) has contributed to the identification and systematics of the insect-parasitic nematodes.

In the United States, contemporary workers include Johnson (*12*), Nickle (*13, 14*), and Poinar (*15*), who recently wrote a book on insect-parasitic nematodes.

BIOSYSTEMATICS OF THE INSECT-PARASITIC NEMATODE GROUPS

The name of a parasite is important for several reasons. The most obvious, of course, is that we should know what organism we are talking about. Future work on the same parasite needs a designation under which information can be accumulated. Catalogers can provide this useful service only if they have the scientific names and key words needed to do their work properly. Several entomological and other scientific journals as standard policy reject papers submitted to them if the species of the parasite discussed has not been determined. Graduate students' dissertations often do not meet with approval when the principal parasite under study is not completely identified. Sometimes these parasites are new to science, thus requiring additional work.

The biosystematics of nematodes that parasitize insects is in an accelerating growth period, undergoing many changes and additions. Catalogers are not in a position to keep track of new species and synonymies or to weigh the relevance of complicated nomenclatural changes. The useful life of most keys is short and they are normally used only for preliminary identifications. The instability of these keys is mostly the result of incomplete knowledge of the systematic organization of nematodes. This must be overcome by additional work.

Recognizing a potentially important insect-parasitic nematode, and understanding its life cycle, pathology, and possible host range are important contributions of the systematist. The mosquito mermithid, found naturally in only about five ponds in Louisiana and Maryland, and the *Autographica californica* virus, found only once in an individual insect in California, are good examples of useful biological control agents that had a limited distribution.

About half of the nematode specimens I receive for identification are not important parasites. Often the insect larva died for reasons unrelated to its internal fauna, and bacterial-feeding, saprophytic nematodes entered the cadaver and built up in sufficient numbers to be recognizable to the investigator. Identification of these nematodes as saprophytes and not parasites saves money for the field researcher as he can then disregard the saprophytes and continue to look for true parasites of the insects.

Knowledge of the morphology, systematics, habitat, host specificity, and geographic distribution of effective and proven insect-parasitic nematodes is basic to the research for new parasites. I have been asked whether there are any nematode parasites of a specific tipulid pest that was introduced into the northwestern part of the United States and has caused damage to turf on golf courses. One existed on that species in England and that parasite was placed on a list of priority needs for the European Beneficial Insect Introduction Laboratory located in Paris, France. Recently, a mermithid was found to kill up to 60% of the gypsy moth caterpillars in the Soviet Union (*1*). Attempts are now being made to introduce it into the northeastern United States where it is not yet present. A mermithid nematode parasite of the corn rootworm is now also being studied in an attempt to use it as a biological control agent.

MERMITHIDAE

Mermithidae. The mermithids have taken first place as the group of insect-parasitic nematodes having the greatest potential as biological control agents. They are effective, safe, sometimes easily produced and stored, and often become established and self-perpetuating in the new site, killing a high percentage of the pest insects from then on. The Environmental Protection Agency of the U.S. Government is leaning toward classifying mermithids as parasites and exempting them from the regulations of the

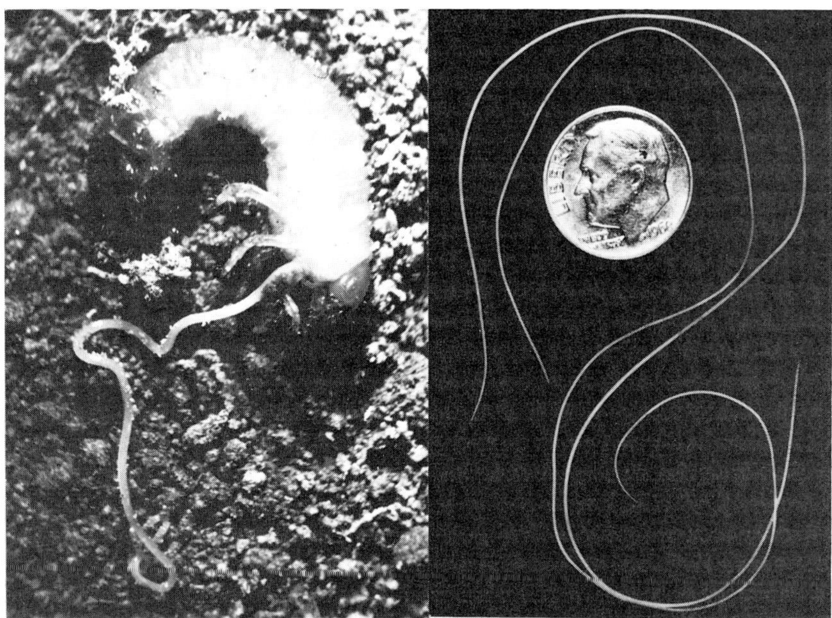

Figures 3.1–3.3. Members of the mermithid group of insect-parasitizing nematodes. *3.1 (above left).* Mermithid nematode parasite *Psammomermis* emerging from Japanese beetle grub. *3.2 (above right). Hexamermis albicans* mermithid taken from a gypsy moth caterpillar from Europe. *3.3 (bottom).* Grasshopper, *Schistocerca gregaria,* with emerging *Mermis nigrescens.* (Courtesy of R. Gordon and J. M. Webster.)

Federal Insecticide, Fungicide, and Rodenticide Act. Mermithids are roundworms. They vary in length from 1 to 50 cm, according to species of parasite and size of host. The sex of the mermithid nematode is often determined by the number of nematodes present in the body cavity of the insect. If there is only one nematode parasite in the insect, the nematode becomes a female. If there are three or more mermithid nematode parasites in the insect body cavity, these nematodes all become males. Mermithids kill their insect hosts by making large exit holes in their body walls. The insect loses its essential body fluids through this hole and dies (Fig. 3.1–3.3).

The systematics of the Mermithidae is currently dominated by the Soviet workers, who have the family split into two distinct groups: aquatic forms, and terrestrial forms. Their reasons for adopting this sytematic division are not clear to me.

More than 60 genera of mermithids are described in the literature. Some of these descriptions were of larval forms or of only one sex and these "genera" are now unidentifiable. The genera in Table 3.1 are recognizable and they are listed with the type of host insect in which they are most likely to be found.

Mermithids parasitizing mosquitoes. Here we are dealing with an ideal situation in an aquatic environment where mermithid eggs hatch, the preparasites penetrate young mosquito wrigglers, and the postparasitic

Table 3.1. Classification of the Mermithoidea.

Nematode	Host
Mermithoidea (Weinland, 1858) Wülker, 1924	
Tetradonematidae Cobb, 1919	
Tetradonema Cobb, 1919	Fungus gnat type
Aproctonema Keilin, 1917	Fungus gnat, biting midge
Corethrellonema Nickle, 1969	Chaoborid fly
Mermithonema T. Goodey, 1941	Scavenger fly
Mermithidae (Weinland, 1858) Braun, 1883	
Mermis Dujardin, 1842	Grasshoppers
Agamermis Cobb, Steiner, Christie, 1923	Grasshoppers
Allomermis Steiner, 1924	Unknown edaphic host
Amphimermis Kabaraki & Imamura, 1932	Rice borer, moths, damselflies
Bathymermis Daday, 1911	Unknown aquatic host
Capitomermis (Frostia) Rubtsov, 1968	Chironomids
Diximermis Nickle, 1972	Anopheline mosquitoes
Gastromermis Micoletzky, 1925	Many aquatic insects
Hexamermis Steiner, 1924	Many lepidopterans
Hydromermis Corti, 1902	Several aquatic insects
Lanceimermis Artyukhovsky, 1969	Aquatic host
Limnomermis Daday, 1911	Chironomids
Neomesomermis Nickle, 1972	Black flies
Octomyomermis Johnson, 1963	Chironomids
Perutilimermis Nickle, 1972	Saltmarsh mosquitoes
Psammomermis Polozhentsev, 1941	May beetles and Japanese beetles
Romanomermis Coman, 1961	Mosquitoes

Figure 3.4. Larvae of the mosquito *Culex pipiens* with the "Skeeter Doom" (see text) mermithid *Romanomermis culicivorax* visible in their thoracic regions.

mermithids emerge from the late instar mosquitoes, causing their death (Fig. 3.4). A few years ago several vials of mosquito mermithids were submitted to me for identification by an Agricultural Research Service entomologist, J. J. Petersen, in Louisiana. The material contained adult specimens of the mermithid parasites and included good host data. After a thorough study, these parasites were separated into five distinct genera, three being new to science. One of the species, *Romanomermis culicivorax (Reesimermis nielseni),* was found to have a wide mosquito host range. After I found the infective stage and elucidated the life cycle, the parasite was mass-reared and eventually became the first nematode-based biological control agent on the market. Its trade name is *Skeeter Doom.** I take pride in pointing out that Agricultural Research Service personnel found, researched and developed, mass-reared, and turned over to private industry this biological control agent. The other mosquito nematodes are either already well distributed, difficult to mass-rear, or have not been thoroughly studied. *Diximermis peterseni* Nickle, 1972, appears to be specific to the members of the mosquito genus *Anopheles* (Meigen), and *Perutilimermis*

*Mention of a trademark or proprietary product does not constitute a guarantee or warranty of the product by the U.S. Department of Agriculture, and does not imply its approval to the exclusion of other products that may also be suitable.

culicis Nickle, 1972, seems to be host-specific to the saltmarsh mosquito, *Aedes sollicitans* (Walker). *Hydromermis churchillensis* Welch, 1960, is an effective parasite of *Aedes communis* (DeG.).

Mermithids parasitizing black flies. The natural habitat of black fly larvae and their associated mermithid parasites seems cold and inhospitable to us. The larvae are attached to the undersides of rocks and also to leaves and other vegetation in rapidly flowing streams in the very early spring when the water temperature ranges between 0° and 15°C. Often we find high rates of parasitism (80 to 95%) by mermithid nematodes (Fig. 3.5). Our studies on a small, springfed tributary of Beaver Dam Creek on the Beltsville Agricultural Research Center show that nematodes of the genus *Isomermis* can kill 85% of the larvae of the black fly *Cnephia mutata* (Malloch) once the water temperature reaches about 12°C. *Neomesomermis flumenalis* (Welch, 1962) Nickle, 1972, has a wide black fly host range and a large geographical distribution, whereas species of *Gastromermis* are more host-specific and are not widely distributed. A classical biological control release of *Neomesomermis flumenalis* from upper New York state to Maryland was attempted in the spring of 1976 and studies to determine the extent of establishment will be carried out in 1977.

Mermithids parasitizing grasshoppers. In moist, agricultural river bottom land and in areas where rainfall exceeds 100 cm, we find that the mermithid nematode *Mermis nigrescens* Dujardin, 1842 is parasitic in

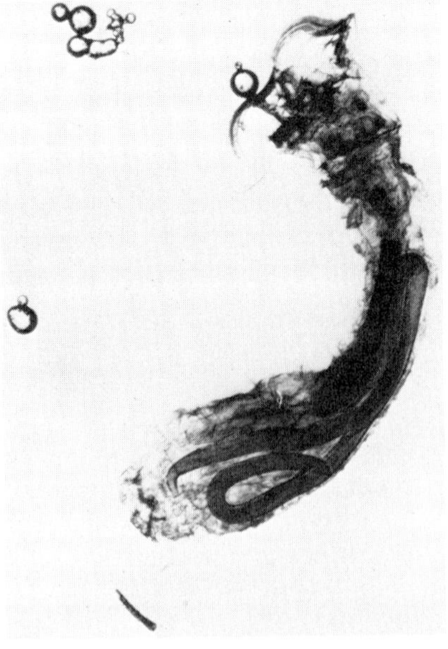

Figure 3.5. Larva of the black fly *Cnephia mutata* parasitized by an *Isomermis* nematode.

several genera and species of grasshoppers on a worldwide basis. The female worm, about 10 cm long, climbs up onto vegetation wet with the early morning dew and lays thousands of eggs having filamentous appendages. She later returns to the soil. The eggs adhere to the plant surfaces upon drying. When a grasshopper eats a leaf with an egg it may become parasitized (Fig. 3.3). I visited the Soviet Union in 1968 and was told that the workers there "milk" the female *Mermis nigrescens* of their eggs and spray the eggs on cabbage plants as a grasshopper control measure.

Mermithids parasitizing gypsy moths. One of the main pest insects of the forests in the northeastern United States is the gypsy moth. It was accidentally introduced into this country from Europe without its natural parasites and predators. A Soviet report *(1)* asserts that up to 60% of the gypsy moth caterpillars south of Moscow are parasitized by a mermithid, *Hexamermis albicans* (Siebold, 1848); and as we do not have this parasite in gypsy moth caterpillars in our infested areas, I requested that some be introduced to allow us to institute a classical release and establishment program here. Since my request was made, a few specimens were brought in from the Soviet Union, Austria, and Japan (Fig. 3.2). So far, I have been unable to maintain a living culture. One of our problems is that the life cycle of this parasite is not known, and we are unable to infect the caterpillar. *Hexamermis albicans* has a wide host range of lepidopterans and is found all over the world. This species complex is currently under study in our laboratory and I anticipate receiving several hundred nematode parasites from Japan later this year (1977).

Mermithids parasitizing Japanese beetles. Recently, research work on the Japanese beetle in Vermont, upper New York, and Connecticut revealed that up to 60% of the beetle grubs were parasitized fatally by a long (20 cm) mermithid (Fig. 3.1). Mermithids usually emerge from insects as last-stage nematode larvae and require one more molt to become adult worms. This one from Japanese beetle larvae required 66 days in the soil before it became adult. About 95% of all insects spend some time in the soil, which is also a good habitat for insect-parasitic nematodes. Interestingly enough, it turned out to be a *Psammomermis* sp. known up to now to kill up to 60% of the May and June beetle grubs in the Soviet Union. Both of these hosts are important pests to agriculture. I hope that we can spread and establish this parasite throughout the northeastern United States.

SPHAERULARIIDAE AND ENTAPHELENCHIDAE

The Sphaerulariidae is a family of insect-parasitic nematodes in the superfamily Tylenchoidea, which contains most of the important plant-parasitic nematodes (Table 3.2). The evolutionary progression of insect parasitism appears to have developed from the fungus-feeding forms. Identification of insects parasitized by these nematodes is usually easily made because dis-

Table 3.2. Classification of the Sphaerulariidae and Entaphelenchidae.

Nematode	Host
Sphaerulariidae Lubbock, 1861	
Sphaerulariinae (Lubbock, 1861) Pereira, 1931	
Sphaerularia Dufour, 1837	Bumblebees, wasps
Sphaerulariopsis Wachek, 1955	Bark beetles, weevils, braconids
Tripius Chitwood, 1935	Fungus gnat, gall gnat
Allantonematinae Pereira, 1931	
Allantonema Leuckart, 1884	Weevils, bark beetles, muscoid fly
Aphelenchulus Cobb, 1920	Hickory borer
Bovienema Nickle, 1963	*Pityogenes* bark beetles
Bradynema zur Strassen, 1892	Dung beetle, bibionid fly, Hemipterans
Contortylenchus Rühm, 1956	Bark beetles
Heterotylenchus Bovien, 1837	Muscoid flies, fleas, ground beetles
Howardula Cobb, 1921	Cucumber beetle, flies, thrips, mites
Metaparasitylenchus Wachek, 1955	Various beetles
Neoparasitylenchus Nickle, 1967	Bark beetles
Parasitylenchoides Wachek, 1955	Various beetles, rove beetles
Parasitylenchus Micoletzky, 1922	Bark beetles, fruit flies
Proparasitylenchus Wachek, 1955	Rove beetles
Protylenchus Wachek, 1955	Mud-loving beetles
Sulphuretylenchus Rühm, 1956	Bark beetles
Scatonema Bovien, 1932	Scavenger fly
Fergusobiinae J. B. Goodey, 1963	
Fergusobia Currie, 1937	Eucalyptus gall fly
Iotonchiinae T. Goodey, 1953	
Iotonchium Cobb, 1920	Unknown, found in fungi
Entaphelenchidae Nickle, 1970	
Entaphelenchus Wachek, 1955	Rove beetles
Peraphelenchus Wachek, 1955	Burying beetles
Praecocilenchus Poinar, 1969	Weevils
Roveaphelenchus Nickle, 1970	Rove beetles

section reveals 5,000 to 10,000 worms of two to five size groups in the body cavity of the insect. Such dissection should take place in saline water, otherwise the nematodes will burst and become unidentifiable. During the last 15 years, I have studied representative specimens from 13 of the 20 genera that have been placed in the family Sphaerulariidae. The parasite does not often kill the insect, but it usually causes debilitation, decreased egg production, or sterility. At present, there are four subfamilies, 20 genera, and about 120 species in the family.

The Face Fly Nematode, Heterotylenchus. Musca autumnalis DeGeer, the face fly pest of cattle and horses, was introduced into the northeastern part of the United States and has since moved across the country to the west coast. The front edge of this infestation, moving westward, was free of parasites and predators, and there were no reliable control measures. This resulted in a maximum number of flies and a maximum amount of distress to our cattlemen and dairymen, because the flies laid their eggs in the eyes of the cattle. The maggots then caused irritation, rubbing of eyes on barbed wire fences, and weight loss.

A nematode parasite of the face fly was found at Cornell University and specimens were sent to us for identification. Even though this parasite was in one of my specialty groups, it took two weeks to determine that it was a new species in the genus, *Heterotylenchus,* and it took several months to ascertain its life cycle. The parasitized flies are sterilized by the nematodes which pack their ovaries. These flies oviposit balls of infective-stage nematodes which infect the healthy maggots in the cow dung. The nematode would tend to follow the healthy wave of flies about two summers later. Agricultural Research Service scientists mass-reared the parasite on flies. At the request of the states of Montana and California, the parasites were introduced into the healthy waves of flies, resulting in sterilization of 25 to 50% of them.

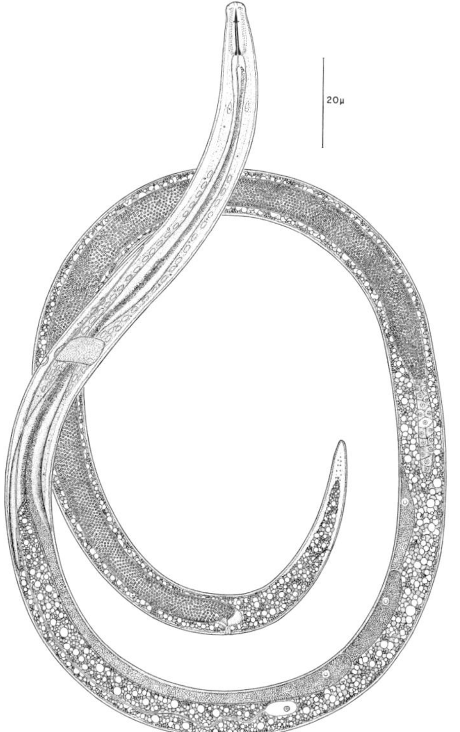

Figure 3.6. Infective stage female *Contortylenchus elongatus,* which parasitizes the bark beetle *Ips confusus.*

Bark Beetles. These insect pests of forests, orchards, and suburban trees are often parasitized by sphaerulariid-type nematodes. In fact, bark beetle frass is usually teeming with nematodes of many shapes and sizes. One or more types of parasitic nematodes from 3 to 4 mm in length, along with their progeny of up to 7,500 nematode eggs and larvae, may be found in the body cavity of a single adult bark beetle. Nematodes of the genera *Parasitylenchus* and *Contortylenchus* cause a reduction in the number of beetle eggs produced, if not total sterility. Another group of a few hundred nematodes may be located in the hind gut of the beetle, feeding on the epithelial layer. One may also find up to 200 larval *Parasitaphelenchus* in the body cavity of the beetle and hundreds more in a sticky mass attached to the underside of the elytra. *Cryptaphelenchus* nematodes may be found attached to the Malpighian tubules or located in the gut of the beetle grubs. Members of the phoretic diplogasterid group can be found attached to the genitalia, body segments, folds, or elytra of bark beetles.

The infective stage of *Contortylenchus elongatus* (Massey, 1960) (Fig. 3.6) was discovered by this writer. It was the first sphaerulariid infective-stage nematode known in the United States. It penetrates into the body cavity of the bark beetle grub in the gallery, swells, and lays 5,000 to 10,000 eggs inside the insect host.

The family Entaphelenchidae consists of members of the fungus-feeding aphelenchs that have evolved to parasitize insects. Biologically they behave the same as the sphaerulariids but are more rare and therefore, so far at least, are less important as biological control agents.

Table 3.3 Classification of the Genera *Neoaplectana* and *Heterorhabditis*.

Nematode	Host
Neoaplectana dutkyi Jackson, 1965, (DD-136)	Wide host range
Neoaplectana glaseri Steiner, 1929	Japanese beetle
Neoaplectana menozzi Travassos, 1931	Weevils
Neoaplectana feltiae Filipjev, 1934	European cutworm
Neoaplectana bibionis Bovien, 1937	Bibionid fly
Neoaplectana affinis Bovien, 1937	Bibionid fly
Neoaplectana chresima Steiner, 1942	Corn earworm
Neoaplectana leucaniae Hoy, 1954	Tussock moth
Neoaplectana carpocapsae Weiser, 1955	Codling moth
Neoaplectana janickei Weiser and Kohler, 1955	Sawfly
Neoaplectana bothynoderi Kirjanova and Puckhova, 1955	Weevil
Neoaplectana melolontha Weiser, 1958	June beetle
Neoaplectana georgica Kakulia and and Veremchuk, 1965	June beetle
Neoaplectana hoptha Turco, 1970	Japanese beetle
Heterorhabditis bacteriophora Poinar, 1975	*Heliothis*

NEOAPLECTANA AND HETERORHABDITIS

These genera of insect-parasitic nematodes have a life cycle that includes vectoring a bacterium that actually causes the death of the insect (Table 3.3). The ensheathed nematode larva is ingested by the insect grub. After entering the gut, it penetrates to the body cavity of the insect and releases its bacteria into the insect's blood, thus causing a septicemia which kills the

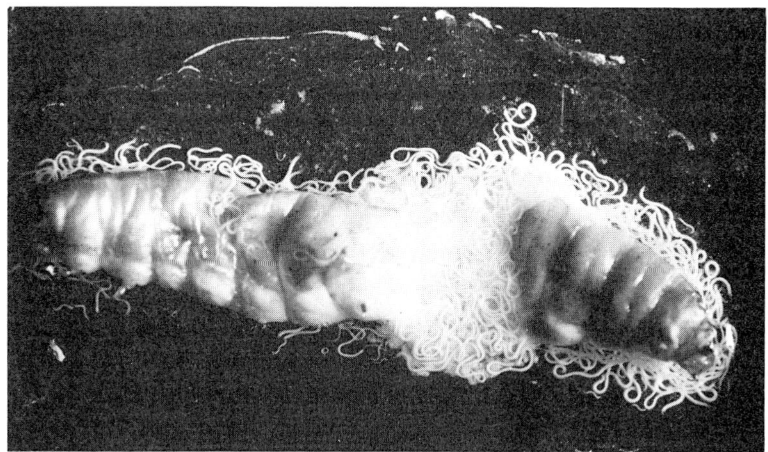

Figure 3.7. *Neoaplectana dutkyi* nematodes on the greater wax moth.

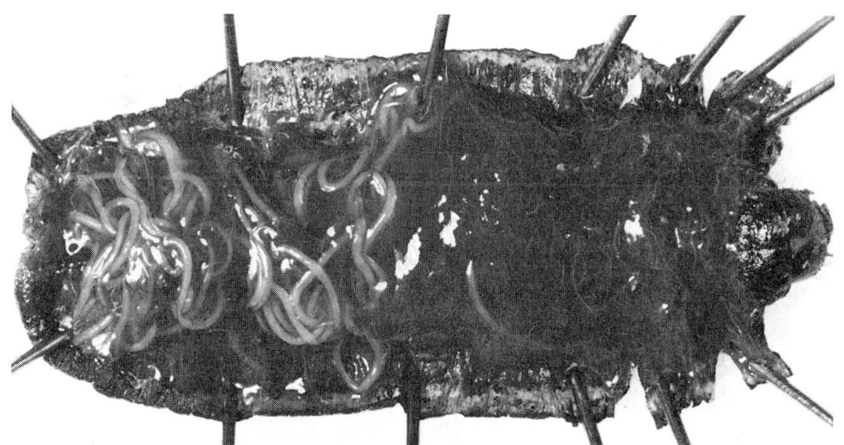

Figure 3.8. Nematodes of the species *Heterorhabditis bacteriophora* inside the greasy cutworm. (Courtesy of W. M. Wouts)

insect. In a test that I ran two years ago, the Colorado potato beetle and the Mexican bean beetle stopped feeding within 36 hours after ingestion of the parasites. After death of the insect host, the nematode feeds on the bacteria in the cadaver and builds up to larger numbers (Fig. 3.7). The Environmental Protection Agency and some research workers, myself included, will be concerned about the widespread field use of this nematode until more research has been done to show that it is not capable of vectoring bacterial pests of man, animals, and agricultural plants through contamination of cultures.

The systematics of the genus *Neoaplectana* is complicated by the wide host range of *Neoaplectana dutkyi* (DD-136), which has over 1,000 host insects. Also, *N. carpocapsae* Weiser, 1955, has been installed as another name for this nematode.

The only known species in the genus *Heterorhabditis* vectors a bacterium and kills the insect host within 48 hours. A photograph showing the *Heterorhabditis* parasites after dissection of the host, the greasy cutworm, is presented as Fig. 3.8.

CONCLUSIONS

In the last few years, researchers have found and studied important new nematode parasites of major pest insects such as mosquitoes, black flies, gypsy moths, Japanese beetles, face flies, bark beetles, Colorado potato beetles, Mexican bean beetles, and corn rootworms. Alpha taxonomy of these nematodes has either been done or is currently underway. A few larger revisions and life cycle studies have been undertaken and appear to be most useful and in demand because they provide information regarding the nematode groups that are effective biological control agents. New genera and species are described each year as more workers enter the field.

The mermithids appear to be the group of insect-parasitic nematodes that show the most promise for field application. A mosquito mermithid preparation is now for sale under the "Skeeter Doom" label. Much additional effort on a worldwide scale is being expended on this nematode and other similar mermithids.

The sphaerulariids debilitate, reduce the egg-laying capacity, or sterilize a wide range of insects that are rarely parasitized by mermithids. The systematics of this family and its genera appears to be set up on a firm base, as changes have been minimal.

The neoaplectanids and heterorhabditids kill many kinds of insects quickly. The ease with which they are mass-reared allows their use as biological control agents in larger-scale field applications. However, they have the stigma of being known vectors of bacteria and this must be researched more conclusively.

LITERATURE CITED

1. Artyukhovsky, A. K. 1953. *Infection of Porthetria dispar caterpillars with the nematode Hexamermis albicans* (Siebold, 1848) Steiner, 1924. Pages 31-33 *in* K. I. Skrjabin 75th Anniv. Comm. Vol. *Contributions to Helminthology.*

2. Artyukhovsky, A. K. 1963. *On the taxonomic characters of the genus Amphidomermis Filipjev, 1934; Melolonthinimermis gen. nov. and Spiculimermis gen. nov. (Mermithidae, Nematoda).* Pages 195-197 *in* K. I. Skrjabin 75th Anniv. Comm. Vol. *Helminths, Human, Animal and Plant, and the Control of Them.*

3. Artyukhovsky, A. K. 1969. *New genera of the family Mermithidae (Enoplida, Nematoda).* Zool. Zh. 48: 1309-1319.

4. Bovien, P. 1937. *Some types of association between nematodes and insects.* Vid. Meddel. fra Dansk Naturh. Forening, Kobenhavn 101: 1-114.

5. Bovien, P. 1944. *Proatractonema sciarae n. g., n. sp., a parasitic nematode from the body cavity of a dipterous larvae.* Vid. Meddel. fra Dansk Naturh. Forening, Kobenhavn 108: 1-14.

6. Christie, J. R. 1936. *Life history of Agamermis decaudata, a nemic parasite of grasshoppers and other insects.* J. Agric. Res. 52: 161-198.

7. Fuchs, G. 1914. *Tylenchus dispar curvidentis m. and Tylenchus dispar cryphali m.* Zool. Anz. 45: 195-207.

8. Fuchs, G. 1938. *Neue Parasiten und Halbparasiten bei Borkenkäfern und einige andere Nematoden.* Zool. Jahrb. (Syst.) 71: 123-190.

9. Goeze, J. A. E. 1782. *Versuch einer Naturgeshichte der Eingeweidewurmen thierischer Korper.* Blankenburg 471 pp.

10. Gould, W. 1747. *An account of English ants.* London. 109 pp.

11. Hagmeier, A. 1912. *Beiträge zur Kenntnis der Mermithiden. 1. Biologische Notizen und systematische Beschreibung einiger älter und neuer Arten.* Zool. Jahrb. (Syst.) 32: 521-612. (Abstr.)

12. Johnson, A. A. 1963. *Octomyomermis itascensis gen. et sp. nov. (Nematoda, Mermithidae), a parasite of Chironomus plumosus (L.).* Trans. Amer. Micro. Soc. 82: 237-241.

13. Nickle, W. R. 1967. *On the classification of the insect parasitic nematodes of the Sphaerulariidae Lubbock, 1861 (Tylenchoidea: Nematoda).* Proc. Helm. Soc. Wash. 34: 72-94.

14. Nickle, W. R. 1972. *A contribution to our knowledge of the Mermithidae (Nematoda).* J. Nematol. 4: 113-146.

15. Poinar, G. O., Jr. 1975. *Entomogenous nematodes.* E. J. Brill, Leiden. 317 pp.

16. Polozhentsev, P. A. 1941. *On the fauna of mermithids from May beetles, Melolontha hippocastani Fabr.* Proc. Bashkirskoi Sci. Res. Vet. Stat. 111: 301-441.

17. Polozhentsev, P.A. 1952. *New Mermithidae of sandy soil of pine forests.* Studies of the Helminthology Laboratory, Academy of Sciences, USSR 6: 376-382.

18. Rubtsov, I. A. 1972. *Aquatic Mermithids (Part I).* Academy of Sciences of the USSR, Zoology Institute, Leningrad. 253 pp.

19. Rubtsov, I. A. 1974. *Aquatic Mermithids (Part 2).* Academy of Sciences of the USSR, Zoology Institute, Leningrad. 222 pp.

20. Rühm, W. 1956. *Die Nematoden der Ipiden.* Parasitolog. Schriftenreihe 6: 1-437. (Fischer, Jena).

21. Steiner, G. 1929. *On a collection of Mermithids from the basin of the Volga River.* Zool. Jahrb. (Syst.) 57: 303-328. (Abstr.)

22. Taylor, A. L. 1935. *A review of the fossil nematodes.* Proc. Helm. Soc. Wash. 2: 47-49.

23. Wachek, F. 1955. *Die entoparasitischen Tylenchiden.* Parasitolog. Schriftenreihe (Fischer, Jena) 3: 1-119.

24. Welch, H. E. 1959. *Taxonomy, life cycle, development, and habits of two new species of Allantonematidae (Nematoda) parasitic in drosophilid flies.* Parasitology 49: 83-103.

25. Welch, H. E. 1960. *Hydromermis churchillensis n. sp. (Nematoda: Mermithidae) a parasite of Aedes communis (DeG.) from Churchill, Manitoba, with observations of its incidence and bionomics.* Can. J. Zool. 38: 465-474.

4] Ecosystematics

by JAMES A. DUKE*

It is . . . taxonomists, not taxa, that make distinctions between the formal disciplines of ecology, evolution, and systematics. [Kruckeberg, 1969]

We are engaged in the construction of a framework on which to hang or arrange the total available biological information, the data about life whether on the molecular or the organismal or the population levels. . . . [Fosberg, 1972]

ABSTRACT

Ecosystematics (here defined as integrated studies of the ecology, evolution, and systematics of ecosystems) is a logical but difficult integration of disciplines because the environment (subject of ecology), acting on the biotic genotypes, selects (subject of evolution) those taxa (subject of systematics) which persevere in a given ecosystem. Narrower definitions of ecosystematics in the literature are not too far removed from broader definitions of biosystematics. Ecosystematics is vitally important to agriculture since it is not isolated taxa that will feed and clothe the world, rather it is the integrated group of taxa in a functional ecosystem, be it a cornfield or a crayfish pond. The Plant Taxonomy Laboratory uses a four-digit code (giving annual precipitation and mean annual temperature) for retrieving names and yields of economic plants in various ecosystems around the world. If crop pests such as fungi, insects, nematodes, and weeds were entered into the same matrix, one could make crop recommendations for specific ecosystems in which the economic plant, by virtue of a broader ecological amplitude, might grow without being exposed to certain pests with narrower ecological amplitudes. The ecosystematic codes are useful in matching germplasm donors with germplasm recipients, in predicting where alien weeds are liable to find hospitable ecological niches in the United States, and in comparing crop amplitudes. Knowledge of the ecosystematic amplitudes of economically important organisms can dictate logical choices in the selection of new introductions.

*Plant Taxonomy Laboratory, Plant Genetics and Germplasm Institute, Beltsville Agricultural Research Center, U.S. Department of Agriculture, Beltsville, Maryland 20705.

DEFINITIONS

When I first used the word ecosystematist (*10*), I had in mind one who performs, or participates in integrated studies of the ecology, evolution, and systematics of an ecosystem. Ecosystematics is a logical but difficult integration of disciplines because the environment (subject of ecology), acting on the biotic genotypes, selects (subject of evolution) those units which systematists and biosystematists describe as taxa.

Several other definitions have been proposed. Lloyd Knutson, convenor of this symposium, first called my attention to the proceedings of "A Symposium on Ecosystematics," (held in 1971 in Fayetteville, Arkansas) edited by Robert T. Allen and Frances C. James (*2*). I failed to find a clearly labelled definition of ecosystematics in those proceedings. The stated objective of that symposium was to convene scientists doing research in which ecological theory and techniques had been applied to problems of systematics. On the premise that "ecology is the study of interrelationships between organisms and their environment," and that "systematics is the study of diversity among organisms," the planners of that symposium hoped to discover to what extent the diversity among organisms is adaptive and an ecosystematic consideration of biological phenomena will lead to a better understanding of the evolutionary process.

I wrote to all speakers and many discussants in "A Symposium on Ecosystematics." Unanimously they concurred that Drs. Allen and James had been responsible for the choice of the word ecosystematics. Dr. James said that she was responsible for the defintion of the term as used in that symposium. "The term," she said, "was used to mean: how the study of variations between organisms (systematics) can be understood in terms of the interactions of animals and plants with their physical and biotic environment (ecology)." (Personal communication from F. C. James, 1977.) In 1975 Allen stated that "a new term 'ecosystematics,' was coined by Allen and James (1972) to signify the union of some areas of study encompassed by ecology and systematics but studies not intended to be as broad as biosystematic studies." Allen (1) contended that ecosystematics is narrower than biosystematics.

According to Davis and Heywood (*8*), "Biosystematics, in its broadest sense, is that part of systematics concerned with the variation and evolution of species. It is often more concerned with the process of evolution than it is with the classification itself." As a botanist, I accept this definition. During the planning of this symposium, it became obvious that definitions of biosystematics varied widely. According to Seigler (*36*), biosystematics attempts to study as much as possible of the biology of living populations of organisms and to utilize these studies to clarify the taxonomic and phylogenetic relationships of the taxa involved. Seigler (*36*) said: "The phylogeny of the taxon is its evolutionary history." To me, a biosystematist is one who studies a group of related organisms from an ecological and taxonomic point of view, examining the phylogenetic relationships between them. When ecosystem, ecology, and evolution are tied to systematics, it becomes my *ecosystematics.* To me ecosystematics is the study of the components of the ecosystem (*systematics*), their functions (*ecology*), and origins (*evolution*).

When I was a student under Prof. A. E. Radford at the University of North Carolina at Chapel Hill, we studied and cataloged the plant species in various North Carolina ecosystems such as the sandhills (*9*), the maritime forests, and the grassy balds. This is an interesting but limited ecosystematic approach. It becomes more intricate when the functional relationships among taxa of plants and animals within the ecosystem are studied, as in the sea level canal surveys in Panama and Colombia (*11, 12, 13, 17*). Dozens of scientists of many disciplines were trying to determine the roles of all major taxa, animal, plant, and, egocentrically, man, in various ecosystems of eastern Panama and adjacent Colombia. Apparently there was some parallelism of thought between Prof. Radford as he studied the flora of the ecosystems of the Carolinas and his first student (myself) studying the flora and fauna of the ecosystems of isthmian America.

In 1972, Radford (*34*) defined ecosystematics as one of three fields of systematics; ecosystematics was that field "dealing with floristics and plant communities, primarily observational and descriptive, and mostly treating the species, genus, and family taxa." His most recent definition of plant ecosystematics, scheduled to appear in the 1977 University of North Carolina catalog, is: "A study of definition and classification of elements of the ecosystem for ecological characterization, classification, nomenclature, and identification of plant species, communities, and habitats." (Personal communication from A. E. Radford, 1977.)

Radford's plant-oriented definition, while more detailed, parallels Allen's summary (*1*): "The term *ecosystematics* is used to designate a multipurpose approach to the study of certain questions in insect systematics and ecology." I propose that we subtract the insect from Allen's definition, the plant from Radford's, and bring in all bios. Why should we not add evolutionary interpretations, since species and ecosystems are good examples of coevolution? That is my formula for arriving at a broader definition of ecosystematics, the ecology, evolution, and systematics of ecosystems. As Kruckeberg (*26*) noted, ecology, systematics, and evolution have usually been practiced as separate disciplines, each with its own cadre of practitioners.

The materials of the systematist are products of interactions between organisms and their biotic and abiotic environments. The moment the systematist goes beyond the classification of organic diversity to seek causal relationships, he is drawn into ecological-evolutionary questions. Evolution occurs through the interplay of heredity and environment. Therefore the objects of systematic studies (individuals, populations, taxa, etc.) are shaped by their past and present ecologies. Genecology constitutes to date the greatest point of contact between ecology and plant systematics. Ecotypic differentiation adds a complicating but essential component for evaluating community composition. Local and regional climate, soils, other organisms—all can produce ecotypic responses and must inevitably do so in one and the same population (*26*).

I reiterate Dansereau's plea of 25 years ago (*6*) that the description of a given taxon include a measure of its ecological adaptation as well as its morphology and its genetics. Such inclusion would contribute much to our understanding of the species. Still agreeing with Dansereau, I urge all collectors to report, where possible, the long-term record of annual precipitation (dm) and annual mean temperature (°C) as the climatic half of an ecosystematic code for their collecting site. A unified soil classification, numerical or otherwise, should be used as the edaphic half of the code.

The ecologist should no longer stop short at the study of natural selection, the former domain of the evolutionist; the ecologist should no longer hesitate to look systematically and comparatively at environmental tolerance limits within species, genera, or families, former domain of the systematist. Teams of broad-minded ecosystematists should ignore the boundaries or walls between disciplines, answer the challenging questions at the interfaces, synthesize and give us the truly twentieth-century Natural History of Kruckeberg (*26*).

In my most detailed ecosystematic study (*13*), I noted how the paucity of systematic information on Panama and the difficulty of identifying tropical forest dominants caused many ecologists to consider the forest from a physiognomic and geographic, rather than a taxonomic, point of view. Even so, one ultimate aim of ecological studies is ecosystematic, to specify the function of each taxon in the overall processes of the ecosystem. Without the all-important name, how do we refer to the taxon?

Perhaps more than others, we in agriculture should strive toward an ecosystematic approach, viewing predator and prey, pollen producer and pollinator, host and parasite, commensals, symbionts, etc., as integral parts of that bigger whole, the agricultural ecosystem, rather than as isolated morphological descriptions or museum specimens. Personally I believe that the word *systematist* has a better connotation to the non-taxonomic scientist in agriculture than the word *taxonomist*. The isolated *taxon* will not feed the world; the world is fed by integrated groups of taxa in functional ecosystems, be they cultivated corn fields, catfish ponds, crab cays, or codfisheries. These ecosystems, functionally inviolate, will feed today's world.

Heywood (*22*) expressed his surprise that conservation had been regarded chiefly as the concern of the ecologist while the taxonomist's role has remained quite vague in our environmental crises. "Taxonomists in fact occupy a key position in relation to our knowledge of the environment and this is only somewhat belatedly being recognized."

One major goal of practical taxonomy is to prepare classifications to serve a range of different interests, including those major users of classifications, the ecologists. Automated data processing (ADP) may eventually provide alternative, more flexible means of classification. Meanwhile, ecologists and taxonomists should engage in dialogues so that those in each discipline will come to understand the reference terms, working concepts, and requirements of the other. This is particularly important because both use primarily the same data and raw material and even employ the same terms, for example, population, species, sampling, etc., although with different meanings (*22*). Guinochet (*18*) stated (though I disagree; others agree) that phytosociology is not a branch of ecology, but is essentially concerned with floristic lists, with ecology invoked only to interpret phytosociological phenomena. I prefer Daubenmire's conclusions (*7*): "(1) Vegetation reflects the sum of all the elements of the environment which are important to plants. (2) The species with highest competitive powers are the best indicators . . . (5) Many characters of vegetation have potential significance as ecologic indicators. (6) Types of environment ('habitat' types) are the most basic ecologic units of landscapes." I agree with Guinochet (*18*) that inventories (from a given region) listing different species may indicate the existence of different conditions. Conversely two similar lists often indicate similar conditions. Such thinking led to a computerized approach to determine ecological parameters of remote areas by analyses of floristic checklists.

When a species occurs in several different associations, different genotypes may be involved. Comparing the distribution and floristic composition of phytosociological units confirms the link between the origin and evolution of taxa and of plant associations (*18*). Species have not evolved in a vacuum; they have evolved ecosystematically in concert with other evolving species of different phyla and kingdoms, each having some influence on the other.

MacArthur and Wilson (*32*) were unable to see any real distinction between biogeography and ecology. According to Snaydon (*38*), there is interdependence between taxonomy and ecology, and a need for cooperation. Do taxonomic groupings, which are mostly defined morphologically, reflect ecological similarities, which are mostly determined physiologically? The emphasis on environmentally stable characters in plant taxonomy has tended, although not deliberately, to exclude ecologically important characters. The resulting classifications therefore have a smaller content of ecological information and less predictive value in ecology than might be the case if plastic, but more functional, characters were considered. Should the aim of taxonomy be to classify abstract genotypes or concrete phenotypes? According to Davis and Heywood (*7*) *phyletic* emphasizes "belonging to a particular line of descent (phyletic lines)," while *phenetic* reflects the "overall similarity as it is perceived through the senses. Phenetic similarity is assessed without reference to ancestors or a time dimension." I sense a strong parallel between the words *genotypic* and *phyletic* (or perhaps *phylogenetic*) on the one hand and *phenotypic* and *phenetic* (and perhaps *artificial*) on the other. Contrasted to *phyletic* similarities, *phenetic* is a horizontal concept, a look only at the tips of the remaining branches of the phyletic tree.

Agriculture, medicine, and all phases of the human economy ultimately depend on natural living organisms of which no real understanding can be had without a basic inventory and knowledge of spatial and environmental relationships (*37*). Man gets all his carbohydrates and nearly 75 percent of his protein from plant sources. Of course, nearly all the food we get from animals is in turn derived from plants (*21*). Hence man is dependent on the study of ecosystematics for his very survival.

There is need for the word, and for the study of, ecosystematics. After reading the introduction to *Annual Review of Ecology and Systematics* (*24*) and Heywood's *Taxonomy and Ecology* (*22*), I sympathize with the coining of the word "ecosystematics" to give substance to the long overdue marriage of disciplines.

> Some ecologists and systematists will wonder at the combination of ecology and systematics . . . others will recognize the unity of the disciplines but wish that some unit term had been used for the title. . . . Because ecologic work proceeds from a systematic baseline and any informed systematic work is also ecologic, we think we deal with a viable concept. (*24*)

EVOLUTION

Lewis (*31*) hinted that speciation by chromosomal reorganization may be induced by environmental extremes: "Ecological differentiation functions as a barrier to gene exchange when hybrids or their derivatives are unlikely to become established. . . . Whenever intermediate conditions occur such that hybrid derivatives are as well adapted as either parental type, the lines between taxa become indistinct. Unfortunately for the taxonomist but not for the plants, this is a very frequent situation. . . . Ecological differentiation has special significance as a barrier to gene exchange because it is the essence of most evolutionary change and to a far greater extent than spatial separation may set the stage for the development of reproductive barriers."

Tropical alliances are adapted to warm frostless climates and they probably have not greatly changed their climatic requirements during time because climatic tolerances of larger taxa are genetically controlled, just as are their morphologic features (*5*). Kruckeberg (*26*) (1969) maintains that all characters or character complexes used by systematists somehow result from selection by prevailing ecological conditions.

Hybridization, more frequent in hybrid habitats, more frequent in disturbed "telescoped" montane ecosystems, has a mutagenic effect in itself. The late Edgar Anderson (*3, 4*) was the eloquent spokesman for the importance of introgression in evolution. Introgression is "the process by which the genes of one taxon are mixed with the genes of another by hybridization of the two taxa followed by backcrossing of the hybrid plant with either of the two parents. Even when hybrids are not significant in relative numbers, they can allow gene flow and mixing, producing increased variability of the two parental types." (*36*) Dr. Anderson, not a strong advocate of mutation as an evolutionary force, concluded (*3*): "the natural variability of wild populations in North America must be almost exclusively the *immediate* result of introgressions . . . rather than of mutation. Mutation either must be the primary or secondary result of introgression, or else it must take place at such a slow rate as to be comparatively unimportant either for theories of population dynamics or for programs of plant and animal improvement." Anderson (*4*) added that it "not only supplies the heightened variability which is the raw material for selection to work upon; the variables which it introduces into the breeding population *have already been selected to work well together,* thus greatly enhancing the possibility that something practicable might be evolved from the complex."

Lewis (*30*) expounded on catastrophic selection in relation to speciation as follows:

1) During wetter years, populations of mesic species would disperse beyond the normal range of the species into outlier situations.
2) Xeric-tolerant mutant alleles, and especially chromosomal rearrangements, might occur anywhere in the population and might on occasion arise in an outlier population.
3) Changes to the xeric part of the climatic cycle would annihilate the outlier colonies on the edge of the range unless they contained some individuals bearing xeric-adapted genetic constituents.
4) These latter favored individuals would survive and produce a population in a xeric environment that had a new adaptive relationship.

Further, Lewis marshalled evidence to show that this had happened more than once in *Clarkia.* Thus there can be production of new species associated with the occupation of, and restriction to, a new ecological zone. First shifts in ecological adaptation might involve only physiology or behavior in animals that would not show up morphologically in the fossil record (*35*).

Dealing with insects, Ross (*35*) outlined how frost tolerance might evolve:

1) In a tropical species occurring adjacent to the temperate zone, outlier populations become established seasonally in areas that are temperate.
2) In such outlier populations, mutations occur, conferring a certain amount of frost tolerance to individuals bearing them.
3) Freezing in the temperate outlier kills off the individuals devoid of frost tolerance.

If these insects were associated with plant hosts, might not similar evolution take place in the plants? If the insects were pests on a crop species, and that crop species evolved the frost tolerance while the insect pest did not, would we not be one step ahead? Evasion could be a battle stratagem in biological control. We hope to develop ecological amplitude tables for crop pests, using the same parameters we have used for crops (*15*) and weeds (*14*). Such ecological amplitude data, entered into a computer, could be retrievable to determine where ecological amplitudes of a crop exceed those of a serious pest. We could then make cropping recommendations accordingly.

Species that occupy two or more ecological zones are in a position to give rise to a species restricted to a derived zone. Snaydon (*38*) noted that morphological and physiological differences between populations are frequently associated with environmental differences and appear to be adaptive. Ecological factors may play a paramount role in species divergence, perhaps more important than reproductive isolation.

If a species occupies ecologically divergent regions and these regions contain different arrays of parasites and predators, we might expect the species to diverge in defensive strategy to adjust to regional differences in selective forces. Similarly, if groups of species occupy ecologically divergent regions, we might expect species sharing a common region to have similar defensive compounds among themselves, but to differ from species in another region. Intra- and interspecific differences in phenolic profile sometimes correlate with geography and ecology (*29*). If different congeneric plant species growing in one habitat solve their defense problems in a similar fashion, and if the nature of the pest pressure differs among habitats, one might expect some ecological associates from different families to have similar defensive compounds, that is, convergent defensive evolution. Levin (*29*) raised the question: If convergence occurs for attraction (red-flowered hummingbird attractors, for example), may it not also occur for repulsion?

The important role of ecological factors in species divergence may be underrated. It is easy to forget that reproductive isolation simply maintains, at any one time, the status quo; some other mechanism must be invoked to account for the divergence between species. Occasionally, differences between populations in response to soil and climatic factors are as great as those between species that are widely diverse ecologically, yet these populations usually are fully interfertile. Many related allopatric, even sympatric, species are reproductively compatible, and may produce ecologically successful offspring, provided there are suitable environmental conditions. Such hybridization has been responsible for the evolution of many plant species.

The capacity of phenotypes to respond to environmental influences is itself under genetic control (*38*), suggesting that there is no clear-cut distinction between genetic and environmental control of the phenotypes we classify. Environmental modifications are frequently adaptive and of evolutionary and ecological importance. Phenotypic plasticity and genetic differentiation are alternative strategies in adaptation.

Coevolution. According to Leppik (*28*), "It is no wonder that the highest evolutionary levels of flowering plants with stereomorphic and zygomorphic flower types are rich in carbohydrates, proteins, volatile oils, vitamins, seed oils, and other nutritional substances for man and higher animals. . . . Hence a well correlated coevolution between different groups of organisms and their environments takes place in which man is an important factor."

Sneath (*39*), noting evidence that vegetative characters tend to vary independently of other characters, questioned which should be used in constructing classification. Clearly the flowers may be evolving under selection of anthophilous insects and other biotic factors whereas the vegetative evolution may respond relatively more to abiotic factors. Despite reticulate evolution in flowering plants, I suspect that there will be better correlations between ecosystematic codes and vegetative characters than floral characters. In Australia, to the surprise of many, xeromorphy and sclerophylly were shown to be due to low phosphate, not aridity. Kruckeberg (*26*) suggested that biotic interactions, both vegetative and reproductive, may underlie major evolutionary events in higher organisms, namely insects, birds, herbivorous mammals, and angiosperms. An examination of pollination ecology reveals at least three reproductive barriers: seasonal differences, spatial limitations (within or between habitats), and/or pollen vector specificities. Pollination conditions such synecological phenomena as species diversity, pattern, dominance, and successional (seral) and climax community composition (*26*).

Symbio-taxonomy. One case of coevolution of particular interest to agriculture is that of legumes and their nitrogen-fixing symbionts. Studies of the legume-*Rhizobium* symbiosis were plagued by the fact that historically they were made by specialists interested in only one aspect of the subject. They were begun by agricultural chemists, followed by bacteriologists, followed by agronomists. The fact that a *symbiosis* was being studied received scant attention. Taxonomic botanists then, as now, tended to ignore the presence of nodules on the roots of the legumes; it occurred to no one that the evolutionary history of the legume might have significance (*33*). Norris uses the term "symbio-taxonomy" for this phenomenon. As species diverge, the ability to symbiose effectively with each other's *Rhizobium* is lost before the ability to form nodules. Ecosystematically, there is a better chance that the rainforest legume of acid soils will have the "cowpea-type" *Rhizobium* than some of the derived species. Derived legume species might fail in such an environment, because of calcium deficiency effects on the legume, not the *Rhizobium*. An acid-producing *Rhizobium* strain will respond better to lime application on an acid soil than will an alkali-producing *Rhizobium* (*33*). A clover inoculated with *Rhizobium* showing minimal acid production will do better than one inoculated with an alkali-producing *Rhizobium*. The taxonomic position of an unstudied legume is all important in predicting which *Rhizobium* will do best. In pelleting legume seeds, the seeds are inoculated, pelleted in gum and coated with lime or superphosphate. Those species with acid-producing *Rhizobium* should be pelleted in lime, for example, *Lathyrus, Lens, Leucaena, Medicago, Melilotus, Pisum, Trifolium, Trigonella, Vicia, Phaseolus coccineus* L., and *Phaseolus vulgaris* L. Those species with alkali-producing *Rhizobium* should be pelleted with rock phosphate, for example, *Arachis, Cajanus, Calopogonium, Canavalia, Cassia, Centrosema, Clitoria, Crotalaria, Cyamopsis,* and other species of *Phaseolus*. The plant dominates the symbiosis; its ability to accept or reject the bacterium is hereditary.

In nature a non-nodulating or shy-nodulating habit is recessive because it functions virtually as a lethal factor. There is an hereditary shy-nodulating character in unselected strains of *Centrosema pubescens* Benth. *Glycine wightii* (R. Grah. ex Wight & Arn.) Verdc. is a poor and tardy nodulator. The uncautious geneticist may derive a strain with fine agronomic features unable to fix sufficient nitrogen (*33*). We have talked largely about the evolution and coevolution of species. As species coevolve with other species, do they not coevolve with the ecosystem containing them? "Man and his domesticates have been bound together for some millenniums in an adaptive coevolution." (*19*)

ECOSYSTEM

Recently the Plant Taxonomy Laboratory at Beltsville developed a broad-based numerical system for codifying certain ecological attributes of economically important angiosperms and for plotting their distribution in various ecosystems. I have two reasons for using numbers instead of words, for example, 2525 instead of Tropical Moist Forest Life Zone. The numbers are more precise in definition and economical of space, and the numbers are more readily assimilated by computers than the words. Sneath (*39*) discussed organisms rather than ecosystems when he said: "Numerical methods can yield acceptable classifications, and they much increase taxonomic understanding of the groups that are examined. . . . Some way is now needed to integrate voluminous new data now being accumulated from all sources into the body of taxonomy. Numerical methods offer the only satisfactory way of doing this." (*39*)

The numbers we gather in our Laboratory will be useful in modeling certain agricultural ecosystems. We hope to superpose yield and input matrices onto the ecological data already in our computer. My ignorance prevents me from discussing the taxonomy of ecosystem models. "Simulation modeling of ecosystems is a young subdiscipline even when compared to the relatively recent origins of ecology itself." (*41*) We are gradually assigning numerical values to species in ecosystems and the attributes of the ecosystems in which the species occur. These ecosystematic values, like chemosystematic values, will enable us to tell more about evolution of the species and the ecosystems containing them. One of our goals is to determine from a floral checklist the ecosystematic code of a remote area, then make crop recommendations based on our knowledge of the crop's performance in other areas with similar ecosystematic codes.

In a paper being processed now by the American Society of Agronomy (unpublished data) I elaborate on the use of ecosystematic and ecosystematic amplitude codes. The ecosystematic code, which I urge be included on specimen labels, is a four-digit number, the first two digits representing the long-term figures for annual precipitation in decimeters and the second two digits, the annual mean temperature in degrees Celsius. A four-letter code for soils is used as a suffix to the four-letter climatic code. The ecosystematic amplitude code for a species, on the other hand, is an 8-digit number, the first two digits representing the minimum annual rainfall, the second two the maximum annual rainfall (long-term averages), the third two the minimum, and the fourth two the maximum annual temperature the species is reported to tolerate in our data base. The ecosystematic amplitude code of the European Gooseberry (*Ribes uva-crispa* L.) is 05170511, indicating that it is reported from regions with annual rainfall ranging 05 to 17 dm, annual temperature, from 05 to 11°C. The code for *Zoysia japonica* Steud. is 07410927. If our data base represented the true world rather than our small sample of the world, the coexistence of *Ribes uva-crispa* and *Zoysia japonica* in some remote area would indicate that the annual mean temperatue of that area was between 9 and 11°C. Dr. T. C. Davidson is programming our computer so that eventually we can deduce the climate of remote areas by an analysis of the ecosystematic amplitudes of the local flora. Dr. A. A. Atchley and I are trying to locate lists of economic plants associated with climatic and soil data so that we can expand our data base on economic plants and weeds to the point at which we can ecosystematically determine the climate, perhaps even the soil types, of remote areas. Dr. E. E. Terrell has shouldered the huge responsibility of looking into the nomenclature of the economic plants, and Dr. C. R. Gunn is developing a system for the computerized identification of legume seeds (see Chapter 14, this volume). If all goes well, our laboratory should be able to deduce the climate of some remote area fairly accurately from study of seeds of several species collected there.

SYSTEMATICS

Most ecologists now recognize the ecological importance of ecotypic diversity within species, thanks to genecologists, but these "ecotypes" still lack taxonomic description, nomenclature or classification. There is wide variation within samples, high genetic variability, susceptibility to environmental modification, and lack of correlation with conventional taxonomic markers, making them "bad" taxonomic characters.

Much such variation is physiological or biochemical, with or without correlated morphological expression. Variables correlated with morphological factors might be considered "good" taxonomic characters, uncorrelated, "bad." Would morphological characters, without physiological or biochemical correlates, then be "bad" ecological or chemical characters, respectively? It is the correlation among characters from different subsystems of ecosystematics that allows predictability.

Many problems faced by taxonomists describing and classifying infraspecific variation may be similar to those faced by ecologists describing and classifying vegetation. Careful unbiased sampling by taxonomists at obscure specific boundaries, like careful sampling by ecologists, may show continuous variation rather than clear-cut entities. Using a few floral characters in plant taxonomy is rather like using dominant species in vegetative classification: both practices may suggest that units are more discrete than if more characters were used. In both disciplines the "characters" may be expressed qualitatively or quantitatively, with the choice greatly affecting the resulting classification.

Snaydon (*38*) viewed the problems of describing and classifying infraspecific variation as one of the most challenging and rewarding areas of interrelationships between taxonomy and ecology, requiring new techniques in biometry. An ecologist, faced with the paradox of requiring both stable nomenclature and changes in the ground rules of taxonomy, Snaydon hints at separating the two classical subdivisions of taxonomy, nomenclature and classification.

Species diversity. Janzen (*23*) reiterated his hypothesis that the high diversity of lowland tropical forest is largely maintained by the herbivore community. Still, he provisionally explained higher diversity of herbivorous insects (and associated parasites) at intermediate elevations (900 to 1,300 m) as resulting from more photosynthate due to lower metabolism of plants during cool montane nights. I attribute it also to the increased diversity in habitat, as I expect more species in a cove, slope, and ridge than in any one ecosystem. I expect more species between 900 and 1,300 m than I do at 1,300 m or at 900 m or at a monotonous sea level, in a given unit area in the tropics. Montane evolution is facilitated by the multitude of coexisting ecological niches and by patchy distribution, leading to genetic drift and adaptive radiation (*20*). "Genetic variability should be greater in heterogeneous environments." (*38*)

Chemotaxonomy. We should not divorce ecological characters from our description of taxa; neither should we divorce the chemical characters. The slug selecting a low-nicotine tobacco seedling, the beetle a low-saponin alfalfa, the greenbug a low-phenolic barley, the bollworm a low-gossypetin cotton seedling, the bark beetle a low-juglone hickory, the vole a low-cyanogenic lotus, like the harvester selecting the seed of a flax-mimicking weed, etc.—all contribute to selective ecosystematic pressures resulting in one ecotype being favored at the expense of another. Substantial evidence indicates that secondary substances serve as passive defense mechanisms in plants. Still, many of these same secondary plant substances are used by parasitic and grazing animals as the labels by which they recognize their food (*25*). Exocrine substances excreted by plants may exert allelopathic pressures favoring one taxon at the expense of another, again contributing ecosystematic selectional forces to determine what succeeds in a given ecosystem.

Onion varieties resistant to the fungus *Colletotrichum circinans* Curzi accumulate flavones, anthocyanins, and simple phenolics (all water soluble) which diffuse to infected spots and inhibit fungal germination and penetration. Pea varieties, resistant to root diseases, contain more phenolics in their seed coats than do susceptible varieties. Phenolics in the waxy coverings on apple leaves inhibit the fungus *Podosphaera leucotricha* (Ell. & Ev.) Salmon. High chlorogenic acid contents lower *Verticillium* infection, and differential solubilities play a role in resistance to *Cronartium ribicola* (J. C. Fisch) (*29*).

Phytoalexins are specific antifungal substances of relatively low molecular weight produced in living cells of some plants. They appear in appreciable amounts after inoculation or infection with any one of a wide, nonspecific range of fungi, but in healthy plants the compounds are at trace or low levels. The phytoalexin defense reaction is confined to the attack zone of the fungus, and is toxic to many fungi, and nonspecific in fungicidal spectrum. Some reported phytoalexins are capsidol, chlorogenic acid, hircinol, ipomeamarone, loroglossol, lubimin, medicarpin, orcinol, phaseollin, phytuberin, pisatin, rishitin, scopoletin, and scopolin.

Citrus root nematodes are more damaging to susceptible citrus rootstocks, which experience a reduction in phenolics following infection. Tolerant rootstocks exhibit an increase in phenolics (*29*).

Plant phenolics interest plant physiologists, biochemists, pathologists, ecologists, geneticists, and systematists. Hydroxylated or methylated derivatives may offer much of utility in disease resistance and for ecosystematic studies (*27*). Some methylation steps are under single gene control.

I suspect that few chemicals have a neutral value, that is, are of no adaptive significance in prey-predator and in allelopathic struggles for survival. Recent work in plant stress studies indicates the role of some "neutral" chemicals in favoring or disfavoring the ecotype that possesses them in competing with other ecotypes under abiotic stress. Isolated mitochondria from populations of diverse habitats may help explain different respiratory rates in different populations. Variables in cytochrome, enzyme, and DNA systems will no doubt show some ecophysiological significance in adapting or preadapting certain ecotypes to certain ecosystems. To the agricultural ecosystematist, chemosystematics is important in defining taxa tolerant and nontolerant to various biotic and abiotic stresses, stresses annually costing billions of dollars. As they provide drugs, flavors, and perfumes, secondary plant metabolites are also responsible for the flavor, taste, and smell of most of our plant foods.

"Similarities in the chemistry of plant taxa . . . may reflect an evolutionary or phyletic similarity, but may also be the result of convergent evolutionary processes." (*36*) Since sequences of amino acids in genera of the same family are more similar to each other than to those in unrelated families, phylogeny permits the prediction of amino acid sequences in unstudied species. Turner (*40*) concluded that micromolecular data (essentially monomeric compounds, for example, flavonoids, alkaloids, terpenoids, amino acids, etc.) will prove most useful in resolving taxonomic problems at the generic level and lower, while macromolecular data (polymeric compounds such as RNA, DNA, and proteins) will prove most useful in resolving phyletic problems at the generic level and higher. Resolution of these pure taxonomic problems will enable the agricultural ecosystematist better to predict the behavior of untested taxa, based on the assumption that closely related taxa are more likely to share attributes than are distantly related taxa.

CONCLUSIONS

Work in our laboratory will better enable us to 1) map the agricultural (actual and potential) ecosystems of the world, giving good estimates of potential yields of usual and unusual crops in years with usual weather and unusual weather; 2) make cropping recommendations so that certain economic plants can climatically evade certain economic pests; 3) predict the amount of biomass and potential energy available in various agroecosystems; 4) make crop recommendations based on analyses of local floras in remote areas; 5) deduce the climate of remote areas; and 6) predict where in the U.S. and elsewhere exotic crops and pests, even endangered species, are most likely to gain an ecological foothold. We are approaching these goals, whether we call our methods biosystematic, ecosystematic, or systematic.

ACKNOWLEDGMENTS

The continuing support of Dr. Quentin Jones, Staff Scientist, National Program Staff, Plant and Entomological Sciences, and Dr. John G. Moseman, Chairman, Plant Genetics and Germplasm Institute at the Beltsville Agricultural Research Center of the U.S. Department of Agriculture, is gratefully acknowledged.

LITERATURE CITED

1. Allen, R. T. 1975. *Ecosystematic entomology*. Bull. Entomol. Soc. Amer. 21: 13–17.

2. Allen, R. T., and F. C. James. 1972. *A Symposium on Ecosystematics*. Univ. Arkansas Museum, Occasional Paper No. 4. 235 pp.

3. Anderson, E. A. 1956. *Character association analysis as a tool for the plant breeder*. Pages 123–140 in *Genetics in Plant Breeding*. Brookhaven Symposia in Biol. No. 9. Brookhaven National Lab. 396(C-23), Upton, New York.

4. Anderson, E. A. 1961. *The analysis of variation in cultivated plants with special reference to introgression*. Euphytica 10: 79–86.

5. Axelrod, D. I. 1972. *Ocean-floor spreading in relation to ecosystematic problems.* Pages 15-72 *in* R. T. Allen and F. C. James, eds. *A Symposium on Ecosystematics.* Univ. Arkansas Museum, Occasional Paper No. 4. 235 pp.

6. Dansereau, P. 1952. *The varieties of evolutionary opportunity.* Rev. Can. Biol. 11: 305-388.

7. Daubenmire, R. 1976. *The use of vegetation in assessing the productivity of forest lands.* Bot. Rev. 42: 115-143.

8. Davis, P. H., and V. H. Heywood. 1963. *Principles of angiosperm taxonomy.* Van Nostrand, New York. 556 pp.

9. Duke, J. A. 1961. *The psammophytes of the Carolina fall-line sandhills.* J. Elisha Mitchell Sci. Soc. 77: 3-25.

10. Duke, J. A. 1970. *Ecosystematist.* BioScience 20: 580.

11. Duke, J. A. 1970. *Ethnobotanical observations on the Choco Indians.* Econ. Bot. 24: 344-366.

12. Duke, J. A. 1975. *Ethnobotanical observations on the Cuna Indians.* Econ. Bot. 29: 278-293.

13. Duke, J. A. 1975. *Plant species in the forests of Darien, Panama.* Pages 189-229 *in* F. B. Golley, J. T. McGinnis, R. G. Clements, G. I. Child, and M. J. Duever, eds. *Mineral cycling in a tropical moist forest ecosystem.* University of Georgia Press, Athens.

14. Duke, J. A. 1976. *Perennial weeds as indicators of annual climatic parameters.* Agric. Meteorol. 16: 291-294.

15. Duke, J. A., S. J. Hurst, and E. E. Terrell. 1975. *Ecological distribution of 1000 economic plants.* Información al Día Alerta. IICA-Trópicos. Agronomía No. 1. Turrialba, Costa Rica, 32 pp.

16. Fosberg, F. R. 1972. *The value of systematics in the environmental crisis.* Taxon 21: 631-634.

17. Golley, F. B., J. T. McGinnis, R. G. Clements, G. I. Child, and M. J. Duever. 1975. *Mineral cycling in a tropical moist forest ecosystem.* University of Georgia Press, Athens. 248 pp.

18. Guinochet, M. 1973. *Phytosociologie et systematique.* Pages 121-140 *in* V. H. Heywood, ed. *Taxonomy and ecology.* Academic Press, London.

19. Harlan, J. R. 1976. *The plants and animals that nourish man.* Scientific American 235: 88-97.

20. Hedberg, O. 1973. *Adaptive evolution in a tropical-alpine environment.* Pages 71-92 *in* V. H. Heywood, ed. *Taxonomy and ecology.* Academic Press, London.

21. Heiser, C. B., Jr. 1973. *Seed to civilization.* W. H. Freeman, San Francisco. 243 pp.

22. Heywood V. H. (ed.) 1973. *Taxonomy and ecology.* Academic Press, London, New York. 370 pp.

23. Janzen, D. H. 1973. *Comments on host-specificity of tropical herbivores and its relevance to species richness.* Pages 201-211 *in* V. H. Heywood, ed. *Taxonomy and ecology.* Academic Press, London.

24. Johnston, R. F., P. W. Frank, and C. D. Michener (eds) 1970. Ann. Rev. Ecol. & System. 1. Annual Review Inc., Palo Alto. 406 pp.

25. Jones, D. A. 1973. *Co-evolution and cyanogenesis.* Pages 213-242 *in* V. H. Heywood, ed. *Taxonomy and ecology.* Academic Press, London.

26. Kruckeberg, A. R. 1969. *The implications of ecology for plant systematics.* Taxon 18: 92-120.

27. Lane, F. E. 1972. *Discussion.* Pages 191-193 *in* R. T. Allen and F. C. James, eds. *A Symposium on Ecosystematics.* Univ. Arkansas Museum, Occasional Paper No. 4. 235 pp.

28. Leppik, E. E. 1974. *Phylogeny, hologeny and coenogeny, basic concepts of environmental biology.* Acta Biotheoretica 23: 170-193.

29. Levin, D. A. 1972. *The role of phenolics in plant defense.* Pages 165-185 *in* R. T. Allen and F. C. James, eds. *A Symposium on Ecosystematics.* Univ. Arkansas Museum, Occasional Paper No. 4. 235 pp.

30. Lewis, H. 1962. *Catastrophic selection as a factor in speciation.* Evolution 16: 257-271.

31. Lewis, H. 1969. *Speciation.* Taxon 18: 21–35.
32. MacArthur, R. H., and E. O. Wilson. 1967. *The theory of island biogeography.* Princeton Univ. Press, Princeton. 203 pp.
33. Norris, D. O. 1967. *The intelligent use of inoculants and lime pelleting for tropical legumes.* Trop. Grasslands 1: 107–121.
34. Radford, A. E., W. C. Dickison, and C. R. Bell. 1972. *Vascular plant systematics.* Univ. North Carolina, Student Stores, Chapel Hill. 542 pp.
35. Ross, H. H. 1972. *An uncertainty principle in ecological evolution.* Pages 133–157 *in* R. T. Allen and F. C. James, eds. *A Symposium on Ecosystematics.* Univ. Arkansas Museum, Occasional Paper No. 4. 235 pp.
36. Seigler, D. S. 1974. *Chemists and taxonomy.* Chemistry in Britain 10: 339–342.
37. Smith, A. C. 1969. *Systematics and appreciation of reality.* Taxon 18: 5–13.
38. Snaydon, R. W. 1973. *Ecological factors, genetic variation and speciation in plants.* Pages 1–29 *in* V. H. Heywood, ed. *Taxonomy and ecology.* Academic Press, London.
39. Sneath, P. H. A. 1969. *Recent trends in numerical taxonomy.* Taxon 18: 14–19.
40. Turner, B. L. 1969. *Chemosystematics: recent developments.* Taxon 18: 134–151.
41. Wiegert, R. G. 1975. *Simulation models of ecosystems.* Ann. Rev. Ecol. System. 6: 311–338.

three

RECENT ADVANCES IN BIOSYSTEMATICS

5] Recent Advances in Classification of Protozoa

by NORMAN D. LEVINE*

ABSTRACT

The first protozoa were described by Antony van Leeuwenhoek in 1674. At present, over 64,000 species, of which over half are fossil and about 10,000 are parasitic, have been named. Most parasitic species are not known to be pathogenic. At the turn of the century the phylum Protozoa was divided into two subphyla, Plasmodroma (containing the classes Mastigophora, Sarcodina, and Sporozoa) and Ciliophora. This classification was based primarily on organs of locomotion. In 1964 the Society of Protozoologists introduced a new but fairly similar classification. It is now working on an improved one. The new classification will probably raise the Protozoa to kingdom or subkingdom rank and recognize possibly six phyla: Sarcomastigophora (with flagella, pseudopods, or both), Apicomplexa (with an apical complex of characteristic organelles), Microspora (with unicellular spores having a hollow polar filament), Myxospora (with multicellular spores having a solid polar filament), Haplospora (with spores without a polar filament), and Ciliophora (with cilia and ordinarily having two kinds of nuclei). *Toxoplasma* and *Sarcocystis* have been found only recently to be coccidia. *Toxoplasma* has asexual multiplication in over 200 species of mammals and birds, but sexual multiplication and oocysts only in felids. *Sarcocystis* has asexual stages in prey animals and cysts only in the feces of predators. It is highly host-specific; there are ox-dog, ox-man, sheep-dog, sheep-cat, etc., species. At present 13 species of coccidia are known to have cysts in the feces of the dog, and 14 others in the cat. All named species of *Sarcocystis* with known life cycles are tabulated in an Appendix.

INTRODUCTION

The first protozoa were described by Antony van Leeuwenhoek in 1674. They included the parasitic *Eimeria stiedai* oocysts from the bile of a rabbit. Linnaeus included two species of free-living protozoa (*Volvox globator* L.

*College of Veterinary Medicine and Agricultural Experiment Station, University of Illinois, Urbana, Illinois 61801.

and *V. chaos* L.) in the 1758 edition of his *Systema Naturae,* but he included no parasitic protozoa.

No article on protozoa was included in the 8th edition of the *Encyclopaedia Britannica,* published in 1859–1860. In the 9th (1885) edition, however, protozoa were discussed by E. Ray Lankester in a 36-page article. He divided them into two "grades," the Gymnomyxa and the Corticata—the naked and the corticate—but there is no point in giving more than a sampling of his classification because it is too unfamiliar and too incomplete. To get its flavor, we might mention two of his subclasses of Corticata, the Gregarinidea and the Coccidiidea. Among the gregarines he accepted three genera, *Monocystis, Gregarina,* and *Hoplorhynchus.* We now recognize over 200 genera. Among the coccidia he accepted three orders, each with a single genus. In the order Monosporea was *Eimeria;* in the order Oligosporea was *Coccidium;* and in the order Polysporea was *Klossia.* We now recognize about 57 genera in the group, and we know that *Coccidium* and *Eimeria,* which he put in different orders, are parts of the life cycle of the same species. *Coccidium* has fallen as a synonym of *Eimeria* and other genera, and is merely memorialized in the name of the modern subclass.

Lankester said in 1885, "No disease is known at present due to Sporozoa." He didn't know that Sporozoa caused coccidiosis and malaria—*Plasmodium* wouldn't be recognized for another decade.

The *Zoological Record* began publication in 1864. It included 22 papers on Protozoa in 1870. The number had increased to 167 by 1900, 814 in 1930, 1,682 in 1958, and to 3,228 in 1970.

NUMBER OF PROTOZOAN SPECIES

In 1962 I estimated that 44,250 species of Protozoa had been named between 1758 and 1958. Of these 6,775 were parasitic (*15*). I have now brought these estimates up to date. In Table 1 are given the numbers of species of Protozoa listed in the *Zoological Record* from its beginning in 1865 through 1971, which is the last year for which I have information. By 1971, a total of 46,336

Table 5.1. New species of Protozoa listed in *Zoological Record.*

	1870	1900	1930	1960	1969–71 Mean	1865–1971 Total
Sarcodines	17	139	240	869	425	26,702
Foraminifera[a]		(58)	(231)	(802)	(829)	(21,207)
Flagellates		4	136	420	241	8,081
"Sporozoa"[b]	1	14	92	57	139	4,592
Ciliates	6	9	163	109	54	5,875
Incertae sedis			2	66	52	1,086
Total	24	166	633	1,521	911	43,336

[a]Also included under sarcodines.
[b]Includes Apicomplexa, Microspora, Myxospora, and Haplospora.

Table 5.2. Estimated number of named species of Protozoa, 1758-1976.

Protozoa	Fossil	Parasitic	Free-living	Total
Sarcodines	30,008	250	11,338	41,596
Foraminifera[a]	(26,699)	0	(4,597)	(31,296)
Flagellates	2,496	1,829	5,050	9,375
"Sporozoa"[b]	0	5,582	0	5,582
Ciliates	95	2,176	3,876	6,147
Incertae sedis	1,278	54	202	1,534
Total	33,877	9,891	20,466	64,234

[a]Also included under sarcodines
[b]Includes Apicomplexa, Microspora, Myxospora, and Haplospora.

species, of which something over 21,000 were Foraminifera, had been recorded.

Table 2 gives my estimates of the numbers of named species of Protozoa that existed at the end of 1976. These estimates are based on the information in the *Zoological Record,* on the assumption that the rate of increase during the five years 1972 through 1976 was the same as during the previous three years (1969 through 1971), on my previous estimate for the years 1758 through 1958, and on a personal judgment factor that I shall not explain. I estimate that something over 64,000 species of Protozoa, of which over half are fossil and about 10,000 are parasitic, have been named.

This is not the true number, of course. I have undoubtedly made errors in counting, some species have certainly been missed by the *Zoological Record,* and some of the names are undoubtedly synonyms. Far more important, however, is the fact that only a small percentage of the total existing protozoan species has been named. I once estimated (*15*) that there might be 3,500 different species of the genus *Eimeria* in mammals alone and perhaps 34,000 in chordates. There are about 900,000 species of insects and 73,000 of arthropods exclusive of insects (*22*), yet gregarine protozoa have been named from only 978 named species of insects (0.34%). And they have been reported from only 94 of the approximately 3,100 species of oligochaetes. I would hesitate to estimate how many species of Protozoa there actually are. A million? We don't know. There will certainly be plenty of work for future generations of taxonomists.

PROTOZOAN CLASSIFICATION

The scheme of protozoan classification used during the first half of the present century and even later (*6, 10, 13*) was established in its essentials almost 80 years ago at a time when less than 1/20 as many papers were being published annually on Protozoa as at present. That classification (*3*) separated the Protozoa into two subphyla, Plasmodroma (with a single, vesicular type of nucleus, containing the classes Mastigophora, Sarcodina, and

Sporozoa) and Ciliophora (with two non-vesicular types of nucleus—macronucleus and micronucleus—containing the classes Ciliata and Suctoria, classes which were shortly combined). This is probably the classification that most of you were taught. It is based primarily on organelles of locomotion. The Mastigophora move by means of flagella, the Sarcodina by means of pseudopods, and the Ciliophora by means of cilia. The Sporozoa move, too, though by unknown means since they have neither flagella, cilia, nor pseudopods.

These groups still exist, but they have changed, the Sporozoa probably more than the others. The Society of Protozoologists recognized this, and brought out a revised classification of the phylum in 1964 (*8*). But this classification was soon found to be unsatisfactory, and the Society is now working on a new one. Parts of it have already been published and new revisions are appearing continually. The extent of the flux is illustrated by the fact that J. O. Corliss was an author of three different classifications of ciliates within a year (*2*).

The new classification of Protozoa will raise the group to kingdom or subkingdom instead of phylum rank, and will possibly recognize the following phyla:

Sarcomastigophora (with flagella or pseudopods, or both), divided into the superclasses Mastigophora (with flagella), Sarcodina (with pseudopods and sometimes flagella), and Opalinata (with cilia-like organelles);

Apicomplexa (with an apical complex of characteristic organelles visible only with the electron microscope; with or without spores);

Microspora (with uninucleate spores provided with a hollow polar filament through which the sporoplasm emerges);

Myxospora (with multinucleate spores provided with solid polar filaments through which the sporoplasm does not emerge);

Haplospora (with simple spores without polar filaments); and

Ciliophora (with cilia, subpellicular infraciliature, and two types of nucleus (macronucleus and micronucleus).

The turn-of-the-century Sporozoa have now been split into four—the Apicomplexa, Microspora, Myxospora, and Haplospora—and I feel sure that there will be still more changes in the next decade or so.

What systematics tries to do is to compress the wild exuberance of Nature into a simple scheme. It's impossible. What we have are some large main groups (which are the successful ones), sometimes the descendants of some transition forms, and the surviving descendants of some of Nature's less satisfactory experiments. One could document this statement with data from almost any group.

Actually, and forgetting purely taxonomic considerations, there are seven main groups of protozoa of significance to agriculture. These are:

1. Mastigophora, movement by means of flagella;
2. Sarcodina, movement by means of pseudopods;

3. Apicomplexa, movement by gliding or by undulation of longitudinal ridges, with or without spores;
4. Microspora, which I have already defined;
5. Myxospora, which I have already defined;
6. Haplospora, which I have also already defined; and
7. Ciliophora, movement by means of cilia.

Now, what about recent advances in the classification of protozoa concerned with agriculture? About half the named species are fossils, so we can assume that we have only about 30,000 named species left to consider. Most of them are free-living, and this group includes some that are extremely important to us. Marine flagellates and sarcodines form the base of the ocean's food chain; the white cliffs of Dover and other chalk deposits are composed of their skeletons; and protozoa teem in the soil. One can say that we could not live without them, but one can say the same thing of many components of the ecosphere—energy, oxygen, plants, other animals, for example.

We still don't know what role some protozoa play. I said that they teem in the soil. What do they do there? Do they help make nutrients available to crop plants? Do they help eliminate pathogenic microorganisms? Do they improve the texture of the soil? What would the soil be like without them? We don't know. This is a subject that someone should study.

Disease-causing protozoa markedly affect agriculture. I cannot give a complete inventory here, but I do want to say something about some of them.

TRYPANOSOMES

Protozoa are the main reason for the widespread protein deficiency in natives of Africa. Flagellates of the genus *Trypanosoma* cause sleeping sickness in man there. We've all heard of this disease. In the days of the slave trade, slavers rejected people with Winterbottom's sign—swollen lymph nodes at the back of the neck—because they knew that those people wouldn't live long. The slavers didn't know why, but the cause was trypanosomosis.

Human sleeping sickness is now fairly well controlled, but other species of trypanosomes affect domestic animals and have made it impossible to raise them in about three million square miles of tropical Africa. Where livestock can't be raised, sources of animal protein for the human diet are considerably diminished; one is left with rats, mice, bird eggs, termites, grasshoppers, fish, etc. And quite a few Africans won't eat fish, because they regard them as unclean. Control of the trypanosomes pathogenic to domestic animals would allow an improvement in the human condition in large parts of Africa.

Trypanosomes once presented a taxonomic problem, but they don't seem to nowadays. At one time people gave a new name to every trypanosome they found in a different host, but we now know that most of the African ones (of livestock, at least) aren't host-specific enough to warrant this. As a result, there are many synonyms in the genus *Trypanosoma*. There is one exception. Trypanosomes of the subgenus *Herpetosoma* (many of which occur in rodents) are very host-specific even though they look much alike. One cannot even transmit *T. lewisi* (Kent, 1880) Laveran and Mesnil, 1901, of the rat to the mouse, or *T. musculi* Kendall, 1906, of the mouse to the rat, at least not under normal circumstances. But members of this subgenus are not important pathogens.

African trypanosomes have an ingenious way of counteracting their host's immune response. They have a layer of antigen on their surfaces against which the host makes antibody, and most of the trypanosomes are then destroyed. However, a few survive and these have a new surface antigen which permits a second wave of multiplication. This process is repeated. As many as 22 of these waves of multiplication, each with a different surface antigen, have been counted. Often, of course, the host animal dies before that many cycles have occurred. But if the trypanosome goes back into its vector, the tsetse fly, it reverts to the original surface antigen (*24*).

THE APICOMPLEXA

As I indicated earlier, the old Sporozoa are gradually getting untangled, and they have now been split into four groups. The first and most important is the Apicomplexa. This group gets its name as a result of the introduction of the electron microscope. The electron microscope revealed that Apicomplexa all have an apical complex, which, when fully developed, consists of (1) one or more polar rings; (2) a conoid—a hollow cone

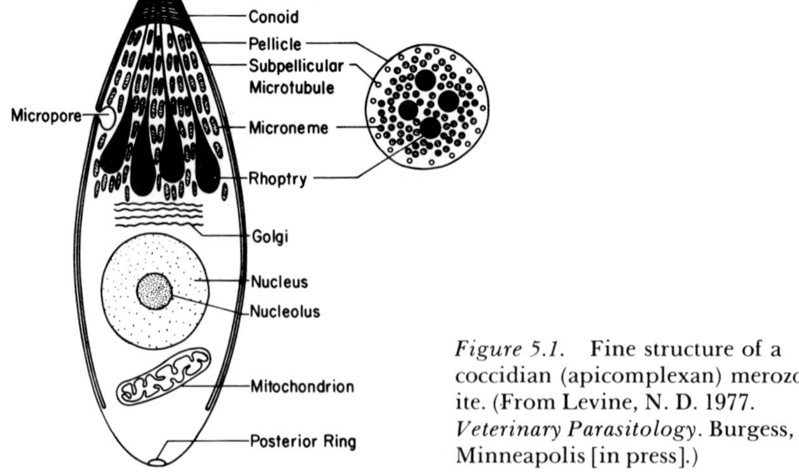

Figure 5.1. Fine structure of a coccidian (apicomplexan) merozoite. (From Levine, N. D. 1977. *Veterinary Parasitology*. Burgess, Minneapolis [in press].)

formed by several spirally wound microtubules; (3) several sac-like rhoptries (from *rhoptron*, the Greek word for club) of unknown function (perhaps secretory); (4) a variable number of rod-like micronemes, which may or may not (depending on the authority) be attached to the rhoptries; and (5) a number of subpellicular microtubules that run posteriorly just underneath the pellicle from a polar ring (Fig. 5.1). In addition, they have one or more micropores (used for ingestion), a pellicle, Golgi apparatus, mitochondria, rough endoplasmic reticulum, a nucleus and nucleolus, and a variety of cytoplasmic granules.

MALARIA

The malaria parasites and their relatives are apicomplexans. They are not important in domestic animals, and some people would say that they are not important in agriculture. But a sick man is not a good farmer, and malaria can make people very sick. Malaria was once common in this country, but it is uncommon now. At the turn of the century there were some millions of cases a year; in 1975 there were only 477 cases, and only two of these were acquired in the United States, one as the result of blood transfusion, and the other congenitally. But there are still probably 200 million malaria cases and two million malaria-caused deaths a year in the world.

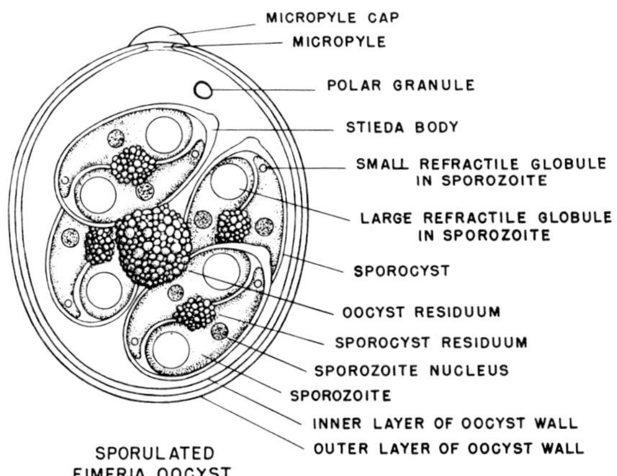

Figure 5.2. Structures of sporulated *Eimeria* oocyst. The oocysts of other genera of coccidia differ in the numbers of sporocysts and/or sporozoites that they contain. (From Levine, N. D. 1973. *Protozoan Parasites of Domestic Animals and of Man*. 2nd ed. Burgess, Minneapolis.)

COCCIDIOSIS

Coccidiosis is the most important of the protozoan diseases of livestock in the United States and it is becoming even more important. It holds about the same place in these animals that malaria does in man, and malaria parasites are closely related to coccidia. The two most important classical genera of coccidia are *Eimeria* and *Isospora*. Both genera occur mostly in the intestinal epithelial cells, and neither has an intermediate host, whereas malaria parasites occur in the red blood cells and are transmitted by mosquitoes. Coccidia are transmitted by cysts (oocysts), which are passed in the feces and are later ingested by the next host. They are highly host-specific. The oocysts of *Eimeria* contain four sporocysts, each with two sporozoites, for a total of eight sporozoites per oocyst (Fig. 5.2), while the oocysts of *Isospora* contain two sporocysts, each with four sporozoites, and thus also a total of eight sporozoites per oocyst. The oocysts of about 20 other genera of coccidia contain from one to *n* sporocysts per oocyst, one to *n* sporozoites per sporocyst, and one to *n* sporozoites per oocyst.

I don't know how many species of coccidia there are. I recently counted almost 1,000 of *Eimeria* and about 170 of *Isospora*, but these are only the named species, and there are many more yet to be named. In each of the last few years about 100 species have been newly named, and the process is far from completed.

TOXOPLASMOSIS

It was only quite recently that the taxonomic position of *Toxoplasma gondii* (Nicolle and Manceaux, 1908) Nicolle and Manceaux, 1909, was clarified. This protozoon was discovered in the gondi *Ctenodactylus gundi* Gray, 1830, (a North African rodent) in 1908 by Nicolle and Manceaux (*17, 18*). Since that time it has been found in over 200 species of mammals and birds, including man. Perhaps a third of the people in the United States are serologically positive for *Toxoplasma gondii*. It causes disease (toxoplasmosis) and death in newborn human babies (and sometimes in adults), and also in lambs, calves, pigs, etc. A great deal of research has been done in *T. gondii*. Jira and Kozojed (*11*), for instance, published a two-volume bibliography with 7,763 entries dated from 1908 to 1967, and many more papers have appeared since.

Toxoplasma was formerly placed variously among the plants, flagellates, amoebae, and Sporozoa by different taxonomists. Its most popular position was *incertae sedis*. More recently, electron microscope studies revealed that the structure of the merozoites (zoites, trophozoites, "spores," etc.) of *Toxoplasma* was similar to that of the merozoites of *Eimeria* and *Isospora*. And in 1969–1970 it was found simultaneously in the United States, Scotland, Denmark, Holland, and Germany (*5, 9, 23, 25, 26*) that *T. gondii* produces a cyst in cats that is indistinguishable from that of the "small race" of *Isospora bigemina* (Stiles, 1891) Lühe, 1906. It is an oocyst with two sporocysts each with four sporozoites, and is passed in the feces in an unsporulated condition.

The coccidia have alternation of sexual and asexual reproduction, so *Toxoplasma gondii* does not differ from *Isospora* in this respect. However, it does differ in that it is heteroxenous, with two types of host and with sexual reproduction in only one of them. On the other hand, *Isospora* species are homoxenous, with both sexual and asexual reproduction in the same animal, and without a vector. With few exceptions, *Isospora* is confined to a single host genus (i. e., it is stenoxenous). The sexual stages of *T. gondii* are quite stenoxenous also—they have been found experimentally only in felids— but the asexual stages occur in a wide variety of hosts (that is, they are euryxenous). These differences and similarities have forced us to change our thinking, and our classification. I, myself, now place *Toxoplasma* in the coccidian family Eimeriidae along with *Isospora*, but not all protozoologists agree with me.

SARCOCYSTOSIS

We now come to the protozoan genus, *Sarcocystis*, in which no sexual stages were known until recently. Over 70 *Sarcocystis* species have been named *(12)*. They are common in the ox, sheep, horse, pig and various herbivores, but not in meat-eaters. They produce cysts known as Miescher's tubules in the muscles (hence the name *Sarcocystis*—muscle cyst), and these contain merozoites (formerly called spores, or Rainey's corpuscles). But the electron microscope revealed that these merozoites are very similar in structure to the merozoites of *Toxoplasma, Eimeria, Isospora,* etc. Attempts at transmission of *Sarcocystis* had not been satisfactory, and no one knew how it was transmitted from one animal to another or why it was so common.

The similarity of the merozoites of *Sarcocystis* to those of *Toxoplasma* and the discovery of the complete life cycle of *Toxoplasma* made it inevitable that someone would try to learn whether it had a similar life cycle. Rommel, Heydorn and their associates in Germany *(7, 20, 21)* did just that. But first Fayer *(4)* found that a *Sarcocystis* of grackles (*Quisculus quiscula*) in Maryland produced a cyst in tissue culture that resembled that of a coccidium.

Sarcocystis is now known to undergo alternation of sexual and asexual reproduction, with asexual reproduction in the muscles of a prey animal of some type *(7, 20, 21,)* and sexual reproduction in the intestinal cells of a meat-eating predator. This alternation of generations has been amply confirmed.

The oocyst of *Sarcocystis* is similar to that of *Isospora*, with two sporocysts, each containing four sporozoites. There is one important difference, however. In *Toxoplasma* and ordinarily in *Isospora* and *Eimeria* the oocysts are unsporulated when they leave the host; they contain a one-celled zygote. In *Sarcocystis*, however, they are completely sporulated, and as a matter of fact the stage generally found in the feces is a sporocyst containing four sporozoites that has broken out of the oocyst. (I said "ordinarily" in *Isospora* and *Eimeria* because the oocysts of some of the species of *Isospora* and *Eimeria* in fish, amphibia, and reptiles are sporulated when they leave their host. But how many Isosporas will eventually turn out to be really *Sarcocystis* we cannot predict.)

A second difference is that, so far as is known, asexual multiplication does not occur in the predator host of *Sarcocystis*, but only in the prey animal. Oocysts are formed in the intestinal cells of the predator, and this is all. Summing it up, however, I now place *Sarcocystis* in the coccidian family Eimeriidae also, but again not all protozoologists agree with me.

It had been thought that each species of prey animal had a single species of *Sarcocystis*, but it has now been found that each has several, and that each species is restricted to one type of predator animal. Thus, instead of one species, there are at least three species of *Sarcocystis* in the muscles of the ox, each of which produces oocysts in a different species of predator (in this case the dog, cat, and man). So far as is known, the ox-dog *Sarcocystis* is not able to infect the cat or man, the ox-cat *Sarcocystis* is not able to infect the dog or man, and the ox-man *Sarcocystis* is not able to infect the dog or cat.

Not too long ago, we thought that the cat and dog both had the same species of coccidia—all *Isospora*—a big, a medium-sized, and a little one. Not today. Coccidia cannot be transmitted from the dog to the cat or vice versa, which means that they have different species. In addition, the discovery of the oocysts of *Toxoplasma, Sarcocystis*, etc., in the feces of predator animals, and the discovery that many of them look alike, has markedly increased the number of species. At the moment, we recognize 13 species in the dog and 14 in the cat, all of them different. And we can't tell some of them apart, which is always a source of frustration.

MICROSPORA

As I said earlier, the old Sporozoa have been split into four phyla: Apicomplexa, Microspora, Myxospora, and Haplospora. Most Microspora occur in insects and other arthropods, and there is at present considerable interest in them because it is possible that some of them might be useful in the biological control of disease vectors. We shall probably have to wait a few years or decades to find out whether this is true.

MYXOSPORA

The Myxospora are mostly parasites of fish. Their spores are multicellular in origin, and for this reason some taxonomists have removed them from the Protozoa entirely and placed them in the Mesozoa (*16*). However, it is the protozoologists who work on them, so I think that they could stay in the Protozoa as a matter of practicality. At one time, fish and their parasites weren't involved in agriculture. Now with the rise of farm ponds and fish farming, further study of Myxospora may be justified from an agricultural viewpoint.

HAPLOSPORA

Some of the Haplospora are parasites of mollusks. Perhaps they can be used in the biological control of snails, which are crop pests in some areas and also vectors of blood flukes. However so little is known about the Haplospora at present that we can only guess at their possible practical value.

CILIATES

Now, how about the ciliates? There are many species of them. I (*15*) estimated that 4,790 species had been named by 1958, of which 1,700 were parasitic; and Corliss (*2*) said that there were about 7,200. My own present total is about 6,100, so our figures are in the same broad range.

So far as I know, only a handful of ciliates (perhaps four species) are pathogenic. Ciliates are common in ponds and in the soil. They are extremely numerous in sewage treatment plants (*14*). I have no idea what their role there is. They also swarm in the rumen of ruminants and in the cecum and colon of horses, elephants, and their relatives. Again, I don't know what their role is. We know that their hosts can live without them. We also know that these ciliates digest starch and proteinaceous material and make sugars and amino acids available to the bacteria in the rumen, and that they ingest bacteria. Coleman (*1*) said: "The net result of the presence of these protozoa . . . is . . . to increase the rate at which the bacteria grow and are killed, i. e., to increase the rate of turnover of bacterial carbon and nitrogen. In the rumen, under appropriate conditions, 1% of the bacteria can be killed each minute." In other words, the protozoa help increase the nutrients available to the ruminants. The rumen is, in effect, a fermentation vat in which bacteria, yeasts, and protozoa can produce digestible substances from indigestible ones. Will some future research workers develop an artificial fermentation vat so that the somewhat inefficient mammalian ruminant can be eliminated?

PROTOZOAN CLASSIFICATION

The first draft (1961) of the Society of Protozoologists' (*8*) classification was 23 double-spaced, typewritten pages long. The draft of the new classification on which I am now working (1977) is more than twice as long. What is happening? In those 15 years we have increased the number of named protozoan species by about 45%, while more than doubling the length of the classification system. And the end is not in sight.

Several protozoologists have told me that the new protozoan classification is becoming too long and complicated for them to teach. This is certainly true if one tries to include all the details. All that is really necessary is a discussion of the large main groups while leaving out the small intermediate ones which are either transitional or dead-end groups.

Let me give an example. The phylum Ciliophora has been considered to be a fairly uniform group. All its members move by cilia and divide by transverse fission. All other protozoa ordinarily have a single type of nucleus, which is vesicular. The Ciliophora have two types of nucleus—a polyploid macronucleus and a diploid micronucleus; neither is vesicular. Now, there is no difference in fine structure between cilia and flagella. The difference is only in size. So you might think that every protozoon that moves by means of cilia would be a "ciliate." But this isn't what the specialists on the group say. There is a whole family, the Opalinidae (parasites of Amphibia), the members of which have cilia but only a single type of nucleus (vesicular), and which have longitudinal division. The ciliatologists say that these aren't ciliates and have thrown them out of the Ciliophora (a name which means "cilia-bearing"). In addition, there is a genus, *Stephanopogon*, the members of which also have a single type of nucleus (vesicular) but which have transverse division. *Stephanopogon* lives in the sand, and the genus contains only a few named species. It has a mouth, too, which the opalinids lack. The ciliatologists include it in the Ciliophora.

Now, in my opinion, if one wants to be accurate, any protozoon bearing cilia is a ciliate. The real difference between the so-called "ciliates" and the rest of the protozoa is in type of nucleus. But there is that bothersome *Stephanopogon*, and there is that name Ciliophora, hoary with tradition. What can one do about them? My own solution, a purely pragmatic one, is to forget the name Ciliophora and use Heterokaryota (diverse nuclei) instead, and to remove both the opalinids and *Stephanopogon* from the group. I believe that the opalinids are a dead-end group which arose from the flagellates, and that *Stephanopogon* is probably a transitional type between the flagellates and the heterokaryotes. In other words, they are intermediate forms and there is no need to say much about them except in specialized lectures or discussions. The so-called ciliatologists will, I expect, throw up their hands in dismay at this idea.

CONCLUSIONS

What of the future? Obviously, more work should be done, more species of protozoa should be studied, described and named. Animal systematists have now named almost all the vertebrates. We should do the same for the Protozoa, I suppose, but I'm thankful that I shall not be here when the job is done. It's hard enough to handle the 64,000 species already named—we have to divide them up among a whole group of specialists. But it will be still harder when that figure has been multiplied by ten or twenty, as it could be in time. There must be a better way. Should we turn to computers? That might help, but we should always remember the GIGO principle: Garbage In, Garbage Out.

More work needs to be done on many problems: the use of protozoa in the biological control of mosquitoes and other disease vectors; the eradication of disease-causing species; the role of protozoa in the soil; their role in the ocean, in sewage treatment, in rumen physiology, in a myriad of habitats; and on the possibility of using them as food. Systematics is needed for all of these.

The leaden hand of tradition lies more heavily on systematics than on any other science. As in the law, our actions are based on precedent. The names of species, genera and families are determined by the rules of priority; the names of higher taxa are not governed by such rigid rules. There might be some advantage to wiping out everything that has gone before and starting fresh on the basis of our improved knowledge. But such an action would create so much confusion that we would be worse off than we are now.

The protozoa are eukaryotes belonging to the Kingdom Protista. One can, if one wishes, consider them a subkingdom of one-celled animal-like forms, and then divide the subkingdom into two phyla on the basis of type of nucleus—Homokaryota (the former Plasmodroma) and Heterokaryota (the traditional Ciliophora). The Homokaryota can be further divided into the subphyla Sarcomastigophora, Apicomplexa, Microspora, Myxospora, and Haplospora. Another possibility would be to ignore the primary division into Homokaryota and Heterokaryota, and simply to recognize six (or more) phyla. Or one could divide up the phyla on the basis of their resemblance to animals, plants, or fungi. Whatever one does, however, one will run into intellectual difficulties and objections. Classifications are man-made, and the Protozoa know nothing of them.

LITERATURE CITED

1. Coleman, G. S. 1975. *The role of bacteria in the metabolism of rumen entodiniomorphid protozoa.* Pages 533–558 *in* D. H. Jennings and D. L. Lee, eds. *Symbiosis.* Symp. XXIX of the Society for Experimental Biology. Cambridge Univ. Press, Cambridge.

2. Corliss, J. O. 1977. *Annotated assignment of families and genera to the orders and classes currently comprising the Corlissian scheme of higher classification for the phylum Ciliophora.* Trans. Am. Micros. Soc. 96: 104–140.

3. Doflein, F. J. T. 1901. *Die Protozoen als Parasiten un Krankheitserreger, nach biologischen Gesichtspunkten dargestellt.* G. Fischer, Jena. 274 pp.

4. Fayer, R. 1972. *Gametogony of Sarcocystis sp. in cell culture.* Science 175: 65–67.

5. Frenkel, J. K., J. P. Dubey, and N. L. Miller. 1970. *Toxoplasma gondii in cats: Fecal stages identified as coccidian oocysts.* Science 167: 893–896.

6. Hall, R. P. 1953. *Protozoology.* Prentice-Hall, New York. 682 pp.

7. Heydorn, A. O., and M. Rommel. 1972. *Beiträge zum Lebenszyklus der Sarkosporidien. II. Hund und Katze als Überträger der Sarcosporidien des Rindes.* Berl. Münch. Tierärztl. Wochenschr. 85: 121-123.

8. Honigberg, B. M., W. Balamuth, E. C. Bovee, J. O. Corliss, M. Gojdics, R. P. Hall, R. R. Kudo, N. D. Levine, A. R. Loeblich, Jr., J. Weiser, and D. H. Wenrich. 1964. *A revised classification of the phylum Protozoa.* J. Protozool. 11: 7-20.

9. Hutchison, W. M., J. F. Dunachie, J. C. Siim, and K. Work. 1970. *Coccidian-like nature of Toxoplasma gondii.* Brit. Med. J. 1970: 142-144.

10. Jahn, T. L., and F. F. Jahn. 1949. *How to know the Protozoa.* Brown, Dubuque. 234 pp.

11. Jira, J., and V. Kozojed. 1970. *Toxoplasmose—1908-1967.* G. Fischer, Stuttgart. 2 vols. (396 pp. and 464 pp.)

12. Kalyakin, V. N., and D. N. Zashukhin. 1975. *Distribution of Sarcocystis (Protozoa: Sporozoa) in vertebrates.* Fol. Parasitol. 22: 289-307.

13. Kudo, R. R. 1966. *Protozoology.* 5th ed. Thomas, Springfield. 1174 pp.

14. Lackey, J. B. 1925. *The fauna of Imhoff tanks.* N. J. Agr. Exp. Sta. Bull. No. 417. 39 pp.

15. Levine, N. D. 1962. *Protozoology today.* J. Protozool. 9: 1-6.

16. Lom, J. 1973. *Current status of Myxosporidia and Microsporidia.* Prog. Protozool. 4: 254.

17. Nicolle, C., and L. Manceaux. 1908. *Sur une infection à corps de Leishman (ou organismes voisins) du gondi.* C. R. Acad. Sci. (Paris) 147: 763-766.

18. Nicolle, C., and L. Manceaux. 1909. *Sur un protozoaire nouveau du gondi.* C. R. Acad. Sci. (Paris) 148: 369-372.

19. Overdulve, J. P. 1970. *The identity of Toxoplasma Nicolle & Manceaux, 1909, with Isospora Schneider, 1881.* Proc. Konikl. Nederl. Akad. Wetenschappen, Ser. C. 73: 129-151.

20. Rommel, M., and A. O. Heydorn. 1972. *Beiträge zur Lebenszyklus der Sarkosporidien. III. Isospora hominis (Railliet und Lucet, 1891) Wenyon, 1923, eine Dauerform der Sarkosporidien des Rindes und des Schweins.* Berl. Münch. Tierärztl. Wochenschr. 85: 143-145.

21. Rommel, M., A. O. Heydorn, and F. Gruber. 1972. *Beiträge zur Lebenszyklus der Sarkosporidien. I. Die Sporozyste von S. tenella in den Fäzes der Katze.* Berl. Münch. Tierärztl. Wochenschr. 85: 101-105.

22. Ross, H. H. *A textbook of entomology.* 3rd ed. Wiley, New York. 539 pp.

23. Sheffield, H. G., and M. L. Melton. 1970. *Toxoplasma gondii: The oocyst, sporozoite, and infection of cultured cells.* Science 167: 892-893.

24. Vickerman, K. 1969. *On the surface coat and flagellar adhesion in trypanosomes.* J. Cell. Sci. 5: 163-193.

25. Weiland, G., and D. Kühn. 1970. *Experimentelle Toxoplasma-Infektionen bei der Katze. II. Entwicklungsstadien des Parasiten im Darm.* Berl. Münch, Tierärztl. Wochenschr. 83: 128-132.

26. Work, K., and W. M. Hutchison. 1969. *The new cyst of Toxoplasma gondii.* Acta Pathol. Microbiol. Scand. 77: 414-424.

Appendix

SPECIES OF *SARCOCYSTIS* FOR WHICH HOSTS OF BOTH SEXUAL AND ASEXUAL STAGES ARE KNOWN

1. *Sarcocystis bertrami* Doflein, 1901
   Syns., *Isospora rivolta* sporocysts of Gassner (1940), Levine & Ivens (1965) and *auctores* in part
   Isospora bigemina large form of Mehlhorn, Heydorn & Gestrich (1975) and Heydorn, Mehlhorn & Gestrich (1975) in part
   Sarcocystis equicanis Rommel & Geisel, 1975
   Endorimospora bertrami (Doflein, 1901) Tadros & Laarman, 1976
   Intermediate Hosts: Horse *Equus caballus*, ass *E. asinus*.
   Definitive Host: Dog *Canis familiaris*.
2. *S. cernae* Levine, 1977 (in press)
   Syn., Sarcocystis sp. Černá & Loučkova, 1976
   Intermediate Host: Vole *Microtus arvalis*.
   Definitive Host: Kestrel *Falco tinnunculus*.
3. *S. cruzi* (Hasselmann, 1926) Wenyon, 1926
   Syns., *Miescheria cruzi* Hasselmann, 1926
   Sarcocystis fusiformis Railliet, 1897 of Babudieri (1932) and *auctores* in part
   Sarcocystis iturbei Vogelsang, 1938
   Sarcocystis marcovi Vershinin, 1975 in part
   Sarcocystis bovicanis Heydorn, Gestrich, Mehlhorn & Rommel, 1975
   Isospora rivolta sporocysts of Gassner (1940), Levine & Ivens (1965) and *auctores* in part
   Isospora bigemina large form of Mehlhorn, Heydorn & Gestrich (1975) and Heydorn, Mehlhorn & Gestrich (1975) in part
   Endorimospora hirsuta (Moulé, 1888) Tadros & Laarman, 1976
   Intermediate Host: Ox *Bos taurus*
   Definitive Hosts: Dog *Canis familiaris*, coyote *C. latrans*, wolf *C. lupus*, red fox *Vulpes vulpes*, raccoon *Procyon lotor*.
4. *S. dispersa* Cerná, 1977 (in press)
   Syn., *Sarcocystis* sp. Cerná, 1976
   Intermediate Host: House mouse *Mus musculus*.
   Definitive Host: Owl *Tyto alba*.
5. *S. fayeri* Dubey, Streitel, Stromberg & Toussant, 1977
   Intermediate Host: Horse *Equus caballus*.
   Definitive Host: Dog *Canis familiaris*.
6. *S. fusiformis* (Railliet, 1897) Bernard & Bauche, 1912
   Syns., *Balbiania fusiformis* Railliet, 1897
   Sarcocystis blanchardi Doflein, 1901
   Sarcocystis siamensis von Linstow, 1903
   Sarcocystis bubali Willey, Chalmers & Philip, 1904
   Intermediate Host: Water buffalo *Bubalus bubalis*.
   Definitive Host: Cat *Felis catus*.

7. *S. hemionilatrantis* Hudkins & Kistner, 1977
   Intermediate Host: Mule deer *Odocoileus hemionus.*
   Definitive Host: Coyote *Canis latrans,* dog *C. familiaris.*
8. *S. hirsuta* Moulé, 1888
   Syns., *Miescheria cruzi* Hasselmann, 1926 in part
   Sarcocystis fusiformis Railliet, 1897 of Babudieri (1932) and *auctores* in part
   Sarcocystis marcovi Vershinin, 1975 in part
   Sarcocystis bovifelis Heydorn, Gestrich, Mehlhorn & Rommel, 1975
   Endorimospora cruzi (Hasselmann, 1926) Tadros & Laarman, 1976
   Intermediate Host: Ox *Bos taurus.*
   Definitive Host: Cat *Felis catus.*
9. *S. hominis* (Railliet & Lucet, 1891) Dubey, 1976
   Syns., *Miescheria cruzi* Hasselmann, 1926 in part
   Coccidium bigeminum var. *hominis* Railliet & Lucet, 1891 in part
   Isospora hominis (Railliet & Lucet, 1891) Wenyon, 1923 in part
   Lucetina hominis (Railliet & Lucet, 1891) Henry & Leblois, 1926 in part
   Sarcocystis fusiformis Railliet, 1897 of Babudieri (1932) and *auctores* in part
   Sarcocystis bovihominis Heydorn, Gestrich, Mehlhorn & Rommel, 1975
   Endorimospora hominis (Railliet & Lucet, 1891) Tadros & Laarman, 1976
   Intermediate Host: Ox *Bos taurus.*
   Definitive Hosts: Man *Homo sapiens,* rhesus monkey *Macaca mulatta,* baboon *Papio cynocephalus,* (?) chimpanzee *Pan troglodytes.*
10. *S. leporum* Crawley, 1914
   Intermediate Hosts: Cottontail rabbits *Sylvilagus floridanus, S. nuttalli,* and *S. palustris.*
   Definitive Host: Cat *Felis catus.*
11. *S. miescheriana* (Kühn, 1865) Labbé, 1899 (TYPE SPECIES)
   Syns., *Synchrytium miescherianum* Kuhn, 1865
   Sarcocystis miescheri Lankester, 1882
   Coccidium bigeminum var. canis Railliet & Lucet, 1891 in part
   Isospora rivolta sporocysts of Gassner (1940), Levine & Ivens (1965) and *auctores* in part
   Isospora bigemina large form of Mehlhorn, Heydorn & Gestrich (1975) and Heydorn, Mehlhorn & Gestrich (1975) in part
   Endorimospora miescheriana (Kühn, 1865) Tadros & Laarman, 1976
   Cryptosporidium vulpis Wetzel, 1938 (probably)
   Intermediate Host: Pig *Sus scrofa.*
   Definitive Hosts: Dog *Canis familiaris,* (?) red fox *Vulpes vulpes.*
12. *S. muris* (Blanchard, 1885) Labbé, 1899
   Syns., *Miescheria muris* Blanchard, 1885
   Coccidium bigeminum var. *cati* Railliet & Lucet, 1891
   Sarcocystis musculi Blanchard, 1885 of Kalyakhin & Zasukhin (1975) *lapsus calami*
   Endorimospora muris (Blanchard, 1885) Tadros & Laarman, 1976
   Intermediate Host: House mouse *Mus musculus.*
   Definitive Host: Cat *Felis catus.*

13. *S. ovicanis* Heydorn, Gestrich, Mehlhorn & Rommel, 1975
   Syns., *Sarcocystis tenella* Railliet, 1886 in part
      Isospora rivolta free sporocysts of Gassner (1940), Levine & Ivens (1965) and *auctores* in part
      Endorimospora ovicanis (Heydorn, Gestrich, Mehlhorn & Rommel, 1975) Tadros & Laarman, 1976
      Hoareosporidium pellerdyi Pande, Bhatia & Chauhan, 1972 (Probably)
   Intermediate Host: Sheep *Ovis aries*.
   Definitive Host: Dog *Canis familiaris*
14. *S. porcifelis* Dubey, 1976
   Syn., *Sarcocystis miescheriana* (Kühn, 1865) Labbé, 1899 in part
   Intermediate Host: Pig *Sus scrofa*.
   Definitive Host: Cat *Felis catus*.
15. *S. sebeki* (Tadros & Laarman, 1976) nov. comb.
   Syn., *Endorimospora sebeki* Tadros & Laarman, 1976
   Intermediate Host: Long-tailed field mouse *Apodemus sylvaticus*.
   Definitive Host: Tawny owl *Strix aluco*.
16. *S. singaporensis* Zaman, 1977
   Syn., *Sarcocystis orientalis* Zaman & Colley, 1975
   Intermediate Host: Norway rat *Rattus norvegicus*.
   Definitive Host: Reticulated python *Python reticulatus*.
17. *S. suihominis* (Tadros & Laarman, 1976) Heydorn, 1977
   Syns., *Sarcocystis porcihominis* Dubey, 1976
      Endorimospora suihominis Tadros & Laarman, 1976
   Intermediate Host: Pig *Sus scrofa*.
   Definitive Host: Man *Homo sapiens*.
18. *S. tenella* (Railliet, 1886) Moulé, 1886
   Syns., *Miescheria tenella* Railliet, 1886
      Balbiania gigantea Railliet, 1886
      Sarcocystis ovifelis Heydorn, Gestrich, Mehlhorn, Rommel, 1975
      Endorimospora tenella (Railliet, 1886) Tadros & Laarman, 1976
   Intermediate Host: Sheep *Ovis aries*.
   Definitive Host: Cat *Felis catus*.
19. *Sarcocystis* sp. Rzepczyk, 1974
   Intermediate Host: Rat *Rattus fuscipes*.
   Definitive Host: Carpet python *Morelia spilotes variegata*.
20. *Sarcocystis* sp. Janitschke, Protz & Werner, 1976
   Intermediate Host: Grant's gazelle *Gazella granti*.
   Definitive Hosts: Dog *Canis familiaris* (sporocysts 13 to 18 x 8 to 12 μm, mean 16 x 10 μm)
      Cat *Felis catus* (sporocysts 11 to 15 x 8 to 12 μm, mean 13 x 9 μm
   (Sarcocysts from the gazelle were fed to the dog and cat, and sporocysts were found in both; it is likely that two species were involved, but no names can be assigned to them without further investigation)
21. *Sarcocystis* sp. Dissanaike *et al.* (1977)
   Intermediate Host: Water buffalo *Bubalus bubalis*.
   Definitive Host: Dog *Canis familiaris*.

6] Application of Genetics to Insect Systematics and Analysis of Species Differences

BY GUY L. BUSH* and G. BARRIE KITTO†

ABSTRACT

Recent developments in molecular genetics offer extremely useful techniques for resolving systematic and evolutionary problems ranging from the identification and delineation of races and sibling species to the development of accurate and objective phylogenies. Estimates of speciation rates and origination time can also be obtained by applying the molecular clock method. Three methods now in use for establishing genetic variation at the molecular level within and between species of animals and plants are discussed. Examples using one dimensional gel electrophoresis and the more recently perfected two dimensional O'Farrell technique capable of resolving 1,000 or more proteins in a single individual are presented. Applications of microcomplement fixation and DNA hybridization techniques to systematic and evolutionary problems are also considered. Finally, the recent development of gene splicing methods and their significance and potential use by evolutionary biologists is reviewed. It is suggested that incorporation of these tools into the repertoire of the systematist and evolutionary biologist will lead to a better understanding of evolutionary rates, speciation and phylogenic relationships. The importance of these approaches in relation to various pest management problems (for example, the detection of host races and sibling species) is also discussed.

INTRODUCTION

Systematics has been described as the science of organismal diversity (*110*). Its primary objective and contribution has been to order the incredible diversity of life in a way that reflects patterns of evolutionary divergence. Ideally it encompasses the study of all biological relationships among

*Department of Zoology, University of Texas, Austin, Texas 78712.
†Clayton Foundation Biochemical Institute, Department of Chemistry, University of Texas, Austin, Texas 78712.

organisms from the molecular level of organization to the complex interaction of species and their environment (*77*).

In practice, this ideal is seldom realized. For historical and practical reasons, most systematists rely heavily on the comparative morphology of living and fossil organisms to group and rank their taxa. The development of numerical taxonomic techniques, while adding somewhat greater precision and objectivity to the treatment of taxonomic problems, has focused attention more on the deficiencies of contemporary systematics than on resolving problems of systematic or evolutionary importance (*77*). Major questions of phylogenetic relationships, patterns of convergence, parallel evolution and rates of evolutionary change and speciation still remain in all groups of organisms.

The recent advances in these areas have come not from classical systematics, but from molecular and population biologists who are attacking evolutionary and systematic problems at the biochemical and genetic level with startling and exciting results. Their approach sometimes generates interpretations of phylogenetic relationships and evolutionary processes that contrast radically with those established by conventional systematic techniques. If systematists are to contribute significantly to the revolution in biological systematics initiated by recent discoveries in molecular evolution, it will be necessary to incorporate techniques of the molecular biologist into their repertoire of systematic tools, or at least to become familiar with their use and meaning.

This is not to say that morphological studies are no longer needed. Without sophisticated biometric analysis and a clear understanding of evolutionary processes at the organismal level, it is difficult if not impossible to place molecular evidence in a meaningful evolutionary context.

The major obstacle to the acceptance of molecular genetic data as a systematic tool has been the impression among many systematists that phylogenies based on genetic relationships (cladism) disregard divergence that may occur subsequent to branching (*78*). This attitude reflects an erroneous impression widely held by systematists that genetic change and morphological change are synonomous. In their view, large differences in morphology must inevitably be associated with extensive differences at the genetic level. Furthermore, it is considered axiomatic that organismal evolution and speciation cannot originate by a few gene changes, but that they occur primarily as a byproduct of many gene changes accumulated over long periods of time or by way of genetic "revolutions" brought about by stochastic processes and selection (i.e., founder effect) (*30, 41, 77*). All of these basic assumptions must be reexamined in the light of the molecular evidence.

ORGANISMAL AND MOLECULAR RATES OF EVOLUTION

It is now clear that amino acid and nucleotide sequences evolve at relatively constant rates that appear to be independent of phenotypic evolution (*103*,

122, 123) and speciation (*9, 10, 26, 28*). Thus sequence evolution within a group proceeds as rapidly in phenotypically conservative and species-poor phylads (for example, living fossils) as in those that have undergone radical changes in phenotype over short periods of time and are species-rich (*125*). Changes that occur in protein and DNA sequences, therefore, behave like fairly good stochastic evolutionary clocks (*47, 102, 123*). The accuracy of this molecular clock has about twice the variance as one based on radioactive decay (*123*). For this reason, genes represent the best "fossils" available for reconstructing phylogenetic relationships.

Phenotypic characteristics of the type used most frequently by paleontologists and systematists, on the other hand, are somewhat less appropriate indicators of phylogenetic relatedness than molecular genetic criteria. Not only is it difficult to accurately estimate divergence times from the fossil record, but rates of organismal evolution are far from constant. Two species may differ markedly in morphology and other traits (for example, Hawaiian Drosophilidae; man and chimpanzee) (*31, 61*), yet be very closely related genetically and have diverged recently. In other cases sibling species, which are morphologically similar or indistinguishable, may represent ancient species that have diverged extensively in their protein sequences (for example, frogs) (*76*).

This paradox may be resolved when one realizes that two classes of genetic changes are actually involved in adaptive evolution. One involves *structural genes* that encode for proteins and ribonucleic acids and the other represents *"regulatory" genes* involved in regulating structural gene activity (*17, 18, 19*). Recent evidence (*123*) suggests that most adaptively significant mutations, some of which may produce large phenotypic effects and permit the exploitation of new adaptive zones, appear to be regulatory in nature. Changes in structural genes, which are the subject of most genetic studies, are either neutral or more closely associated with minor adaptive adjustments that keep a population in tune with its immediate environment. They are normally not directly involved in speciation or major adaptive shifts (*26*).

In the following section we will consider several biochemical techniques for obtaining estimates of genetic variation within and between populations and for obtaining amino acid and nucleotide sequence data for specific proteins and nucleic acids. Our treatment is not intended to be exhaustive, but should provide enough information to give some idea of the extent to which molecular genetics is being applied to systematics.

METHODS FOR DETERMINING GENETIC DIFFERENCE IN PROTEINS

One-dimensional gel electrophoresis. A technique now widely used to rapidly establish gene frequencies for specific soluble enzymatic and nonenzymatic proteins of the cytoplasm in natural and laboratory populations is one-dimensional gel electrophoresis. The basic procedure is outlined in Fig. 6.1. It was first adapted to genetic and systematic problems in 1966 (see

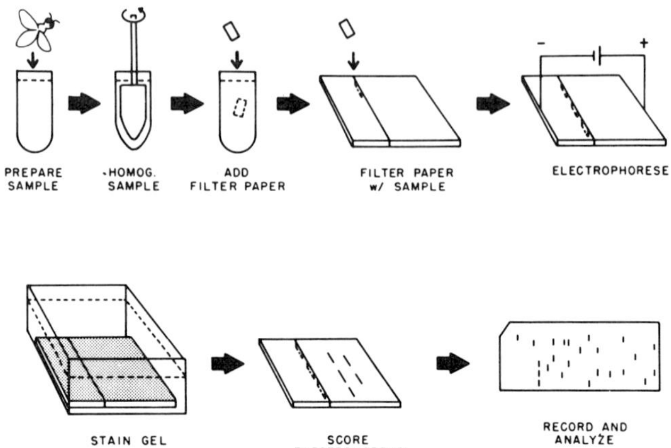

Figure 6.1. Procedure followed in one-dimensional gel electrophoresis. See text for explanation.

Lewontin [*68*] for details). Excellent discussions of the various principles and methods for enzyme staining and electrophoresis have been presented by Harris and Hopkinson (*53*) and by Gordon (*50*). Readers interested in critical reviews on the adaptive function of various proteins and the various ways genetic data can be utilized should consult Lewontin (*68*), Selander and Johnson (*108*), Wagner and Selander (*116*), Ayala (*10*), Nei (*84*), Dobzhansky *et al.* (*42*), Hedrick *et al.* (*54*) and Sarich (*103*).

Electrophoretic separations of genetic variants in proteins are carried out on crude extracts, usually prepared by simply grinding the whole animal or appropriate tissue in a drop of buffer (Fig. 6.1). Enzyme molecules which carry net positive or negative electric charges, according to the pH of the buffer held in the matrix of the supporting gel, will migrate in an electric field. When a sample of an extract is placed in a gel which serves as a supporting medium, its proteins can be separated by zone electrophoresis. The speed and distance of enzyme migration is related to the size of its net charge and the strength of the electric field. The location of enzymes in the gel can be identified by using specific staining techniques. Generally, only one or two enzymes can be stained on a single gel.

Many proteins exist in natural populations in two or more molecular forms called *allozymes,* which differ by one or more amino acid substitutions. If these substitutions involve amino acids that alter the net charge, the multimolecular forms will migrate at different rates. These electrophoretically detectable allozymes are called *electromorphs.* Only about one third of all amino acid substitutions are detectable by gel electrophoresis. Thus the amount of genetic variation is underestimated by this method (*68*).

Due to the processes involved in protein synthesis, individuals that are homozygous or heterozygous for each electromorph can be identified by direct visual inspection of gels. Resolution can be enhanced and the number of detectable electromorphs increased for certain proteins by using variable gel-sieving and heat treatment methods (*58, 59, 111*). For instance, the six electromorphs previously recorded at the esterase locus of *Drosophila pseudoobscura* Frolova have been found to comprise at least 37 distinct allelic forms of the enzyme (*111*).

A measure of genetic differentiation between two or more populations may be obtained by comparing the average degree of genetic differentiation in a random sample of loci of the genome as determined by gel electrophoresis. When several populations or taxa are involved, a dendrogram reflecting genetic distances can be constructed. This dendrogram can be regarded as an expression of phylogeny if one assumes that the amount of genetic divergence averaged over several loci is approximately proportional to the time during which these populations or taxa shared a common ancestor (*42, 84, 103*).

Several methods have been developed for calculating genetic distance (D) or genetic identity (I) between populations and higher taxa (*49, 72, 96, 97*), but the method of Nei (*84*) is currently the most widely used. Cladistic phylogenies based on genetic distance and identity estimates may be constructed using a variety of clustering and ordination techniques such as the unweighted or weighted group pair methods with arithmetic means (*113*) or the maximum likelihood or minimum steps method (*44, 45*). Most give very similar results.

From a practical standpoint, gel electrophoresis can be applied to a wide range of systematic and evolutionary problems. However, it is best used on closely related taxa where divergence times do not exceed 10 to 15 million years. If all proteins have diverged to the point where two taxa differ electrophoretically at every locus, then no estimate of genetic distance can be made. Furthermore, as proteins diverge, certain amino acid substitutions may accumulate that result in orthologous proteins from different taxa having similar mobilities. Other methods described below can be used to study phylogenetic relationships between higher taxa.

As an example of how electrophoretic data can be used to resolve taxonomic problems, we may look at the results of recent studies on the fruit fly genus *Rhagoletis* (Tephritidae). Several members of this group are major pests of fruit in the Holarctic and Neotropical regions.

A previous systematic study, based primarily on morphology, uncovered several sibling species groups whose members could be distinguished only with difficulty or by biological criteria such as host preference, or hybrid sterility (*24, 25*). Other host races were suspected but could not be identified as distinct species in the absence of adequate biological studies. Many of these suspected species were morphologically indistinguishable from major pest species and thus complicated control measures. Furthermore, it had

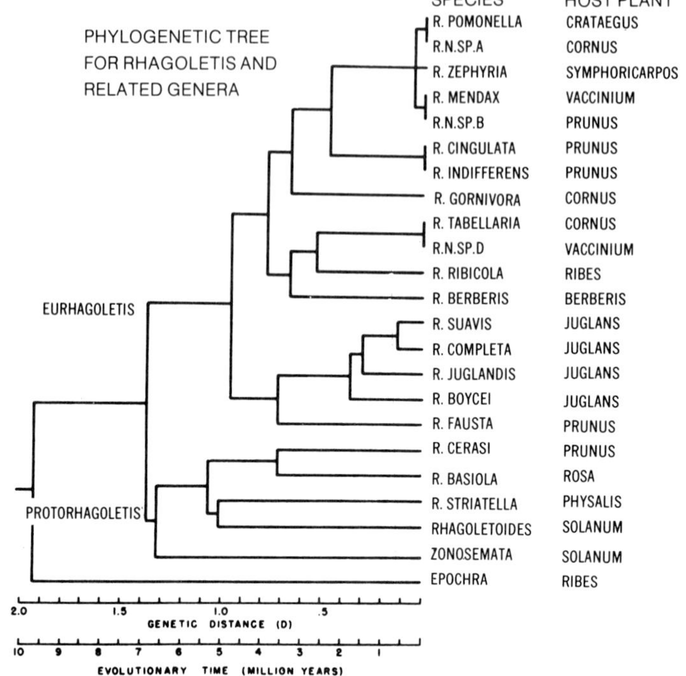

Figure 6.2. Dendrogram of the genus *Rhagoletis* derived from the genetic distance matrix based on 15 enzyme loci. Representatives of related genera are included as outside reference points (from Berlocher, ref. *12*).

been difficult to construct an adequate phylogeny for *Rhagoletis* on the basis of morphology.

An electrophoretic study carried out over the past few years by Berlocher (*12*), Bush and Huettel (*11, 29*) and Berlocher and Bush (in preparation) has resolved many of these problems. Not only do we now have a satisfactory picture of phylogenetic relationships within the genus (Fig. 6.2), but several interesting and unexpected patterns of divergence have emerged.

For instance, members of the *pomonella* species group (Fig. 6.3), with one exception, have all.speciated very recently. In this group, speciation is always accompanied by a major shift in host preference and has occurred sympatrically (*25, 26*). In others, like the *suavis* group (Fig. 6.3) whose species infest only walnuts, speciation has occurred more slowly by geographic isolation and has never involved a host shift.

The same technique has also been applied to the study of recently evolved host races. The European cherry fruit fly, *Rhagoletis cerasi* (Linn.), occurs

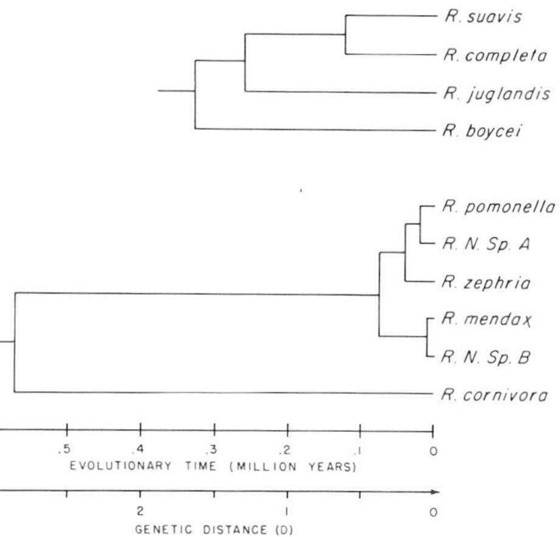

Figure 6.3. Dendrograms of *R. suavis* and *R. Pomonella* group species (from Berlocher, ref. *12*). The time pattern of divergence in the *suavis* group which infests only walnuts (*Juglans*) contrasts with that of the *pomonella* group whose species all infest unrelated host plants and have evolved recently. The former have speciated geographically and the latter sympatrically (*26*).

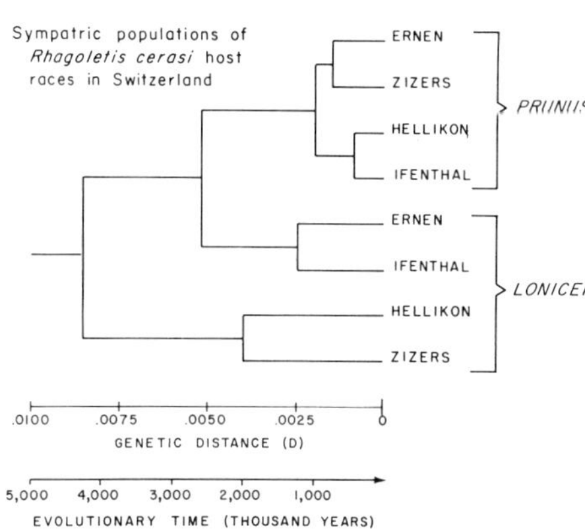

Figure 6.4. Dendrogram of *R. cerasi* host races in Switzerland. Estimates of genetic distances and divergence times are consistent with the known age of the flora in Switzerland which was glaciated until about 8,000 years ago (from Boller *et al.*, ref. *16*).

on both cherry (*Prunus* spp.) and honeysuckle (*Lonicera* spp.) in forms that are biologically distinct in nature but are morphologically and karyotypically indistinguishable (*15, 16, 27*). Small but consistent genetic differences between these morphologically indistinguishable races have been found in sympatric populations (Fig. 6.4) (Bush and Boller, in preparation). Genetic distances between sympatric populations of the two host races are small (Fig. 6.4), suggesting they have diverged recently. However, at least within Switzerland, all the *Lonicera* fly populations are more closely related to one another than they are to members of the *Prunus* fly race. This suggests that the two races probably represent distinct reproductively isolated species and should be regarded as such for control purposes. For further examples of gel electrophoresis applied to insect taxo-

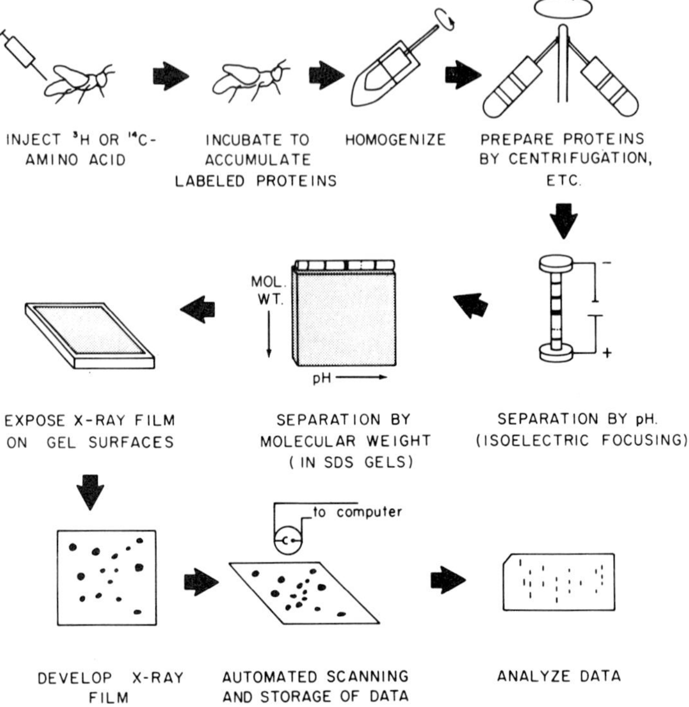

Figure 6.5. Procedure followed in high resolution, two-dimensional gel electrophoresis by the O'Farrell technique. See text for explanation.

nomic problems, Ayala (*10*) and Wagner and Selander (*116*) can be consulted.

Several methods have been suggested for translating genetic distance estimates based on electrophoretic data into evolutionary time (*84, 103*). We have applied the molecular clock using Nei's method (*83*) to calculate time of divergence and to establish an evolutionary time scale. As there is no fossil record of Tephritidae that can be used to set the molecular clock for the 16 proteins studied, we have had to use rates worked out for other organisms and from fossil evidence of host plant distributions. The time scale provided in the figures should therefore be regarded as very tentative.

Two-dimensional gel electrophoresis. Recently a more sophisticated technique has been introduced to study genetic variation in natural and domesticated insect populations. This is the high resolution two-dimensional gel electrophoresis technique of O'Farrell (*86*) (Fig. 6.5). This technique, which greatly increases the number and range of proteins that can be detected simultaneously in a single individual, differs from the conventional one-dimensional gel electrophoresis in several important ways.

Because all classes of proteins can be tested, both enzymatic and nonenzymatic proteins can be studied. This technique, therefore, shows promise as a powerful tool for studying genetic variation in all kinds of proteins, not just those of the cytosol. Proteins are separated according to their isoelectric point by isoelectric focusing in one dimension, and according to molecular weight by electrophoresis in SDS (sodium dodecyl sulfate) gels in the second dimension. The positions of the proteins, which have been labeled with ^{3}H or ^{14}C-amino acids, are established on the gel by autoradiography. Film images can be enhanced and the autoradiographic exposure time greatly reduced by using fluorography techniques developed by Laskey and Mills (*67*). This system can resolve proteins differing in only a single charge and at extremely low concentrations. Consequently, a large number of naturally occurring electromorphs can be detected.

The advantage of the O'Farrell technique over currently used one-dimensional methods is that a very large number of electromorphs can be characterized simultaneously in a single individual on one gel. Over 1100 protein components of *Escherichia coli* have been resolved by this method using appropriate techniques so that specific classes of proteins (e.g., chromosomal, membrane, cytosol, etc.) can be examined separately (*5, 52*). An example is shown in Fig. 6.6.

The fact that the O'Farrell technique resolves several hundred proteins on a single gel, however, makes it very difficult to analyze each gel for

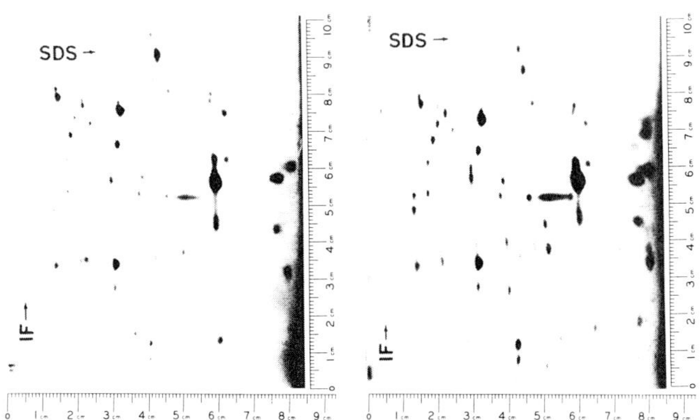

Figure 6.6. Photograph of two-dimensional slab gels of *Escherichia coli* cell envelopes. Proteins from the parent strain have been separated in the upper gel and those from a succinic dehydrogenase mutant in the lower gel (from Ames and Nikaido, ref. *5*).

genetic variants (for example, a position change). A small number of samples can be handled manually, but recording total genetic variation on the scale necessary for population studies is laborious, time consuming, and subject to error. If the technique is to be used for population genetics, a method must be developed for rapidly estimating gene frequencies for each locus.

A computer technique for the scanning and analysing of individual gels for position changes, which reflect charge changes of specific polymorphic proteins, is now under development in several laboratories. The method has not yet been applied to systematic or evolutionary problems, but attempts are now underway to do so.

Microcomplement fixation. Electrophoretic methods provide estimates only of differences in charge and size between proteins. While this information can be very useful for insect systematic studies, it is often desirable to have more quantitative estimates of the degree of amino acid sequence difference between specific proteins from different species. Complete amino acid sequence information from the proteins in question would provide the most direct means of comparison. However, for most practical purposes such comparisons are precluded for insect systematics, not only

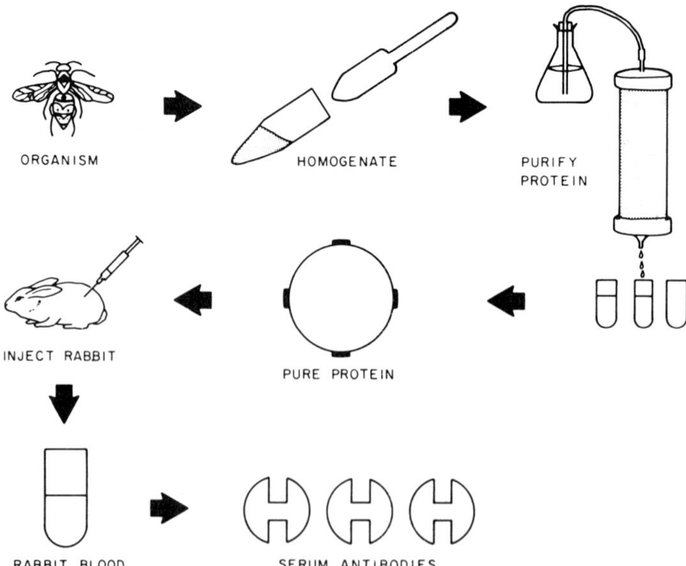

Figure 6.7. Preliminary steps in the preparation of purified proteins by affinity chromatography and serum antibodies for microcomplement fixation procedures.

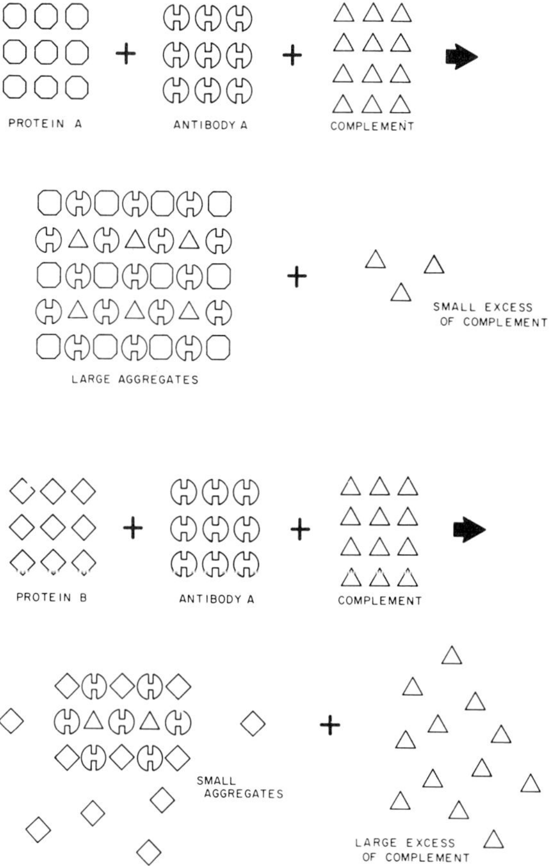

Figure 6.8. Diagrams Illustrating the interactions of proteins and antibodies when incubated with complement. (*Above*) When protein A and its antibody are incubated with complement, large aggregates are formed leaving only a small excess of complement. (*Below*) When an orthologous protein from another species (protein B) is substituted for protein A, it interacts less well with antibody A and complement and forms smaller aggregates. A large excess of complement, therefore, remains.

because amino acid sequence analysis is time consuming and expensive but also because each protein to be compared must be highly purified from each separate species. Immunological techniques offer a less direct but much more rapid means of comparing protein sequences and can still provide the quantitative results required for constructing phylogenetic trees and estimating speciation rates and divergence times.

The most useful immunological procedure for this purpose is that of

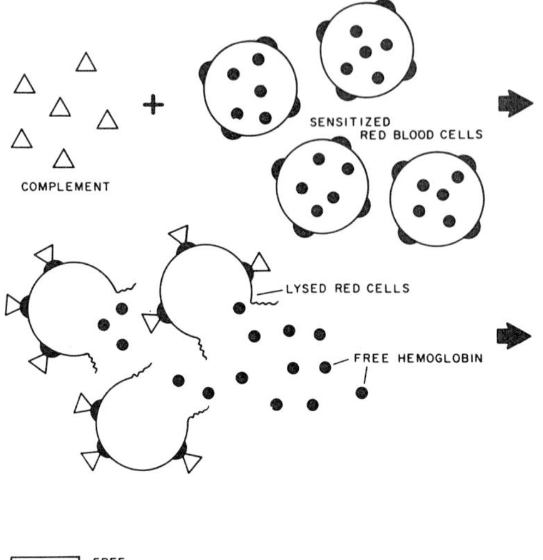

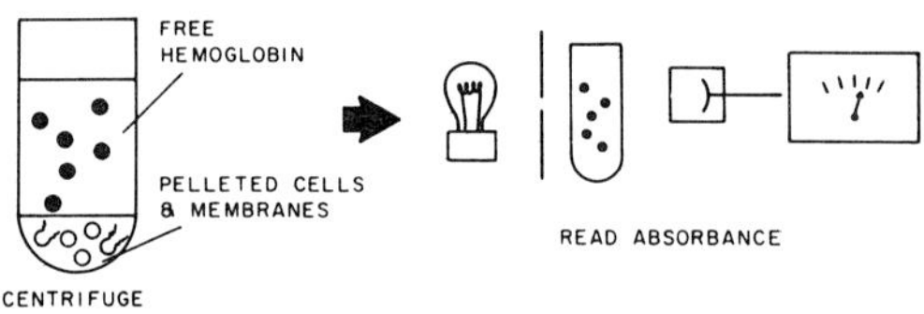

Figure 6.9. Measurement of amount of residual complement by addition of sensitized red blood cells. The amount of hemoglobin released into solution from lysed cells is measured spectrophotometrically and is inversely proportional to the amount of complement fixed in the antigen-antibody-complement complexes.

microcomplement fixation (MC′F), developed in 1961 by Wasserman and Levine (*119*). This technique has been adapted for systematic studies primarily by Allan C. Wilson and his colleagues at the University of California at Berkeley and is now in widespread use. Two advantages of this procedure are that it requires from 100 to 1000 times less material than other common immunological procedures such as macrocomplement fixation and quantitative precipitin tests and that it is not necessary to purify the cross-reacting protein for each species to be compared. The basic MC′F procedure is illustrated in Figs. 6.7 to 6.9.

The first step is to obtain a highly purified sample of the protein to be examined (for example, an enzyme such as α-glycerophosphate dehydrogenase) from a single insect species. The recently developed and highly efficient procedure of affinity chromatography is often used for this purpose (*38, 39, 107*). The highly purified protein is then injected into rabbits in multiple doses over a 2–3 month period, to generate an antiserum specifically directed against this protein (Fig. 6.7). This antiserum is then used in the MC′F test for quantitating cross reactions with the same enzyme in related species. The MC′F technique makes use of two special properties of a complex of proteins found in vertebrate sera and which are collectively termed complement (C′). The first property employed is that of complement to bind irreversibly to antigen-antibody complexes (Fig. 6.8). The

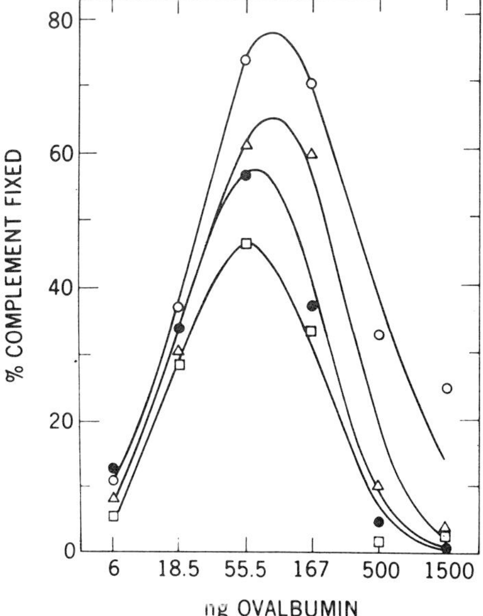

Figure 6.10. The effect of antiserum concentration on complement fixation. The antiserum used was rabbit anti-chicken ovalbumin (M19-11). The concentrations were: 4-1/4,000; △-1/4,600; 4-1/5,200; □- 1/5,800. The test antigen was chicken ovalbumin (from Champion *et al.*, ref. *32*).

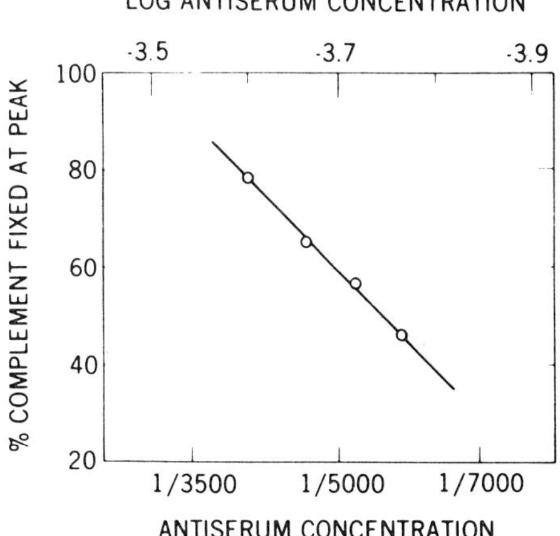

Figure 6.11. The dependence of the peak height of the complement fixation curve on the concentration of antiserum. Each point represents the peak height of a complement fixation curve (obtained from *Figure 6.10*) for a particular antiserum concentration (from Champion *et al.*, ref. *32*).

second is the ability of complement to lyse or break open red cells which have been coated with anti-red blood cell antibody (sensitized RBC).

In the MC′F test a serial dilution of the protein used for immunization of the rabbit (the homologous antigen) is incubated with the antiserum and an excess of complement, resulting in the formation of antigen-antibody-complement complexes. The residual complement is then reacted with sensitized red blood cells, resulting in lysis of the cells and release of hemoglobin. The degree of lysis and amount of hemoglobin released into solution is proportional to the amount of residual complement. The free hemoglobin concentration is measured spectrophotometrically and is inversely proportional to the amount of complement fixed in the antigen-antibody-complement complexes (Fig. 6.9). In such an experiment a plot of the percentage of complement fixed versus the antigen concentration yields a bell-shaped curve. The peak height of the curve varies with antibody concentration as shown in Figs. 6.10, 6.11. The factor by which the antiserum must be diluted to give 75% C′ fixation is termed the titer of that antiserum. The antiserum can now be used to measure quantitatively the crossreaction of orthologous proteins from other species. Orthologous proteins from closely related species crossreact well while there is less crossreaction with proteins from more distantly related species. Experimentally the crossreaction is quantified by increasing the antiserum dilution until the peak of the complement fixation curve is at 75% complement fixed (Fig. 6.12). For these tests it is not necessary to purify extensively the crossreacting protein. In most cases a simple tissue extract will suffice. A

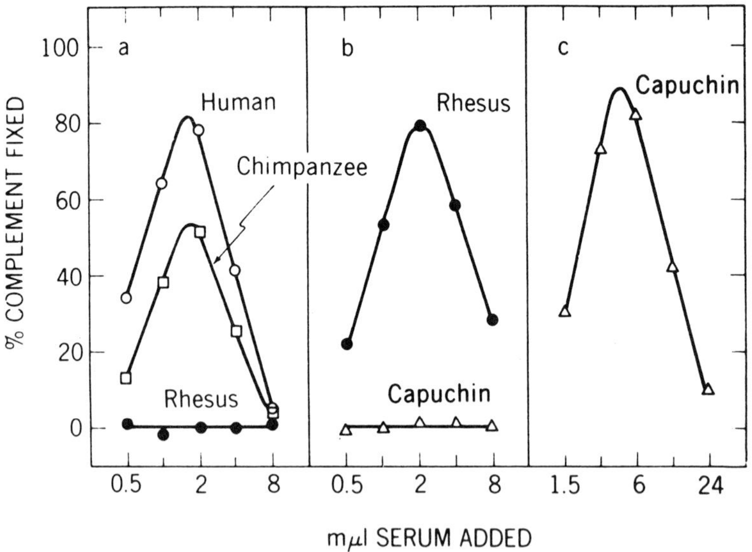

Figure 6.12. The effect of antiserum concentration on complement fixation by primate serum albumins. The antiserum, rabbit anti-human albumin (6-15) was tested at three concentrations (a. 1/11,000; b. 1/4,600; c. 1/2,500. As the concentration of albumin in serum is about 40 mg/ml, the amount of albumin required for maximum fixation for human, chimpanzee, rhesus and capuchin monkey was approximately 70, 70, 80, and 180 ng, respectively (from Sarich and Wilson, ref. *104*).

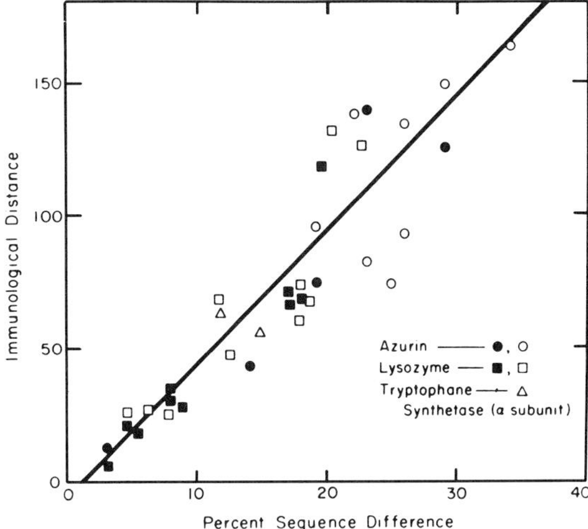

Figure 6.13. Correlation between immunological distance between two proteins and their differences in amino acid sequence (from Wilson, ref *122*)

measure of the immunological relatedness of the homologous and heterologous proteins is termed the *Index of Dissimilarity* (I.D.) and is defined as

$$\text{I.D.} = \frac{\text{Antiserum titer of the homologous antigen}}{\text{Antiserum titer of the heterologous antigen}}$$

The results of MC′F experiments can also be expressed as units of *Immunological Distance* which is defined as 100 × the log I.D. Complete details of the technique are available in a paper by Champion *et al.* (*32*). As mentioned earlier, the technique is very sparing of materials and thousands of cross-reactions can be carried out with but a few milliliters of antiserum. The procedure is also particularly sensitive to small changes in protein structure, and even single amino acid substitutions have been detected (*33, 93, 94*).

Extensive studies by Wilson and his colleagues (*32, 89*) have shown that there is an excellent correlation between the immunological distance between two proteins and their difference in amino acid sequence (up to approximately 25% of difference in sequence) (Fig. 6.13). It is this fact that makes the MC′F test well suited for systematic studies. Moreover, the procedure can be utilized for systematic comparisons across a very broad range by judicious selection of the type of proteins chosen as antigens. With very rapidly evolving proteins, such as insect larval hemolymph proteins, quantitative comparisons at the genus level are most effective (*13*), while with more slowly changing proteins it is possible to detect crossreactions between proteins from different phyla (*8*).

Because the microcomplement fixation procedure provides an estimate of the degree of sequence difference between proteins, rather than the exact sequences, matrix procedures must be employed for the construction of dendrograms. The procedures most commonly used for this purpose are those of Fitch and Margoliash (*48*), Farris (*44*) and Moore *et al.* (*80*). These methods can also be employed for constructing dendrograms from DNA hybridization data.

To date most of the systematic studies utilizing the MC′F technique have been confined to proteins of vertebrate or bacterial origin. For example, serum albumin, lysozyme, carbonic anhydrase, and transferrin have been used to examine the relationships of the primates (*85, 104, 105, 124*); these proteins plus ovalbumin and penalbumin have been examined in birds (*6, 7, 55, 60, 87, 88, 89, 90, 91*), while studies on the relationships of amphibia and reptiles have utilized serum albumins and lactate dehydrogenases (*14, 51, 76, 117, 118, 125*). Relatively few microcomplement fixation studies have been carried out with insect proteins, but those that have been reported show that the technique holds considerable promise for insect systematics. Brosemer and his colleagues (*22*) have utilized rabbit antibodies directed against honeybee α-glycerophosphate dehydrogenase to examine the crossreactivity with this protein from other insect species. The order of decreasing crossreaction was found to be honeybee (*Apis mellifera* Linn.), bumblebee (*Bombus fervidus* Fab.), robust mining bee (*Tetralonia* sp.), leaf cutting bee (*Megachile rotundata* Fab.), sphecoid wasp (*Sphecidae* sp.), yellow jacket (*Vespula maculifrons* Buyss.), syrphid fly (*Syrphus* sp.) and flesh fly (*Sarcophaga bullata* Parker) which is in excellent agreement with the classical taxonomic order. In a second study (*46*) the crossreactivity of the α-glycerophosphate dehydrogenases of bumblebees was investigated. In contrast to the single form of this enzyme found in honeybee species, the bumblebees show complex patterns of α-glycerophosphate dehydrogenase isozymes. Antisera directed against the major α-glycerophosphate dehydrogenase isozyme of *Bombus nevadensis* Cresson gave identical crossreactions with extracts of nine (six *Bombus* and three *Psithyrus* species) of the 11 other bumblebee species examined (ID=8). The enzymes from *Bombus bifarius* Cresson and *B. occidentalis* Greene showed less crossreactivity, while honeybee extracts crossreacted very weakly.

Marquardt *et al.* (*74*) have examined the immunological relatedness of honeybee (*Apis mellifera*) glyceraldehyde phosphate dehydrogenase to this enzyme in other insect species. While the results were generally in accord with classical taxonomy, anomalous results were obtained with extracts of the leafcutting bee (*Megachile rotundata*). The authors suggest that this anomally may be due to the relatively conservative changes in this slowly evolving protein.

In a more recent study, Collier and MacIntire (*36*) have prepared rabbit antisera against purified α-glycerophosphate dehydrogenases from *Droso-*

phila melanogaster Meigen, D. *virilis* Sturtevant, and D. *busckii* Coquillett. Crossreactions were carried out on the enzymes from a total of 34 Drosophilid species, and the results, in general, show good agreement with the accepted phylogeny of the *Drosophila*. Of note is the finding that the antiserum against the *D. melanogaster* enzyme could distinguish between allelic variants of α-glycerophosphate dehydrogenase. Because this enzyme evolves relatively slowly, it was possible to test the crossreactivity of the *Drosophila* α-glycerophosphate dehydrogenases with those from other Dipteran families, and again the results were in accord with phylogenies proposed from morphological criteria.

Beverly and Wilson (*13*) have recently given a preliminary report on the purification and characteristics of an insect larval hemolymph protein that appears to hold particular promise for quantitative comparisons at the species level. This rapidly evolving protein is readily purified from larval extracts and shows strong antigenic properties. The protein has already been used to explore *Drosophila* systematics using MC′F, and the results to date are in excellent accord with conventional taxonomy (*13* and personal communication from S. M. Beverly and A. C. Wilson).

In summary, the MC′F technique shows great promise for use in insect systematics though it has as yet been relatively little exploited. The procedure is rapid compared with protein sequencing, yet provides a good estimate of the degree of amino acid sequence resemblance between proteins. The tests are very sparing of both the specific proteins to be examined and of antibody. The procedure does require that the primary proteins which serve for immunization be highly purified but crossreactions can be carried out with crude cell extracts. For highly detailed phylogenetic studies it is desirable that several proteins be examined from each species and that at least a few of the comparisons be carried out in reciprocal fashion. It is now recognized that for obtaining antisera for use in systematic studies the immunization of rabbits should be carried out over a period of two to three months (*32*).

DNA HYBRIDIZATION

Not long after the basic complementary double-stranded, helical structure of DNA was established by Watson and Crick (*120, 121*), it was found that the strands of DNA could be separated by heat treatment (100° C), and if the strands were then incubated at a moderate temperature (60°C) they would renature to the double-stranded form (*43, 72, 73*). If single-stranded nucleic acids from different origins are incubated together, hybrid double-stranded molecules can be formed. In a classical experiment of this type Spiegelman showed that messenger RNA is complementary to its DNA template (*114, 115*). The technique was soon adapted to explore the relatedness of DNAs from different species (*56, 57, 79, 106*). If single-stranded DNAs from two closely related species are allowed to hybridize, the strands will share a great deal of sequence complementarity and consequently highly base-paired,

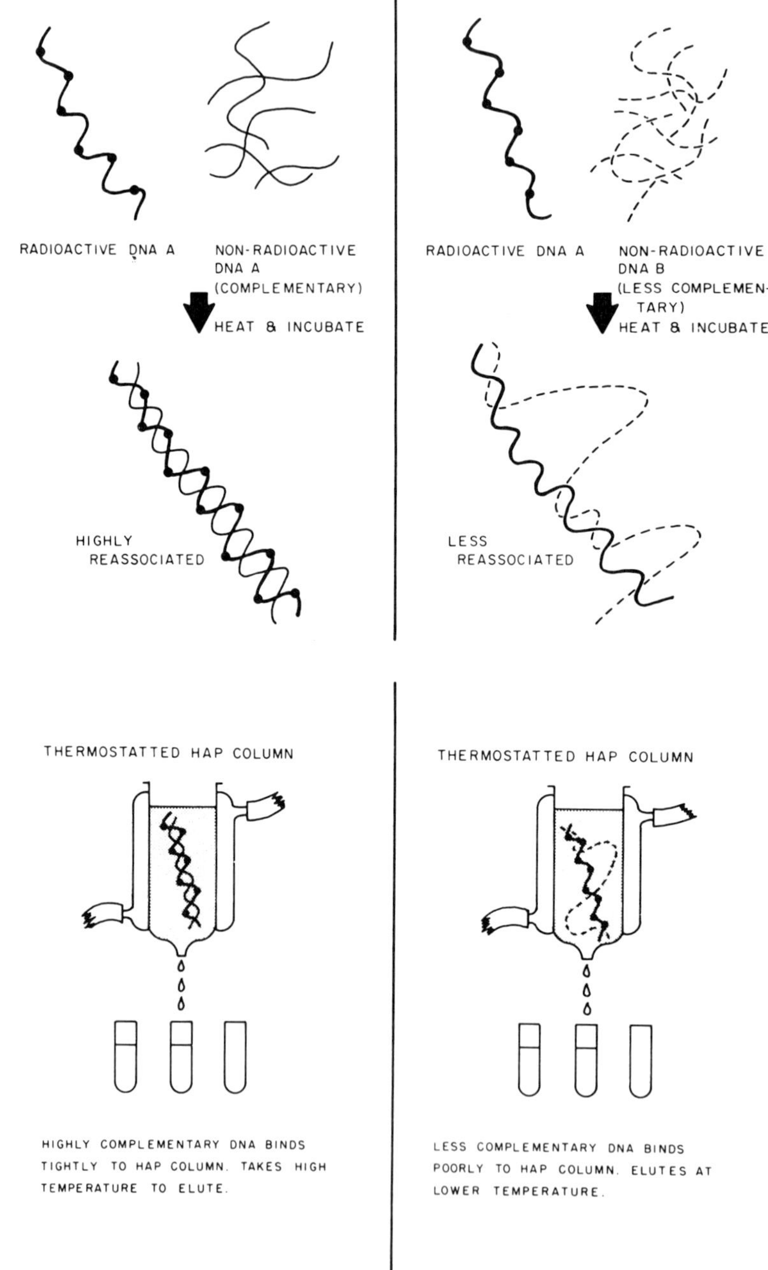

Figure 6.14. Basic procedure for one type of DNA hybridization (see text for details).

strongly interacting, double-stranded hybrids will be formed. With DNAs from distantly related species there will be less complementarity between the nucleotide sequences, and much less tightly structured hybrids will result. The degree of complementarity can be quantitated by determining the temperature required to separate the hybrid DNA strands.

The basic procedure for one type of DNA hybrid analysis is illustrated diagrammatically in Fig. 6.14, and additional details are available in papers by Kohne and Britten (*63*), Shields and Strauss (*109*), and Kohne *et al.* (*65*). The DNA is extracted and purified from cells by treatment with a detergent (often sodium dodecyl sulfate) and sodium perchlorate, and the proteins are removed by shaking with chloroform and isoamyl alcohol. The DNA is precipitated with ethanol (*70, 71, 109*).

For hybridization purposes it is necessary to use DNA fragments of relatively uniform size and these are prepared by homogenization or sonication. One of the DNA strands to be hybridized must carry a radioactive label. In earlier studies this was accomplished by growing cells in tissue culture, but techniques are now available for much more conveniently labelling the DNA in vitro with radioactive iodine (*37, 92*). To prepare the hybrid DNAs a sample of the radioactive DNA is mixed with a sample of unlabelled DNA of different origin (usually in the ratio of one part radioactive to 1000 parts unlabelled); the mixture is heated to 100°C to ensure initial strand separation and the two types of DNA are annealed at 60°C for 3 to 5 days. The hybridized DNA samples are then applied to a thermostatted hydroxylapatite (HAP) column. Double-stranded DNA binds to the hydroxylapatite, while any unreassociated single-stranded DNA can be eluted. The temperature of the HAP column is slowly raised from 60°C as the column is eluted. Those DNA hybrids which show poor complementarity will be dissociated into single strands and be eluted from the column at a relatively low temperature, while those with closely matched sequences will not be dissociated until higher temperatures are reached. The temperature at which the strands are dissociated is monitored by examining the eluted fractions for radioactivity.

DNA hybridization studies are complicated by the fact that in higher organisms there are two general classes of DNA. One type is termed repetitive DNA and consists of multiple copies of the same or very similar DNA sequences in a cell. The amount of repetitive DNA can range from 20% to 80% of the total DNA in cells of higher organisms (*20, 21*). The other type, termed "unique" or "single copy" DNA, is present in only one copy per cell. The presence of variable quantities of these two types of DNA in cell extracts hampered early systematic studies using DNA hybridization and gave rise to some inconsistent results. The "unique" DNA is best suited for systematic studies (*18, 62, 64, 65, 95*). Fortunately, it is now a relatively simple matter to separate the two DNA classes (*109*). Since the repetitive DNA contains many copies of identical or very similar sequences, these recombine very rapidly under reassociating conditions compared with the "unique" DNA. The DNA samples are simply partially hybridized for a short period and the associated repetitive sequences removed by passage through an hydroxylapatite column. The double-stranded repetitive sequences are retained by the column while the "unique" copy DNA, which is still single-stranded, passes through.

While the results of early studies utilizing unfractionated DNA must be treated with caution, more recent studies utilizing the "unique" copy DNA have provided very useful systematic data. A few examples will suffice to demonstrate the utility of the procedure. Kohne *et al.* (*64, 65*) have applied the DNA hybridization technique to an examination of the primate phylogenetic tree. Their results (Fig. 6.15) are entirely in accord with evolutionary trees based on both immunological and morphological evidence. It should be emphasized, however, that while the general configuration of such evolutionary trees shows very good agreement, estimates of divergence times for various lineages vary over wide periods, primarily because of a lack of agreement between research groups on paleontological dating.

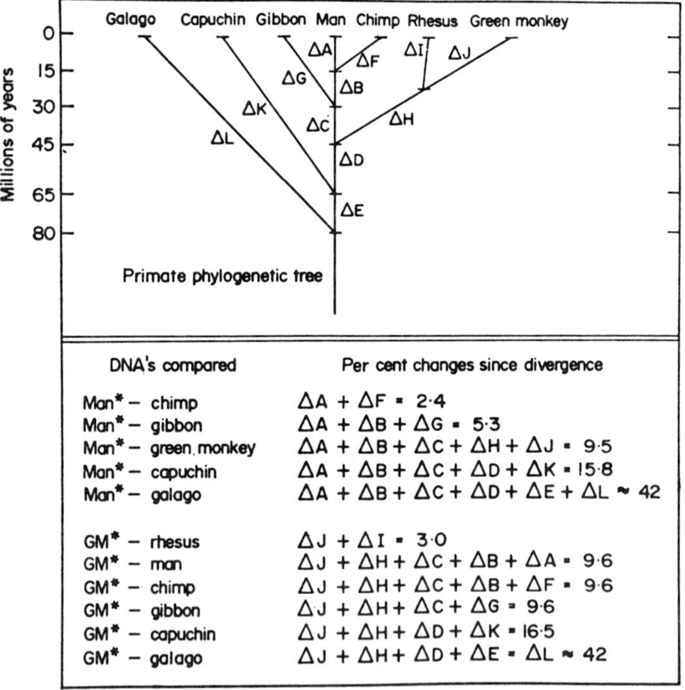

Figure 6.15. Primate phylogenetic tree with symbols assigned to various time periods of the primate evolutionary history. Each value represents the average of two or more thermal stability profiles. Values were reproducible to about ±0.3° C (from Kohne *et al.*, ref. *65*).

As an example of the DNA hybridization technique applied to insects, Sohn *et al.* (*112*) have used the procedure to examine the systematic relationship of six species of black flies. The results they obtained are in general accord with classical procedures (Fig. 6.16), but on a quantitative basis it was observed that some species usually classified in the same genus (e.g., *Simulium venustum* Say and *S. pictipes* Hagen) are genetically as dissimilar from each other as flies classified as species of other genera. As the authors put it, " . . . the present generic limits in black flies bear reexamination." Undoubtedly, as biochemical means of quantitative comparison come into widespread use in systematics, the exact meanings of the terms species, genera, family, etc., as they are applied in different classes or more particularly phyla, will come under increasing scrutiny.

DNA Sequencing. From a biochemical standpoint, the ultimate systematic procedure would be one which allowed the detailed comparison of the total genome of one organism with that of others. Two recent breakthroughs in biochemical techniques have raised the possibility that in the near future this may be much more than an idle dream. The first major advance is the development of extremely rapid procedures for the determination of the nucleotide sequence of DNA molecules (*75*, *99*). Such procedures have already allowed the determination of the complete DNA sequence of the bacteriophage ϕx174, a total of 5,374 nucleotides (*100*). The second, and corequisite development as far as systematics is concerned, is the extraordinary advancement in techniques for specific gene isolation and amplification. These methods, popularly termed genetic engineering, allow for the virtually unlimited propagation of eucaryotic genes in procaryotic hosts (*1*, *34*, *35*, *40*, *81*, *82*).

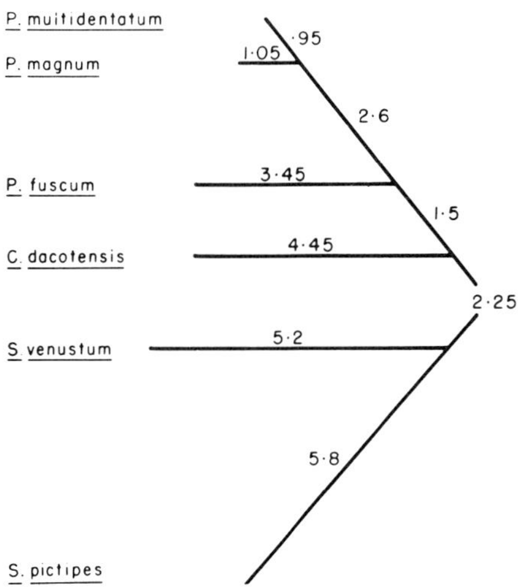

Figure 6.16. Evolutionary tree for six black fly species. The numbers on the tree represent the amount of mutation that has accumulated in each interval—expressed as depression in the melting temperature of hybrids of unique sequences. These values were assigned by assuming the value for *S. pictipes—P. multidentatum* hybrids to be 13.1 and solving for the other values in the algebraic manner used by Kohne *et al.* (*64*) (from Sohn *et al.*, ref. *112*).

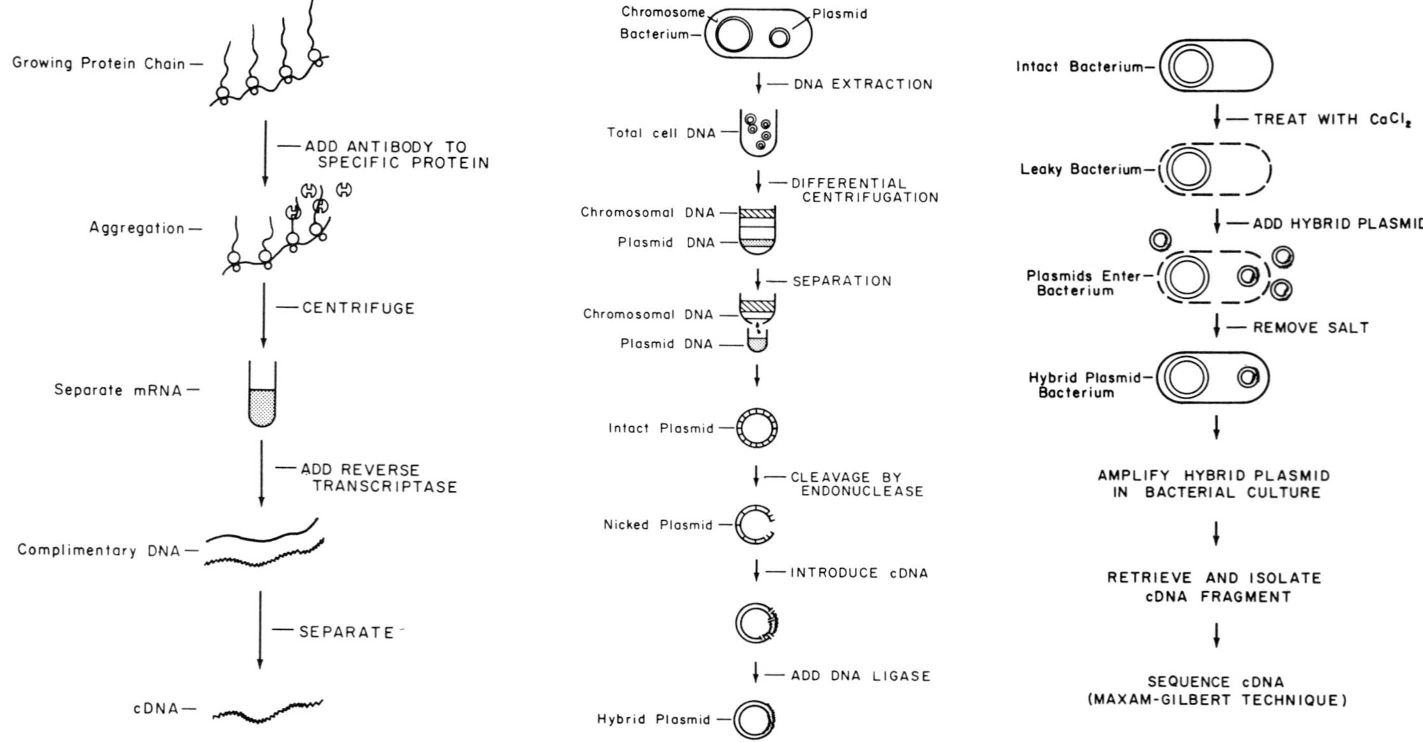

Figure 6.17. Basic procedure for the isolation and cloning of a specific gene for DNA sequencing. *(Left)* Isolation of specific messenger RNA (mRNA) and the production of complementary DNA (cDNA). *(Center)* Separation of plasmid DNA from bacterium and insertion of cDNA to make hybrid plasmid. *(Right)* Insertion of hybrid plasmid into bacterium and subsequent retrieval and sequencing of cDNA.

The necessity of gene amplification, or cloning, procedures as a prerequisite for DNA sequencing should be explained first. Presently available direct gene isolation procedures to provide DNA for sequence analysis are satisfactory for this purpose only when the DNA fragment in question comprises more than approximately 0.005% of the total genome (*23*). While this is adequate for obtaining sufficient quantities of DNA corresponding to the highly reiterated genes such as those for ribosomal RNA and transfer RNA, it is manifestly unsuitable for isolation of sufficient quantities of those genes that are present in only a single copy per genome. For example, a single gene corresponding to a protein of molecular weight 100,000 would comprise only about 1.7×10^{-5}% of the total human genome. Put in other terms, one gram of total DNA would only contain 0.17 micrograms of this gene. Clearly, gene amplification techniques are required in this instance if sufficient quantities of this specific DNA are to be made available for sequencing, and it is here that gene cloning procedures are proving invaluable.

One basic procedure for the isolation and cloning of a specific gene is illustrated in highly schematic form in Fig. 6.17. To cell-free extracts known to be vigorously producing the specific gene product (protein) for which the gene sequence is to be determined are added antibodies to this protein (Fig. 6.17, *left*). This causes aggregation of the combined polysomes, messenger RNA, and antibody. From this mixture the specific messenger RNA is isolated. [In some cases, for example, globin genes in reticulocytes; it may be possible to isolate the mRNA directly (*69*).] A complementary DNA copy (cDNA) of this mRNA is then synthesized by use of reverse transcriptase. This single-stranded DNA is then made double-stranded by use of DNA polymerase. At this stage the DNA (or fragment of it) has to be transferred to a procaryotic host for the generation of multiple copies. This is accomplished by incorporating the double-stranded eucaryotic DNA into a small extra-chromosomal circular DNA element found in many bacteria and termed a plasmid. The incorporation of the foreign DNA into the plasmid is illustrated schematically in Fig. 6.17, *center*.

Once the foreign DNA has been inserted into the plasmid, there remains the problem of reintroducing the plasmid into a viable bacterial cell. This is accomplished by making the bacterial cell walls permeable by treatment with calcium chloride (Fig. 6.17, *right*). Although only a very few bacterial cells incorporate the plasmid (probably only one or two in a million), this is sufficient for further replication of the plasmid since cells containing plasmids can be very selectively isolated. Once the plasmid-containing bacteria are available they can be grown in culture in any quantity desired and the plasmids, now multiplied several millionfold, can be re-isolated. The foreign DNA, now also multiplied in the same ratio, can be isolated by a reversal of the techniques used for its incorporation into the plasmids and can thus be made available in sufficient quantities for sequence analysis.

The determination of protein amino acid sequence has been progressing at an ever increasing pace over the last 20 years with the result that something over 1000 protein sequences have now been established. By contrast, progress in the determination of the nucleotide sequence of nucleic acids moved at a much slower pace until the last few years. This slow pace was largely due to the fact that, while proteins contain 20 different amino acids and this allows the relatively easy overlap of different fragments, DNA contains only four different nucleotide bases, making the alignment of fragments much more difficult. Moreover, no satisfactory enzymatic or chemical procedures existed which could sequentially remove one nucleotide at a time from DNA in a manner analogous to the Edman degradation procedure for proteins. However, the recent development of new sequencing techniques, primarily by Frederick Sanger and his colleagues in Cambridge (*2, 3, 4, 98, 100, 101*), which do not depend upon sequential removal of nucleotides, has made the fact that DNA only has four types of nucleotides a positive advantage rather than a hindrance. These new sequencing methods, together with the use of restriction enzymes to excise small pieces of DNA of interest from large DNA fragments, now makes the determination of DNA sequences less arduous and time consuming than protein sequence analysis. The rapid progress in this field can be gauged from Gilbert's report (*66*) that the determination of a sequence of 20 nucleotides, which previously took two years of work, can now be accomplished in one day.

Two DNA sequencing procedures are now in common use. Though related in basic approach they differ considerably in detail. The technique of Sanger and Coulson (*99*) depends to a large degree on enzymatic steps, while that of Maxam and Gilbert (*75*) depends upon specific chemical cleavage of the DNA. Both procedures yield a series of radiolabelled fragments, differing in length and ending at a specific type of nucleotide. These fragments are separated by electrophoresis on polyacrylamide slabs, the

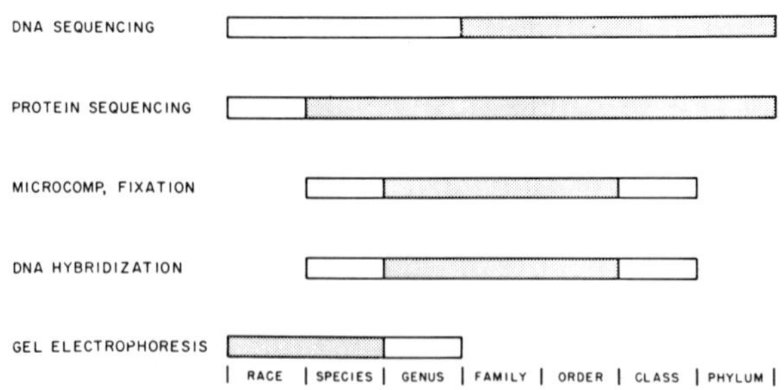

Figure 6.18. Evaluation of various methods for establishing levels of genetic differentiation between taxa with respect to degree of sensitivity and ease of analysis. DNA and protein sequencing should be used on higher categories while gel electrophoresis is the best method for comparing races, species, and closely related genera.

resolution of this procedure being sufficient to cleanly resolve fragments that differ by only a single nucleotide residue. The sequence of the DNA fragment can then be read directly from a radioautograph of the polyacrylamide gel. DNA sequences of 100 nucleotides can now be ascertained routinely on a single gel and the system appears to have the potential for resolving fragments up to ten times that length in one run.

While the whole field of DNA sequencing is still in its infancy, the potential for its application to systematics seems extraordinarily good. A measure of the acceptance of the techniques can be gained from the fact that several research groups are already collaborating in an attempt to define the sequence of the entire genome of a *Drosophila* species.

CONCLUSIONS

Molecular genetics, therefore, provide a very useful approach to biosystematics problems. The method employed (for example, DNA or protein sequencing, MC′F, DNA hybridization, or gel electrophoresis) will depend on what line of systematic delineation is required. As illustrated in Fig. 6.18 gel electrophoresis is the method best suited for studying the genetic structure of populations within and between clines, races, species, and closely related genera. If on the other hand, phylogenetic relationships are to be established among higher taxonomic categories (for example, genera, families, and orders) microcomplement fixation and DNA hybridization are more appropriate. DNA and protein sequencing, however, are still time consuming and expensive. Although they can in theory be applied to all taxonomic categories, their use should be restricted to relationships of high categories such as orders, classes, and phyla.

Also, molecular genetics can help resolve phylogenetic problems and establish a better understanding of the rates of speciation and organismal evolution. We have finally reached the point in systematics and evolutionary biology where critical models of evolutionary significance can be tested with some rigor and precision. It is very possible that in the next decade there will be a veritable revolution in our views on evolutionary processes. This revolution will be less disruptive if the organismal systematists and molecular biologists work closely together.

LITERATURE CITED

1. Abelson, J. 1977. *Recombinant DNA: Examples of present-day research.* Science 196: 159–160.

2. Air, G. M., E. H. Blackburn, F. Sanger, and A. R. Coulson. 1975. *The nucleotide and amino acid sequences of the N(5′) terminal region of gene G of bacteriophage ϕX174.* J. Mol. Biol. 96: 703–719.

3. Air, G. M., E. H. Blackburn, A. R. Coulson, F. Galibert, F. Sanger, J. W. Sedat, and E. B. Ziff. 1976. *Gene F of bacteriophage ϕX174. Correlation of nucleotide sequences from the DNA and amino acid sequences from the gene product.* J. Mol. Biol. 107: 445–458.

4. Air, G. M., F. Sanger, and A. R. Coulson. 1976. *Nucleotide and amino acid sequences of gene G of φX174.* J. Mol. Biol. 108: 519-533.

5. Ames, G. F.-L., and K. Nikaido. 1976. *Two dimensional gel electrophoresis of membrane proteins.* Biochemistry 15: 616-623.

6. Arnheim, N., A. Hindenburg, G. S. Begg, and F. J. Morgan. 1973. *Multiple genes for lysozyme in birds. Studies on black swan egg lysozymes.* J. Biol. Chem. 248: 8036-8042.

7. Arnheim, N., E. M. Prager, and A. C. Wilson. 1969. *Immunological prediction of sequence differences among proteins.* J. Biol. Chem. 244: 2085-2094.

8. Arnon, R., and H. Neurath. 1969. *An immunological approach to the study of evolution of trypsins.* U. S. Natl. Acad. Sci., Proc. 64: 1323-1328.

9. Avise, J. C., and F. J. Ayala. 1975. *Genetic differentiation in speciose versus depauperate phylads: Evidence from the California minnows.* Evolution 29: 411-426.

10. Ayala, F. J. 1975. *Genetic differentiation during the speciation process.* Evolut. Biol. 8: 1-78.

11. Bateman, M. A. 1976. *Fruit flies.* Pages 11-49 *in* V. L. Delucchi, ed. *Studies in Biological Control.* 9. (See especially section by G. Bush and M. Huettel entitled *Population and ecological genetics.* pp. 43-49.) Cambridge Univ. Press, London.

12. Berlocher, S. H. 1976. *The genetics of speciation in Rhagoletis (Diptera: Tephritidae).* Ph. D. Dissertation, University of Texas at Austin. 203 pp.

13. Beverly, S. M., and A. C. Wilson. 1977. *Purification, characterization and immunological cross-reactivity of a hemolymph protein from Drosophila larvae.* Fed. Proc. 36: 927.

14. Bisbee, C. A., M. A. Baker, A. C. Wilson, I. Hadji-Azimi, and M. Fischberg. 1977. *Albumin phylogeny for clawed frogs (Xenopus).* Science 195: 785-787.

15. Boller, E. F., and G. L. Bush. 1974. *Evidence for genetic variation in populations of the European cherry fruit fly, Rhagoletis cerasi (Diptera: Tephritidae) based on physiological parameters and hybridization experiments.* Ent. Exp. and Appl. 17: 279-293.

16. Boller, E. F., K. Russ, V. Vallo, and G. L. Bush. 1976. *Incompatible races of European cherry fruit fly, Rhagoletis cerasi (Diptera: Tephritidae), their origin and potential use in biological control.* Ent. Exp. and Appl. 20: 237-247.

17. Britten, R. J., and E. H. Davidson. 1969. *Gene regulation for higher cells: A theory.* Science 165: 349-352.

18. Britten, R. J., and E. H. Davidson. 1971. *Repetitive and non-repetitive DNA sequences and a speculation on the origins of evolutionary novelty.* Quart. Rev. Biol. 46: 111-133.

19. Britten, R. J., and E. H. Davidson. 1976. *DNA sequence arrangement and preliminary evidence on its evolution.* Fed. Proc. 35: 2151-2157.

20. Britten, R. J., and D. E. Kohne. 1968. *Repeated sequences in DNA.* Science 161: 529-540.

21. Britten, R. J., and D. E. Kohne. 1970. *Repeated segments of DNA.* Sci. Amer. 222(4): 24-31.

22. Brosemer, R. W., D. S. Grosso, G. Estes, and C. W. Carlson. 1967. *Quantitative immunochemical and electrophoretic comparisons of glycerophosphate dehydrogenases in several insects.* J. Insect Physiol. 13: 1757-1767.

23. Brown, D. D., and R. Stern. 1974. *Methods of gene isolation.* Ann. Rev. Biochem. 43: 667-693.

24. Bush, G. L. 1966. *The taxonomy, cytology and evolution of the genus Rhagoletis in North America (Diptera, Tephritidae).* Bull. Mus. Comp. Zool. 134: 431-562.

25. Bush, G. L. 1975. *Sympatric speciation in phytophagous parasitic insects.* Pages 187-206 *in* P. W. Price, ed. *Evolutionary strategies of parasitic insects and mites.* Plenum, New York.

26. Bush, G. L. 1975. *Modes of speciation.* Ann. Rev. Ecol. and System. 6: 339-364.

27. Bush, G. L. 1977. *The chromosome morphology of the Rhagoletis cerasi species complex (Diptera, Tephritidae).* Ann. Ent. Soc. Amer. 3: 316-318.

28. Bush, G. L., S. M. Case, A. C. Wilson, and J. L. Patton. 1977. *Rapid speciation and chromosome evolution in mammals.* U. S. Natl. Acad. Sci., Proc. 74: 3942-3946.

29. Bush, G. L., and R. N. Huettel. 1972. *Starch gel electrophoresis of tephritid proteins. A manual of techniques.* Int. Biol. Program Working Group on Fruit Flies. *Population Genetics Project, Phase I.* 56 pp.
30. Carson, H. L. 1970. *Chromosome tracers of the origin of species.* Science 168: 1414-1418.
31. Carson, H. L., and K. Y. Kaneshiro. 1976. *Drosophila of Hawaii: Systematics and ecological genetics.* Ann. Rev. Ecol. Syst. 7: 311-345.
32. Champion, A. B., E. M. Prager, D. Wachter, and A. C. Wilson. 1974. *Microcomplement Fixation.* Pages 397-416 *in* C. A. Wright, ed. *Biochemical and immunological taxonomy of animals.* Academic Press, New York.
33. Cocks, G. T., and A. C. Wilson. 1969. *Immunological detection of single amino acid substitutions in alkaline phosphatase.* Science 164: 188-189.
34. Cohen, S. N. 1975. *The manipulation of genes.* Sci. Amer. 233(1): 24-32.
35. Cohen, S. N., A. C. Y. Chang, H. W. Boyer, and R. B. Helling. 1973. *Construction of biologically functional bacterial plasmids in vitro.* U. S. Natl. Acad. Sci., Proc. 70: 3240-3244.
36. Collier, G. E., and R. J. MacIntire. 1977. *Microcomplement fixation studies on the evolution of α-glycerophosphate dehydrogenase within the genus Drosophila.* U. S. Natl. Acad. Sci., Proc. 74: 684-688.
37. Commorford, S. L. 1971. *Iodination of nucleic acids in vitro.* Biochemistry 10: 1993-2000.
38. Cuatrecasas, P., and C. B. Anfinsen. 1971. *Affinity chromatography.* Ann. Rev. Biochem. 40: 259-278.
39. Cuatrecasas, P., M. Wilchek, and C. B. Anfinsen. 1968. *Selective enzyme purification by affinity chromatography.* U. S. Natl. Acad. Sci., Proc. 61: 636-643.
40. Curtiss III, R. 1976. *Genetic manipulation of microorganisms. Potential benefits and hazards.* Ann. Rev. Microbiol. 30: 507-533.
41. Dobzhansky, T. 1970. *Genetics of the evolutionary process.* Columbia Univ. Press, New York. 505 pp.
42. Dobzhansky, T., F. J. Ayala, G. L. Stebbins, and J. W. Valentine. 1977. *Evolution.* W. H. Freeman and Co. 572 pp.
43. Doty, P., J. Marmur, J. Eigner, and C. Schildkraut. 1960. *Strand separation and specific recombination in deoxyribonucleic acids: Physical chemical studies.* U. S. Natl. Acad. Sci., Proc. 46: 461-476.
44. Farris, J. S. 1972. *Estimating phylogenetic trees from distance matrices.* Amer. Natur. 106: 645-668.
45. Felsenstein, J. 1973. *Maximum likelihood and minimum-step methods for estimating evolutionary trees from data on discrete characters.* Syst. Zool. 22: 240-249.
46. Fink, S. C., C. W. Carlson, S. Gurusiddaiah, and R. W. Brosemer. 1970. *Glycerol-3-phosphate dehydrogenases in social bees.* J. Biol. Chem. 245: 6525-6532.
47. Fitch, W. M. 1976. *Molecular evolutionary clocks.* Pages 160-178 *in* F. J. Ayala, ed. *Molecular evolution.* Sinauer Assoc., Inc., Sunderland, Mass.
48. Fitch, W. M., and E. Margoliash. 1967. *Construction of phylogenetic trees.* Science 155: 279-284.
49. Flake, R. H., and R. K. Lennington. 1977. *Genetic distances utilizing electrophoretic mobility data.* Biol. Zentralbl. (in press).
50. Gordon, A. H. 1975. *Electrophoresis of proteins in polyacryalmide and starch gels.* 2nd ed. North-Holland Publ., Amsterdam. 213 pp.
51. Gorman. G. C., A. C. Wilson, and M. Nakanishi. 1971. *A biochemical approach towards the study of reptilian phylogeny: Evolution of serum albumin and lactic dehydrogenase.* System. Zool. 20: 167-185.
52. Gurdon, J. B., E. M. DeRobertis, and G. Partington. 1976. *Injected nuclei in frog oocytes provide a living cell system for the study of transcriptional control.* Nature 260: 116-120.
53. Harris, H., and D. Hopkinson. 1976. *Handbook of Enzyme Electrophorphoresis in*

human genetics. North-Holland Publ., Amsterdam.

54. Hedrick, P. W., M. E. Ginevan, and E. P. Ewing. 1976. *Genetic polymorphism in heterogeneous environments*. Ann. Rev. Ecol. Syst. 7: 1-32.

55. Ho, C. Y-K., E. M. Prager, A. C. Wilson, D. T. Osuga, and R. E. Feeney. 1976. *Penquin evolution: Protein comparisons demonstrate phylogenetic relationship to flying aquatic birds*. J. Mol. Evol. 8: 271-282.

56. Hoyer, B. H., B. J. McCarthy, and E. T. Bolton. 1964. *A molecular approach to the systematics of higher organisms*. Science 144: 959-967.

57. Hoyer, B. H., N. W. Van de Velde, M. Goodman, and R. B. Roberts. 1972. *Examination of hominid evolution by DNA sequence homology*. J. Human Evol. 1: 645-649.

58. Johnson, G. B. 1975. *Use of internal standards in electrophoretic surveys of enzyme polymorphism*. Biochem. Genet. 13: 833-847.

59. Johnson, G. B. 1977. *Characterization of electrophoretically cryptic variation in the alpine butterfly Colias meadii*. Genetics 15:665-693.

60. Jollès, J., F. Schoentgen, P. Jollès, E. M. Prager, and A. C. Wilson. 1976. *Amino acid sequence and immunological properties of Chachalaca egg white lysozyme*. J. Mol. Evol. 8: 59-78.

61. King, M. C., and A. C. Wilson. 1975. *Evolution at two levels: Molecular similarities and biological differences between humans and chimpanzees*. Science 188: 107-116.

62. Kohne, D. E. 1970. *Evolution of higher organism DNA*. Quart. Rev. Biophys. 33: 327-375.

63. Kohne, D. E., and R. J. Britten. 1971. *Hydroxyapatite techniques for nucleic acid reassociation*. Pages 500-512 *in* G. F. Cantoni and D. R. Davis, eds. *Procedures in nucleic acid research, Vol. 2*. Harper and Row, New York.

64. Kohne, D. E., J. A. Chiscon, and B. H. Hoyer. 1971. *Nucleotide sequence change in non-repeated DNA during evolution*. Carnegie Inst. Wash. Yearbook 69: 488-501.

65. Kohne, E. D., J. A. Chiscon, and B. H. Hoyer. 1972. *Evolution of primate DNA sequences*. J. Human Evol. 1: 627-644.

66. Kolata, G. B. 1976. *DNA sequencing: A new era in molecular biology*. Science 192: 645-647.

67. Laskey, R. A., and A. D. Mills. 1976. *Quantitative film detection of polyacryalmide gels by fluorography*. J. Biochem. 56: 335-341.

68. Lewontin, R. C. 1974. *The genetic basis of evolutionary change*. Columbia Univ. Press, New York. 346 pp.

69. Liu, A. Y., G. V. Paddock, H. C. Heindell, and W. Salser. 1977. *Nucleotide sequences from a rabbit alpha globin gene inserted in a Chimeric plasmid*. Science 196: 192-195.

70. Marmur, J. 1961. *A procedure for the isolation of deoxyribonucleic acid from micro-organisms*. J. Mol. Biol. 3: 208-218.

71. Marmur, J. 1963. *A procedure for the isolation of deoxyribonucleic microorganisms*. Pages 726-738 *in* S. P. Colowick and N. O. Kaplan, eds. *Methods in enzymology*. Vol. 6. Academic Press, New York.

72. Marmur, J., and P. Doty. 1961. *Thermal renaturation of deoxyribonucleic acids*. J. Mol. Biol. 3: 585-594.

73. Marmur, J., and D. Lane. 1960. *Strand separation and specific recombination in deoxyribonucleic acids: Biological studies*. U. S. Natl. Acad. Sci., Proc. 46: 453-461.

74. Marquardt, R. W., C. W. Carlson, and R. W. Brosemer. 1968. *Glyceraldehyde phosphate dehydrogenase: Crystallization from honeybees; quantitative immunochemical and electrophoretic comparisons of the enzyme in other insects*. J. Insect Physiol. 14: 317-333.

75. Maxam, A. M., and W. Gilbert. 1977. *A new method for sequencing DNA*. U. S. Natl. Acad. Sci., Proc. 74: 560-564.

76. Maxson, L. R., and A. C. Wilson. 1974. *Convergent morphological evolution detected by studying proteins of tree frogs in the Hyla eximia species group*. Science 185: 66-68.

77. Mayr, E. 1963. *Animals, species and evolution*. Harvard Univ. Press, Cambridge, Mass. 795 pp.

78. Mayr, E. 1969. *Principles of systematic zoology*. McGraw-Hill, New York. 428 pp.

79. McCarthy, B. J., and E. T. Bolton. 1963. *An approach to the measurement of genetic relatedness among organisms*. U. S. Natl. Acad. Sci., Proc. 50: 156-164.

80. Moore, G. W., M. Goodman, and J. Barnabas. 1973. *An iterative approach from the standpoint of the additive hypothesis to the dendrogram problem posed by molecular data sets*. J. Theor. Biol. 38: 423-457.

81. Morrow, J. F., S. N. Cohen, A. C. Y. Chang, H. W. Boyer, H. M. Goodman, and R. B. Helling. 1974. *Replication and transcription of eukaryotic DNA in Escherichia coli*. U. S. Natl. Acad. Sci., Proc. 71: 1743-1747.

82. Murray, K. 1976. *Biochemical manipulation of genes*. Endeavour 35: 129-133.

83. Nei, M. 1972. *Genetic distance between populations*. Amer. Nat. 106: 283-291.

84. Nei, M. 1975. *Molecular population genetics and evolution*. North-Holland Publ., Amsterdam. 288 pp.

85. Nonno, L., H. Hershman, and L. Levine. 1970. *Serologic comparisons of the carbonic anhydrases of primate erythrocytes*. Arch. Biochem. Biophys. 136: 361-367.

86. O'Farrell, P. H. 1975. *High resolution two-dimensional electrophoresis of proteins*. J. Biol. Chem. 250: 4007-4021.

87. Prager, E. M., and A. C. Wilson. 1971. *The dependence of immunological crossreactivity upon sequence resemblance among bird lysozymes: Microcomplement fixation studies*. J. Biol. Chem. 246: 5978-5989.

88. Prager, E. M., and A. C. Wilson. 1971. *The dependence of immunological crossreactivity upon sequence resemblance among lysozymes*. J. Biol. Chem. 246: 7010-7017.

89. Prager, E. M., and A. C. Wilson. 1972. *Comparison of multiple duck lysozymes*. Biochem. Genet. 7: 269-272.

90. Prager, E. M., A. H. Brush, R. A. Nolan, M. Nakanishi, and A. C. Wilson. 1974. *Slow evolution of transferrin and albumin in birds according to microcomplement fixation analysis*. J. Mol. Evol. 3: 243-262.

91. Prager, E. M., A. C. Wilson, D. T. Osuga, R. E. Feeney. 1976. *Evolution of flightless landbirds on southern continents. Transferrin comparison shows monophyletic origin of ratites*. J. Mol. Evol. 8: 283-294.

92. Prensky, W. 1976. *The radioiodination of RNA and DNA to high specific activities*. Pages 121-152 *in* D. M. Prescott, ed. *Methods in cell biology*. Vol. 13. Academic Press, New York.

93. Reichlin, M. 1972. *Localizing antigenic determinants in human hemoglobin with mutants: Molecular correlations of immunological tolerance*. J. Mol. Biol. 64: 485-496.

94. Reichlin, M., M. Hay, and L. Levine, 1964. *Antibodies to human A_1 hemoglobin and their reaction with A_2, S, C, and H Hemoglobins*. Immunochemistry 1: 21-30.

95. Rice, N. R. 1973. *Single copy DNA relatedness among several species of the Cricetidae (Rodentia)*. Carnegie Inst. Wash. Yearbook 73: 1098-1102.

96. Richardson, R. H., and P. E. Smouse. 1976. *Patterns of molecular variation. I. Interspecific comparisons of electromorphs in the Drosophila mulleri complex*. Biochem. Genet. 14: 447-465.

97. Rogers, J. S. 1972. *Measures of genetic similarity and genetic distance*. Univ. Texas Publ. 7213: 145-153.

98. Salser, W. A. 1974. *DNA sequencing techniques*. Ann. Rev. Biochem. 43: 923-965.

99. Sanger, F., and A. R. Coulson. 1975. *A rapid method for determining sequences in DNA by primed synthesis with DNA polymerase*. J. Mol. Biol. 94: 441-448.

100. Sanger, F., G. M. Air, G. B. Barrell, N. L. Brown, A. R. Coulson, J. C. Fiddes, C. A. Hutchinson, III, P. M. Slocombe, and M. Smith. 1977. *Nucleotide sequence of bacteriophage ϕX174 DNA*. Nature 265: 687-695.

101. Sanger, F., J. F. Donelson, A. R. Coulson, H. Kossel, and D. Fischer. 1973. *Use of DNA polymerase I primed by a synthetic oligonucleotide to determine a nucleotide sequence in phage fl DNA*. U. S. Natl. Acad. Sci., Proc. 70: 1209-1213.

102. Sarich, V. M. 1972. *Generation time and albumin evolution*. Biochem. Genet. 7: 205-212.

103. Sarich, V. M. 1977. *Rates, sample sizes and the neutrality hypothesis for electrophoresis in evolutionary studies*. Nature 265: 24-28.

104. Sarich, V. M., and A. C. Wilson. 1966. *Quantitative immunochemistry and the evolution of primate albumins: Microcomplement fixation*. Science 154: 1563-1566.

105. Sarich, V. M., and A. C. Wilson. 1967. *Immunological time scale for hominid evolution*. Science 158: 1200-1203.

106. Schildkraut, C. L., J. Marmur, and P. Doty. 1961. *The formation of hybrid DNA molecules and their use in studies of DNA homologies*. J. Mol. Biol. 3: 595-617.

107. Scopes, R. K. 1977. *Purification of glycolytic enzymes using affinity-elution chromatography*. Biochem. J. 161: 253-263.

108. Selander, R. K., and W. E. Johnson. 1973. *Genetic variation among vertebrate species*. Ann. Rev. Ecol. Syst. 4: 75-91.

109. Shields, G. F., and N. A. Straus. 1975. *DNA-DNA hybridization studies of birds*. Evolution 29: 159-166.

110. Simpson, G. G. 1961. *Principles of animal taxonomy*. Columbia Univ. Press, New York. 247 pp.

111. Singh, R. R., R. C. Lewontin, and A. A. Felton. 1976. *Genetic heterogeneity within electrophoretic "alleles" of xanthine dehydrogenase in Drosophila pseudoobscura*. Genetics 84: 609-629.

112. Sohn, U., K. H. Rothfels, and N. A. Straus. 1975. *DNA-DNA hybridization studies in black flies*. J. Mol. Evol. 5: 75-85.

113. Sneath, P. H. A., and R. R. Sokol. 1973. *Numerical taxonomy*. Freeman, San Francisco. 573 pp.

114. Spiegelman, S. 1961. *The relation of informational RNA to DNA*. Cold Spring Harbor Symposia Quant. Biol. 26: 75-90.

115. Spiegelman, S. 1964. *Hybrid nucleic acids*. Sci. Amer. 210(5): 48-56.

116. Wagner, R. P., and R. K. Selander. 1974. *Isozymes in insects and their significance*. Ann. Rev. Entomol. 19: 117-138.

117. Wallace, D. G., and A. C. Wilson. *Comparison of frog albumins with those of other vertebrates*. J. Mol. Evol. 2: 72-86.

118. Wallace, D. G., M-C. King, and A. C. Wilson. 1973. *Albumin differences among ranid frogs: Taxonomic and phylogenetic implications*. Systematic Zool. 22: 1-13.

119. Wassarman, E., and L. Levine. 1961. *Quantitative microcomplement fixation and its use in the study of antigenic structure by specific antigen antibody inhibition*. J. Immunol. 87: 290-295.

120. Watson, D., and F. H. C. Crick. 1953. *Molecular structure of nucleic acid. A structure for deoxyribose nucleic acid*. Nature 171: 737-738.

121. Watson, J. D., and F. H. C. Crick. 1953. *Genetic implications of the structure of deoxyribonucleic acil*. Nature 171: 964-967.

122. Wilson, A. C. 1975. *Evolutionary importance of gene regulation*. Stadler Symp. 7: 117-133. Univ. Mo., Columbia, Mo.

123. Wilson, A. C., S. S. Carlson, and T. J. White. 1977. *Molecular evolution*. Ann. Rev. Biochem. 146: 573-639.

124. Wilson, A. C., and V. M. Sarich. 1969. *A molecular time scale for human evolution*. U.S. Natl. Acad. Sci., Proc. 63: 1088-1093.

125. Wilson, A. C., L. R. Maxson, and V. M. Sarich. 1974. *Two types of molecular evolution. Evidence from studies of interspecific hybridization*. U. S. Natl. Acad. Sci., Proc. 71: 2843-2847.

7] Recent Developments in the Systematics of Pathogenic Fungi

by BRYCE KENDRICK*

ABSTRACT

Since 1968 six large-scale compilations and/or syntheses of almost 600 "good" form-genera of Hyphomycetes have appeared, elicited after a 50-year hiatus by the irruption of developmental concepts onto a scene previously dominated by morphology of mature organisms. But ontogeny is no taxonomic panacea, and the best treatments are now eclectic, since only a developmental process that leaves its mark on the mature organism, or is always associated with other conspicuous features, is taxonomically useful. Of seven ways in which a chain of conidia can develop, most are easily diagnosed retrospectively; in the others we must still rely on unique form, superbly illustrated. Of 600 "good" form-genera of Coelomycetes, only 200 have been ontogenetically investigated, and some important genera of pathogens—*Ascochyta, Diplodia, Septogloeum*—with hundreds of species, are still dismally heterogeneous. *Septogloeum*, for example, has species with solitary, annellidic, annellidic and sympodial, phialidic, and meristem conidium ontogenies. Most such problems should be solved in another decade. In Ascomycetes, lysigenous or schizogenous centrum development is consistently associated with the more easily detected bitunicate asci, and if ascospores are phragmosporous or dictyosporous, odds are greater than 9 to 1 that they occur in bitunicate asci.

Full development of many fungi involves both sexual and asexual states, often separated in time and space. Only when data derived from *all* phenotypic expressions of a fungal genome (the "perfect" fungus) are correlated with comprehensive ecological data can we approach a rational taxonomic system.

INTRODUCTION

The logo designed for this symposium shows a tree of life with three main branches, presumably representing three Kingdoms—Plantae, Protista and Animalia. My own interpretation is closer to the Five-Kingdom scheme

*Department of Biology, University of Waterloo, Waterloo, Ontario, Canada N2L 3G1.

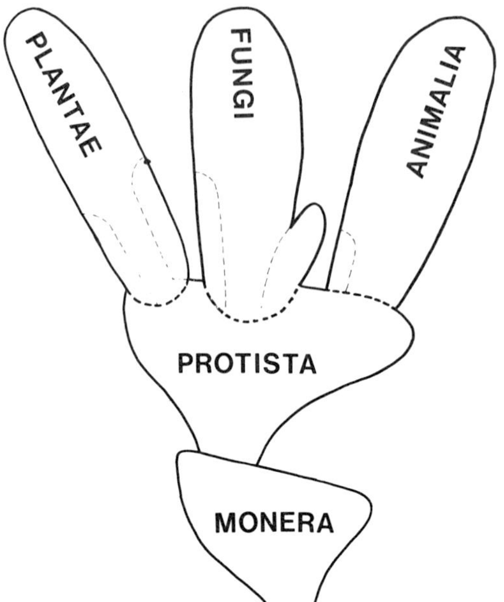

Figure 7.1. Diagram of the Five-Kingdom scheme proposed by Whittaker (*20*).

(Fig. 7.1) proposed by Whittaker (*20*), in that I consider the Fungi to have achieved the status of a separate Kingdom. Perhaps, in deference to my hosts, I can compromise by thinking of the fungi as essentially protistans that have invented a unique form of modular domestic architecture—the tubular hypha—which protects its living contents, and greatly facilitates the exploration and penetration of substrates. This most successful stratagem, combined with their enzymatic versatility, has made the fungi of first-rank importance in the decomposition of the enormous amounts of plant debris—estimated at 85 billion tons per annum—produced in the terrestrial habitat. They are also second only to the arthropods in their depredations on living plants.

Over a third of all fungi (and a majority of plant pathogens) belong to the great Subdivision Ascomycotina. Many members of this group produce asexual mitospores called conidia* that serve as dispersal units, and many of these asexual phenotypes exist independently of the sexual phenotypes. If one finds an unknown insect larva, proper care and feeding will usually result in its arriving at identifiable adulthood. Conidial fungi are not so predictable. The vast majority of them have not yet been connected to their

*A reasonable definition of the term "conidium" is the following: A specialized, non-motile, asexual propagule, usually caducous, not developing by cytoplasmic cleavage or free-cell formation (*9*).

sexual states (indeed, many of them may no longer be capable of producing sexual states), and so we have been forced to classify them independently.

These conidial fungi are known collectively as the Form Subdivision Deuteromycotina. This large group is initially split into two Form Classes: the Hyphomycetes, which produce their conidia from exposed conidiogenous cells, and the Coelomycetes, which form their conidia within some kind of protective structure.

As every plant pathologist knows, many destructive plant pathogenic fungi are found among the Deuteromycotina. Need I do more than mention a few Hyphomycetes such as *Fusarium*, *Helminthosporium*, *Botrytis*, and *Cercospora*, or Coelomycetes such as *Ascochyta*, *Colletotrichum*, and *Septoria*, to make my point? But for over half a century no up-to-date, accurate, comprehensive taxonomic treatment of this group was produced. This deficiency inevitably led to incorrect identifications. Over the years these grew into misleading generic concepts that have had plenty of time to become firmly established.

THE FORM CLASS HYPHOMYCETES

Fortunately, the situation has been considerably ameliorated during the past decade, at least as far as the Hyphomycetes are concerned, by the publication of six large-scale syntheses and compilations: Barron (*3*), von Arx (*1*, *2*), Ellis (*5*, *6*), Kendrick (*9*), and Kendrick and Carmichael (*10*). Further volumes will appear in 1978. What triggered this avalanche of useful books? My guess is that it was some seminal work on the part of Hughes (*7*), who saw clearly that the systematics of the Deuteromycotina was being seriously held back by its excessive reliance on characters of mature morphology, a doctrine enshrined in the monumental works of the great 19th century Italian mycologist, Saccardo, and even inherent in the concept of the *Form* Subdivision Deuteromycotina.

Why, asked Hughes, should we base our whole taxonomic scheme on the frozen moment that is represented by maturity? Why not trace the entire developmental process and see if useful characters emerge? This wasn't exactly a new idea—Hughes didn't *invent* development any more than Darwin invented evolution. The French mycologist Vuillemin (*17*, *18*, *19*) had made some very perceptive observations at the beginning of this century, but he was ahead of his time, and his ideas were effectively buried for nearly 50 years. But after the Second World War had brought increased awareness that molds can be very important to man—not solely as plant pathogens, but as destroyers of textiles, and as producers of antibiotics—the time was ripe. Hughes proceeded to do for Deuteromycete ontogeny what Darwin had done for evolution: he fleshed out the idea with a mass of meticulous observations. These have since been extended, intensified, and refined by many other mycologists. Fungal systematists hoped that, with

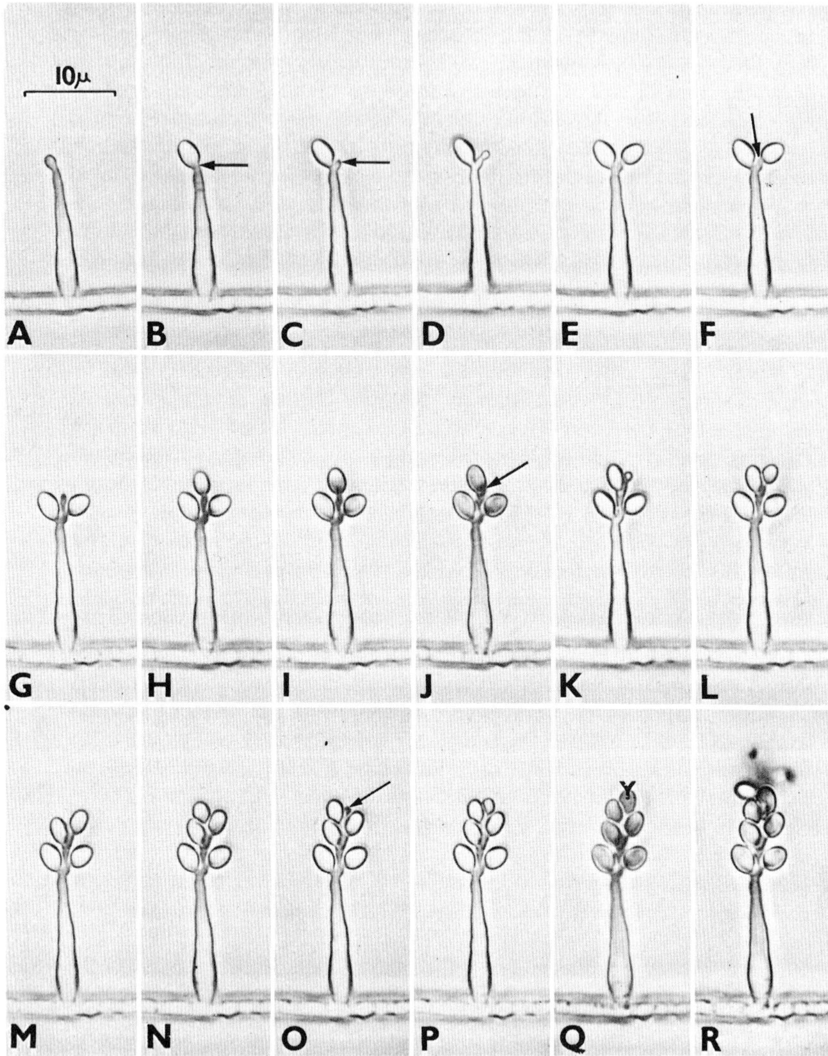

Figure 7.2. Time-lapse photographs of sympodial proliferation of the conidiogenous cell during repeated conidiogenesis in *Tritirachium album* Limber (from Kendrick, ref. *9*).

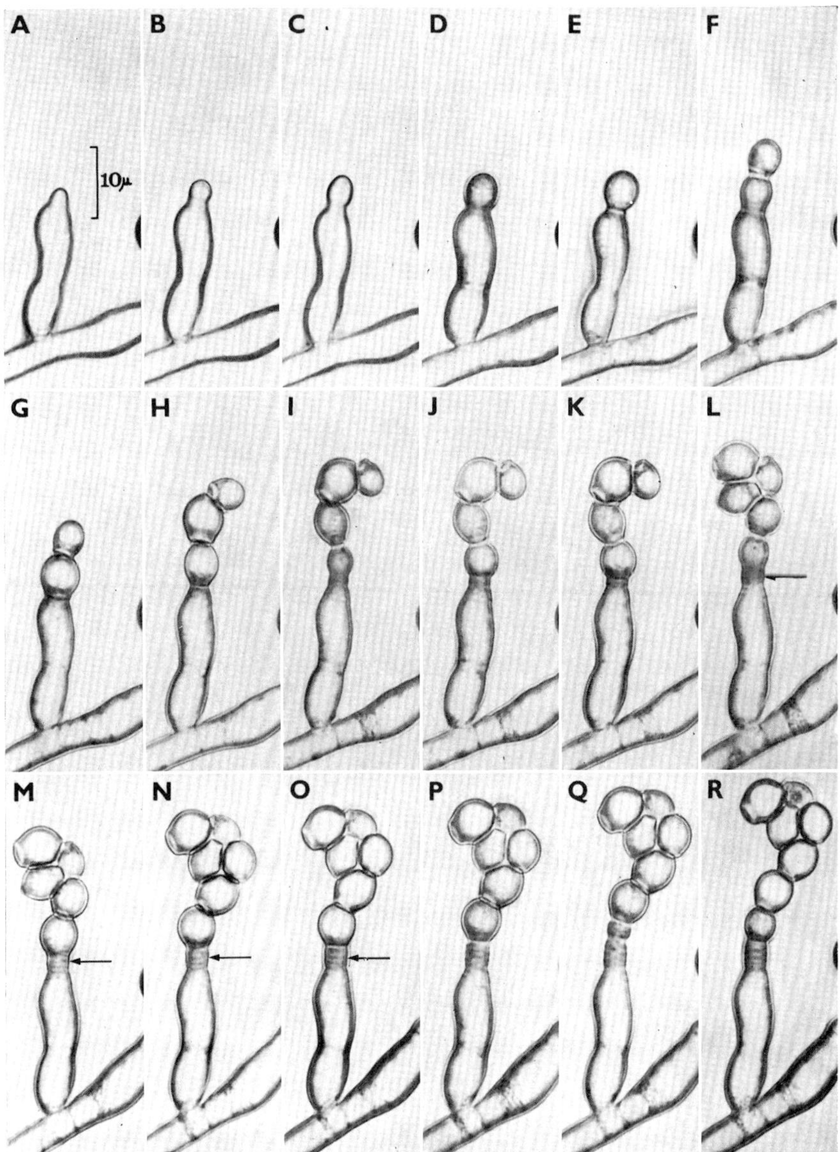

Figure 7.3. Time-lapse photographs of percurrent proliferation of an annellide during repeated conidiogenesis in *Scopulariopsis brevicaulis* (Sacc.) Bainier (from Kendrick, ref. *9*).

the advent and adoption of developmental characters—the recognition of the many different ways in which conidia can be formed singly, synchronously, or successively—a whole new and more 'natural' classification would emerge. I must admit—and this is one of the key points I want to make—that this has *not* happened. Despite the scores of papers published about conidium ontogeny, no workable classification based on development has been proposed, and I am convinced that none will be.

Was the whole developmental concept a chimera, a mirage? Not at all. Despite certain current excesses, the benefits accruing to mycology from the ontogenetic movement have been impressive. Despite disappointments, we have some important new characters to use. In the most extensive compilation of Hyphomycete genera yet attempted, Kendrick and Carmichael (*10*) adopted a very pragmatic philosophy. If developmental characters could be reasonably easily discerned under the light microscope, they would be applied. If mature morphology were diagnostic, then *it* would be used. This eclectic system has already made it easier to identify a Hyphomycete to the genus level than ever before. From the practical point of view, we know that no one will do a time-consuming developmental study of an organism, or examine it with an electron microscope, just to identify it. That is why I have deliberately refrained from using electron micrographs to illustrate this chapter. The only kind of developmental feature fit for general use is

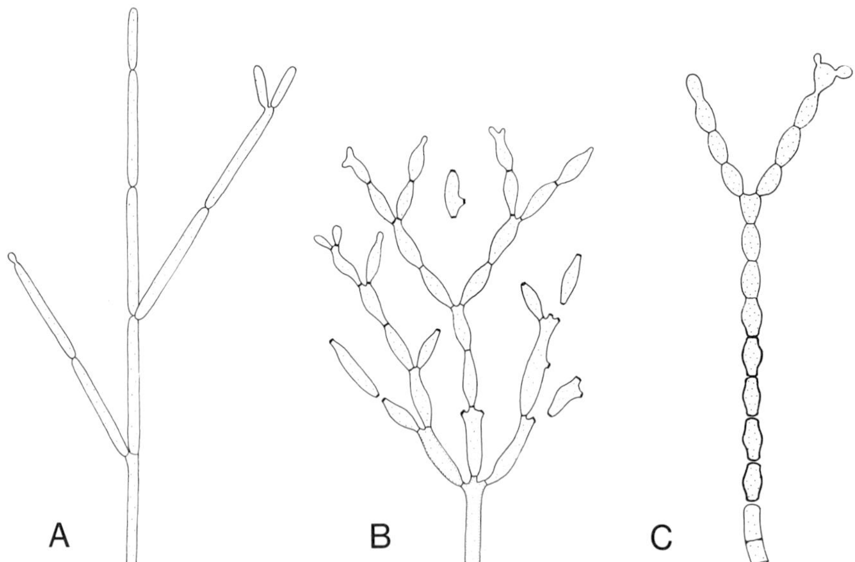

Figure 7.4. Acropetal chains in (A) *Fusidium griseum* Link, (B) *Hyalodendron* sp, and (C) *Monilia cinerea* Bonorden. Note the diagnostic young conidia forming at the tip of some chains.

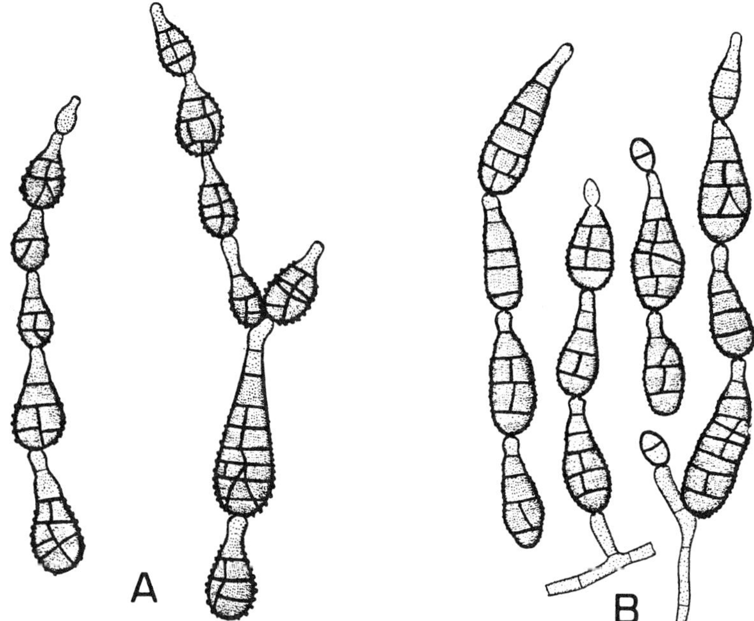

Figure 7.5. Acropetal chains of dictyoconidia in (A) the *Alternaria* state of *Pleospora infectoria* Fuckel and (B) *Alternaria alternata* (Fr.) Keissler (after Ellis, ref. 5).

the kind that leaves its mark on the mature organism or structure, or is invariably associated, for historical (evolutionary) reasons, with other easily recognized features.

Perhaps I should be more specific. If a conidiogenous cell produces a succession of conidia by sympodial proliferation, the appearance of that cell will be permanently changed (Fig. 7.2). The visible changes are frequently diagnostic. If a conidiogenous cell, in producing each new conidium, proliferates vegetatively through the scar left when the previous conidium fell off, the appearance of that cell, too, will be permanently changed. Such a cell is called an annellide (Fig. 7.3). If conidia are produced in chains that grow at the tip (acropetal chains), the presence of a young conidium at the distal end of some of the chains will usually tell us exactly how those chains were formed (Fig. 7.4). Even large, muriformly-septate conidia like those of *Alternaria* (Fig. 7.5) are formed in acropetal chains.

Many molds have been described as having "catenate conidia," or "conidia in chains," and until fairly recently this was thought to be sufficiently diagnostic. But of course those chains were not all produced in the same way. Conidial fungi have evolved no fewer than seven methods of conidial chainmaking. It is convenient to divide them into the four "successive" methods, which add one conidium at a time, and the three "concurrent" methods, in which more than one conidium in the chain is formed at a time. I have already mentioned two of the "successive techniques," the acropetal chain, and the chain produced by some annellides. A third is the chain produced from a conidiogenous cell that doesn't change in length even after producing a large number of conidia (Fig. 7.6). This sausage-machine type of approach is adopted by open-ended conidiogenous cells called phialides, and the youngest conidium in this kind of chain is always at the bottom, having just been extruded by the phialide. Such a chain is called a basipetal chain to distinguish it from the acropetal chain which extends at the top. Phialides and annellides can both produce basipetal chains, and although these two kinds of conidiogenous cell are now considered to be developmentally related, they look different after conidiogenesis, and can therefore be used as bases for form genera.

If the annellide gets longer as it produces conidia, and the phialide stays the same length, could there be a cell that gets shorter as it produces conidia? Indeed there could be, and there is. The so-called retrogressive conidiogenous cell blows out a short segment of itself to produce each conidium, and as the chain gets longer, the conidiogenous cell becomes gradually shorter (Fig. 7.7).

Now let us consider the chains in which two or more conidia develop concurrently. First, there are the meristem conidia. In this kind, several conidia are developing at the same time, though each is at a different stage of development, with the most mature at the top, and very young conidia, hardly differentiated from the cells of the conidiophore, at the bottom (Fig. 7.8). In this case it is hard to tell just where the conidiophore stops and the chain of conidia begins: there is a diffuse meristematic zone which gives these conidia their name. Conidial states of the powdery mildews belong here.

Second, there are the random chains which result when a hypha disarticulates at the septa to form conidia. This disarticulation gives rise to what we call fission arthroconidia (Fig. 7.9). As a variation of this process, every other cell of a hypha or conidiophore degenerates and dissolves, while the alternate cells metamorphose into conidia (alternate arthroconidia, Fig. 7.10).

If we can find ways of recognizing, after the fact, which kind of development has taken place, and show other people how to recognize it, we have considerably increased the taxonomically useful information that can be readily extracted from many fungi.

Sometimes, as with the telltale signs left by sympodial proliferation, or the conspicuous collarette present on many phialides (Fig. 7.11), this information is easy to communicate, if critical illustrations are available, and identification becomes easy. But in other cases, as with the often extremely inconspicuous scars on annellides, or the easily misread phenomena associated with holoblastic-retrogressive conidiogenesis, recogni-

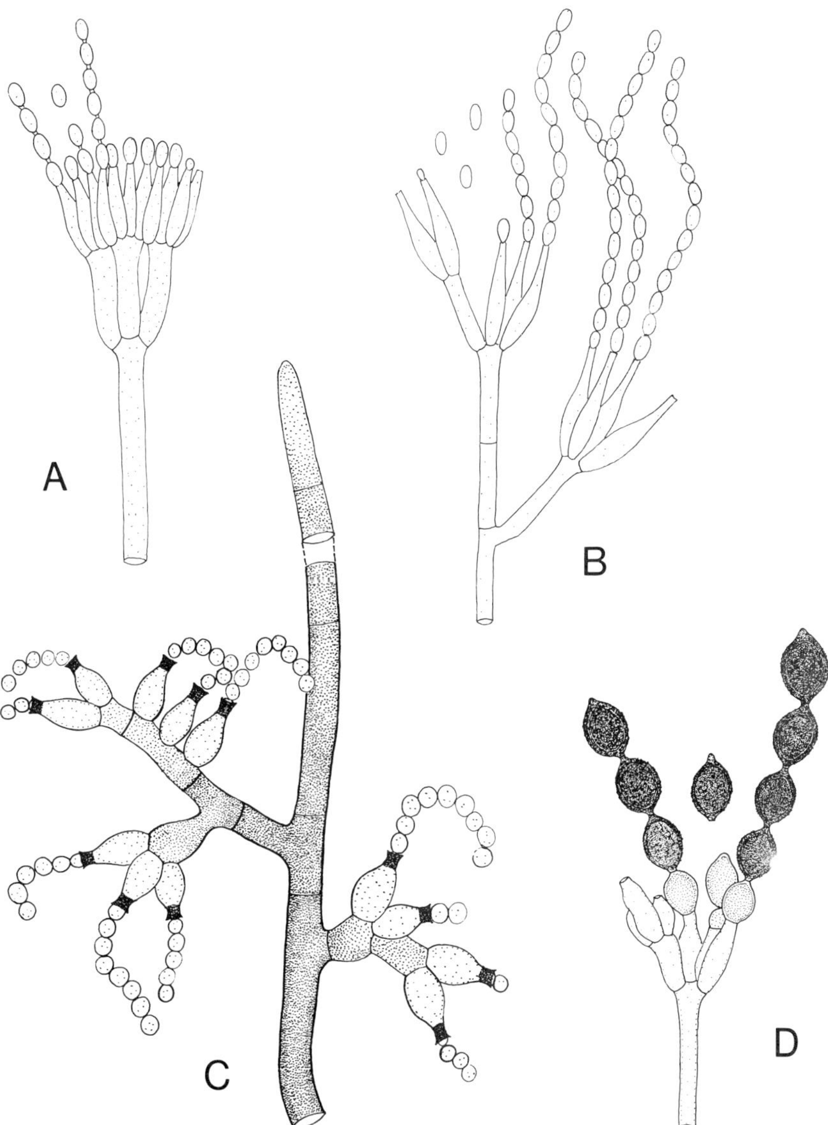

Figure 7.6. Basipetal chains of conidia being produced by phialides in (A) *Penicillium* sp, (B) *Paecilomyces varioti* Bainier, and (C) *Angulimaya sundara* Subram. and Lodha, and (D) *Phialomyces macrosporus* Misra and Talbot (after Kendrick and Carmichael, ref. *10*).

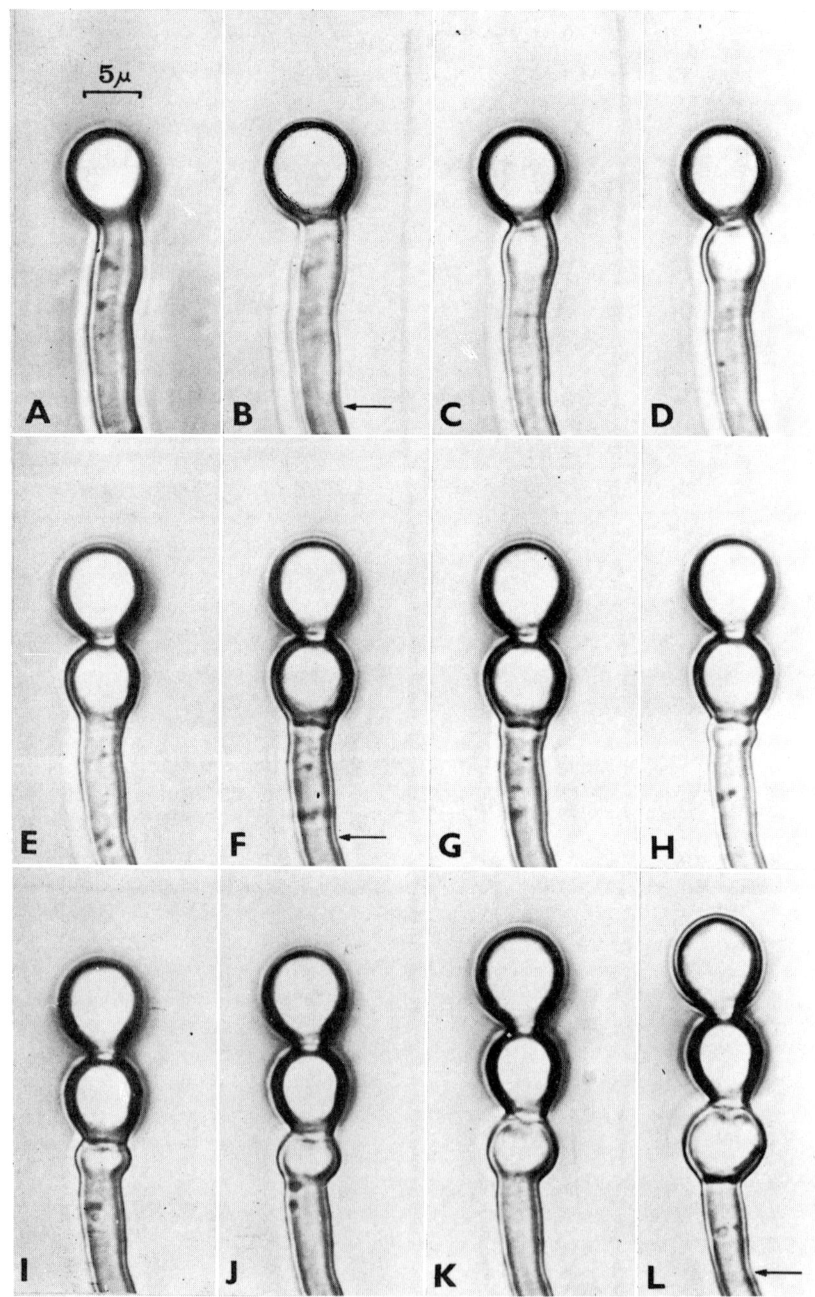

Figure 7.7. Time-lapse photographs of retrogressive conidiogenesis in the *Basipetospora* state of *Monascus ruber* Van Tieghem. The arrows indicate a reference point (from Kendrick, ref. *9*).

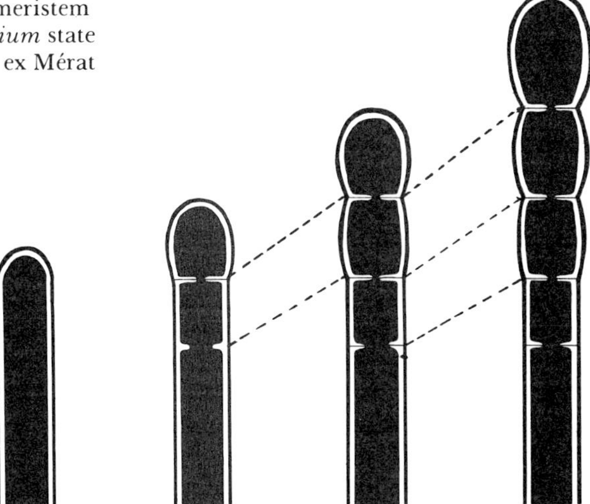

Figure 7.8. Diagram of meristem conidiogenesis in the *Oidium* state of *Erysiphe graminis* DC. ex Mérat (after Kendrick, ref. *9*).

tion of developmental phenomena may call for more critical observations than many practicing plant pathologists have time to make. In such cases we can only hope that truly accurate, life-like illustrations may prove sufficiently diagnostic—that a number of morphological differences, each small in itself, may add up to a unique facies. Certainly, even if one is unaware of their fine developmental criteria, one is unlikely to confuse the annellidic *Spilocaea* state of *Venturia inaequalis* (Cooke) Winter (the apple-scab fungus), or the synchronous blastoconidia of *Botrytis cinerea* Pers. ex Pers., with anything else. Other difficulties confronting those who would establish classifications based on development can be briefly listed as follows:

1) Some ascomycetous genera produce more than one conidial state. Species of *Ceratocystis* produce conidial states assignable to no fewer than 16 form genera (*16*). These conidial states not ony differ in morphology, but also exhibit five different kinds of conidium ontogeny. This may indicate heterogeneity in *Ceratocystis*, but that will not explain all the difficulties inherent in this situation.

2) Members of the same form genus of Hyphomycetes have been reliably connected to widely differing sexual states. This may be explained by convergent evolution of conidial states, but it is a real (if natural) obstacle to the establishment of an integrated classification.

3) Some conidial fungi are themselves pleomorphic, producing two (or occasionally more) morphologically and developmentally different conidial states on one mycelium (for example, *Doratomyces stemonitis* (Pers. ex Fr.) Morton and Smith, and *Echinobotryum atrum* Corda; *Chalara elegans* Nag Raj and Kendrick (*14*) and its unnamed chlamydosporic state—the latter pair formerly known under the single binomial *Thielaviopsis basicola* (Berk. and Br.) Ferraris).

4) Some conidial fungi can switch from one method of conidiogenesis to another

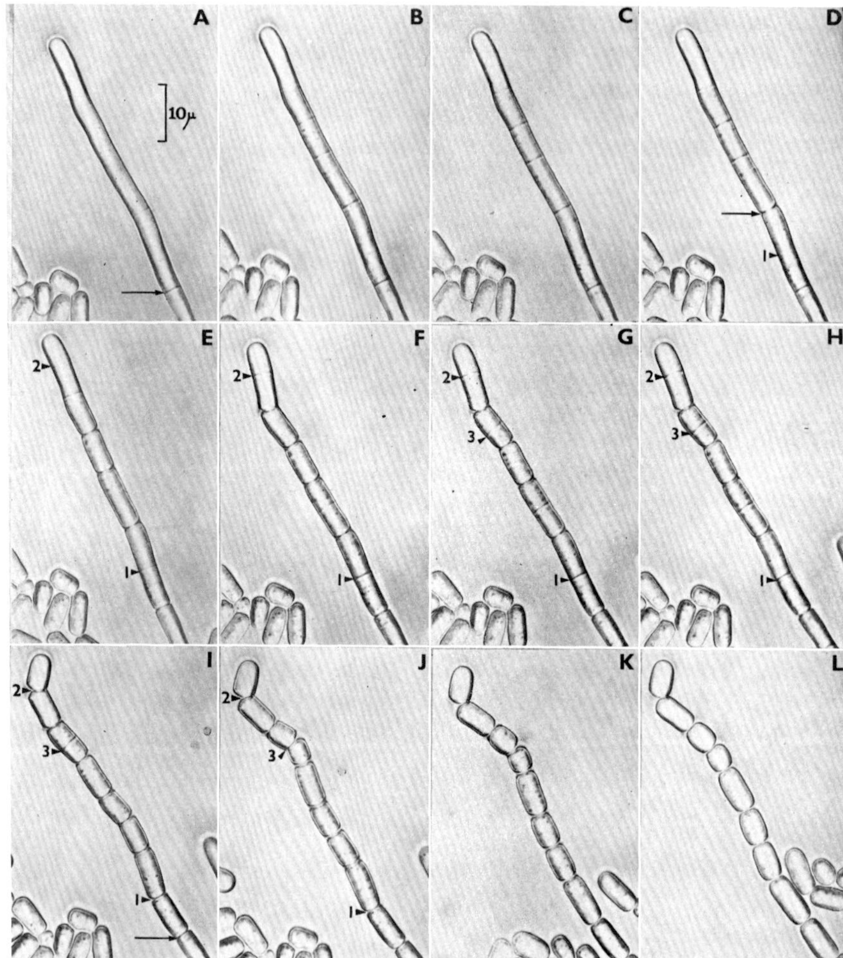

Figure 7.9. Time-lapse photographs of a "random" chain of fission arthroconidia forming in *Geotrichum candidum* Link ex Pers. (from Kendrick, ref. *9*). Note that the sequence of formation of conidia (in some cases indicated by numbers) is neither acropetal nor basipetal.

(for example, *Spiropes melanoplaca* (Berk. and Curt.) M. B. Ellis and *Chionomyces meliolicola* (Ciferri) Deight. and Piroz.—sympodial to annellidic).

I emphasize that developmental characters in most cases have not replaced the more traditional morphological features such as branching, shape, septation, and color, but have supplemented them, and have helped systematists to produce a body of useful and accessible literature on identification. Taxonomists would welcome feedback from the applied mycolo-

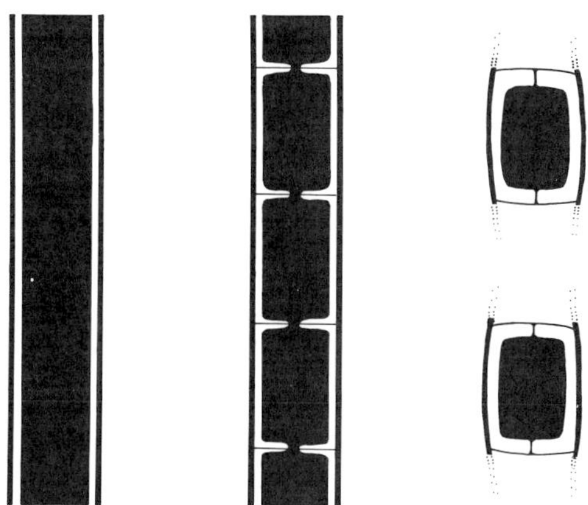

Figure 7.10. Diagram of alternate arthroconidia formation in *Coremiella ulmariae* (MacWeeney) Mason (after Kendrick, ref. *9*).

gists who must now try to use these books, and for whom they are written. Dr. Carmichael and I completed our comprehensive compilation of Hyphomycete genera in 1971. It was published in 1973, and gave coverage of over 1500 generic names, of which we accepted just under 600 as "good." Since 1971 about 240 new generic names have been published. The appearance of our fully illustrated compendium apparently reassured many authors that their "new" taxa were indeed unique. (Sometimes when a genus has been described as new, this has meant only "new to the author"—it *should* have meant that the author knew and could confidently rule out all extant genera.) In addition to this spate of new names, a number of formerly poorly understood older names have now been satisfactorily interpreted. Thus we have undertaken to produce a completely revised, up-dated and expanded version. We hope, but without conviction, that this second edition will last longer than the first.

THE FORM CLASS COELOMYCETES

We formerly divided the Coelomycetes into two Form Orders: the acervular fungi, Form Order Melanconiales, and the pycnidial fungi, Form Order Sphaeropsidales. As the anatomy of more and more Coelomycetes was studied, so many intermediate and aberrant forms were found that we now prefer to call all Coelomycete fructifications "conidiomata." Many genera were published with deficient descriptions and no illustrations. Perhaps this, and the fact that the fifth edition of the *Dictionary of the Fungi** lists over 1,000 generic names for Coelomycetes, daunted taxonomists, because the group was neglected for even longer than the Hyphomycetes. The number of "good" Coelomycete form genera can still be only roughly estimated, but T. R. Nag Raj (personal communication, 1977) suggests that

*Ainsworth, G. C. 1961. Commonwealth Mycological Institute, Kew, Surrey. 547 pp.

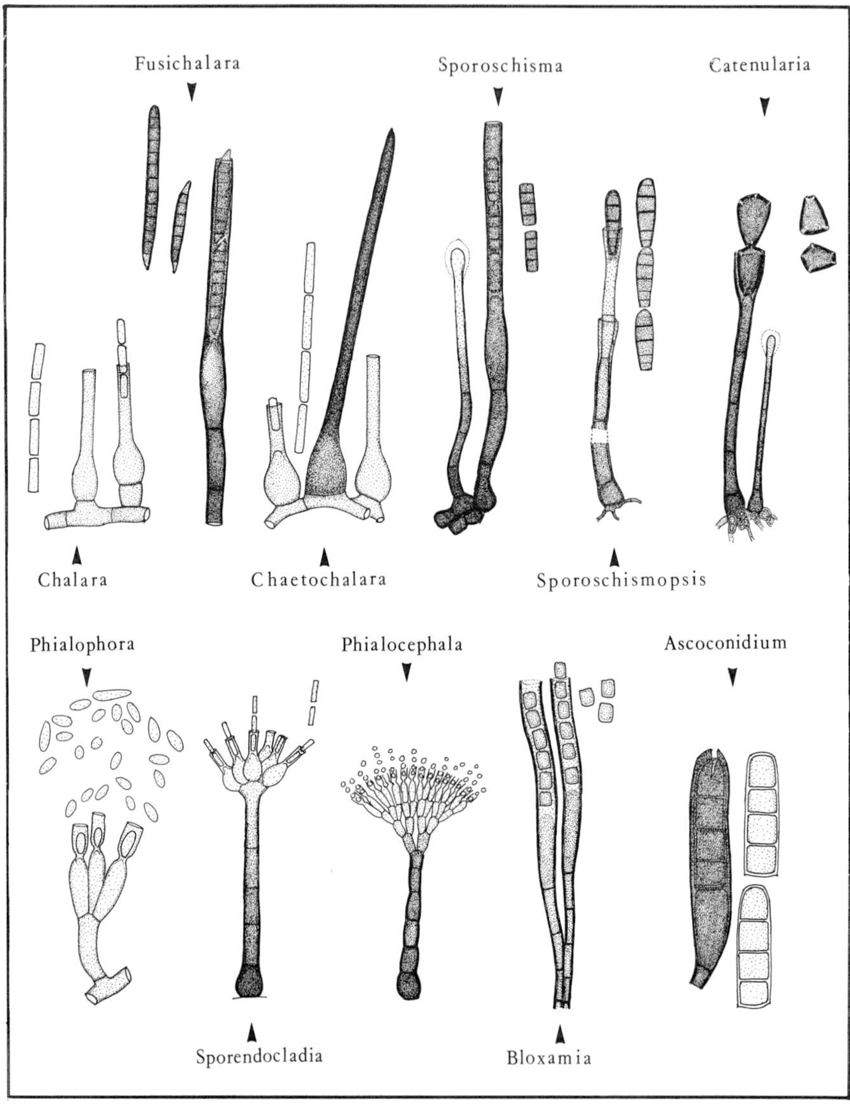

Figure 7.11. Variation in collarette morphology among some phialidic Hyphomycetes (after Nag Raj and Kendrick, ref. *14*).

there may be as many as 600. This makes the group, despite the fact that far fewer mycologists concern themselves with it, of a size and complexity comparable to the Hyphomycetes.

The critical reevaluation of generic concepts in the Coelomycetes is going ahead slowly but surely, and we feel that in about another decade they will rest on a firm developmental-anatomical foundation. But at present, accurate developmental and anatomical data are available for only about a third of the genera. Plant pathogenic genera such as *Ascochyta, Diplodia,* and *Septogloeum* contain several hundred "species" each. They are obviously dismally heterogeneous, and that means bad, genera. *Septogloeum* is known to contain species with five different patterns of conidium ontogeny —annellidic, solitary, sympodial plus annellidic, phialidic, and meristem— and it will be some time before this particular mess is cleaned up. Nag Raj (personal communication) is investigating the genus *Ceuthospora,* and finds that species formerly attributed to this genus should be disposed in *Coleophoma, Dothichiza,* and *Cytospora.*

This kind of information may not be immediately useful to applied mycologists, but I hope they will tolerate our efforts to establish order rather than condemn us because we can't yet produce an over-all classification for them. Clearly, if the process is to be speeded up, more money must be spent on good Coelomycete taxonomists.

THE SUBDIVISION ASCOMYCOTINA

Although many areas of Ascomycete taxonomy are still controversial, some concepts are regarded as established beyond dispute. One of these is that at some time early in Ascomycete evolution a dichotomy occurred between forms producing enclosed ascomata in which the cavity containing the asci was "built-in" during the development of the ascoma, and other forms in which the cavity developed secondarily by dissolution or splitting apart of tissue at the time the asci arose. Again, it is unlikely that anyone trying to name an ascomycetous plant pathogen will do a detailed study to find out which kind is which, because it is much easier to make the decision on the basis of another character altogether—whether the asci themselves are unitunicate or bitunicate.

Painstaking work has shown that ascomata with secondary (lysigenous or schizogenous) cavities always contain bitunicate asci. The constellation of associated characters here is even more extensive—the ascospores in bitunicate asci are usually septate, and they are usually shot out of the ascus one at a time rather than in a single burst. Both of these characters help to differentiate them from the unitunicate Ascomycetes. We can go even further: if ascospores are phragmosporous (having two or more cross-walls), or dictyosporous (with septa running transversely and longitudinally), the odds are at least 9 to 1 that the fungus is one of the bitunicates: ascospore septation is, of course, even easier to observe than the bitunicate ascus.

For some excellent discussions of the implications of developmental characters of the ascoma, I refer you to E. S. Luttrell's accounts published in 1951, 1955, and 1973 (*11, 12, 13*).

THE SYNTHESIS

The Deuteromycotina are not a part of the main taxonomic classification of the fungi. How can they be? They include conidial states of Ascomycetes, Basidiomycetes and, in the opinion of some, Zygomycetes. We group these conidial states into form genera largely for convenience in identification. The species included in any particular form genus must have a high level of morphological and, one hopes, developmental similarity, but they may not be related by descent. There can be no guarantee of that. So it is impossible, or at least misleading, to propose many higher taxa, such as families, for them. Conidial states can only take their place in the main classification when they are linked to their sexual states (if such exist), and we can talk about the "whole fungus."

When I was first introduced to the mysteries of mycology, I gained the impression that the classifications of the Ascomycotina and Basidiomycotina were in good order, while that of the Deuteromycotina was chaotic. Things have changed! Families and Orders of Ascomycetes and Basidiomycetes have been juggled around with amazing dexterity, and even created and abolished with a wave of the pen. Those of us who work with the Deuteromycotina once felt that if we could make connections between the conidial fungi and their sexual states, the sexual classification would help us to put our house in order. Now I think the Ascomycete and Basidiomycete specialists need us almost as much as we need them. In 1971 Kendrick and Carmichael (*10*) noted sexual state connections for about one-quarter of the "good" Hyphomycete form genera—one or more Ascomycete genera were noted as states for 133 form genera, and clamp-connections or named Basidiomycete genera were noted for 25 form genera. Kendrick and Watling (unpublished data) have almost doubled the number of form generic names applied to Basidiomycete conidial states, and our continuing revision of the listing of form-genera shows that Ascomycete connections are now known for about one-third of the "good" form genera.

By the time this book appears, a determined attempt at interpretation of the known connections will have been made. Experts with various backgrounds will have met at the second Kananaskis Conference (Kananaskis, Alberta, September 1977). Here's an example of subject matter: Gary Samuels, collaborating with Amy Rossman at The New York Botanical Garden, concluded that present generic separations of Nectrioid fungi, which are based on ascospore septation, are essentially unnatural, and that well-defined groups (genera?) can be delimited by using a combination of characters, but emphasizing perithecium wall structure and conidial state. Not only have Samuels and Rossman been able to reorganize the perithecial fungi, but they can also define form genera for the conidial states and suggest possible origins and variations of these. Clearly, in this particular exercise each classification—the sexual and the asexual—benefits from the other. In Chapter 11 of this volume E. S. Luttrell ably expounds how this same cross-fertilization process has affected the taxonomy of *Helminthosporium* sensu lato.

THE FUTURE

Morphology. Much remains to be done in the straightforward, classical descriptive mode. The rate of publication of new taxa is still increasing, and I am sure it will be at least a century (all other factors remaining favorable) before it slows down. The general standard of description and illustration has improved in recent years, but much still remains to be done in the way of training mycologists to communicate their taxonomic findings with the completeness that only loving care and consummate skill can achieve.

Development. The development spectrum of the Deuteromycetes is now well understood, but the full integration of ontogeny into the classification will take many more years, since hundreds of genera and thousands of species still await reexamination.

It is now obvious that fungi sharing a particular mode of conidiogenesis are not necessarily closely related. We have had to accept the obvious corollary of this—most, if not all, techniques of producing conidia have arisen more than once. We do not yet know how many times, and I think it will be difficult to find out with any degree of certainty. But we must try.

In the Coelomycetes, we can expect advances in our understanding when detailed studies are made of the mode of origin of the cavities on conidiomata—are they original, or are they lysigenous or schizogenous? We can answer that question for very few genera at present. Yet the relevance of a comparison between conidiomatal cavities and ascomatal cavities is obvious.

Connections. The number of connections established between sexual and conidial states is growing rapidly, and now represents a significant fraction of known fungal taxa. But the process of establishing connections is painstaking and uncertain, and my own belief is that many common conidial fungi have lost the ability to reproduce sexually (probably because it is no longer adaptive, and because they are heterokaryotic and can undergo genetic recombination via the parasexual cycle). In addition, the known connections are concentrated in certain Orders of Ascomycetes, (for example, Hypocreales, Eurotiales, Pezizales) while other Orders have few or no known conidial states (for example, Ostropales, Coronophorales, Lecanorales, Tuberales, Myriangiales). This may be due to taxonomic neglect, or to difficulties involved in culturing certain groups, but to me these observations strongly suggest that our dream of a fully-connected fungal flora may never be realized.

Biochemistry. Bartnicki-Garcia (*4*) has demonstrated that different fungal groups have differing wall compositions and that there are other biochemical markers, whose basis may be well established, as in the various enzyme aggregations involved in tryptophane synthesis as discussed by Hütter and DeMoos (*8*), or obscure, as in many of the chemical color tests outlined by Skalicky and Jechova (*15*). But most of these data apply only at higher levels of classification, and are of little use for segregating genera or species.

DNA base composition and hybridization studies, despite highly optimistic early claims, have been disappointing. The reasons for this are now being elucidated, as L. H. Throckmorton shows in Chapter 13 of this book. We cannot look to the biochemists for solutions to our taxonomic problems.

New characters. Conidium ontogeny arrived in 1953—a quarter of a century ago. It is clearly time for further innovation.

Ecology. In recent years some mycologists have shown that many conidial fungi exhibit considerable phenotypic plasticity in culture when subjected to changes in temperature, pH, radiation, carbon source, etc. These observations are sometimes used to criticize the existing classification, since it appears that species (and even genera) can be transmuted by appropriate manipulations. Yet the species in question often appear to be very stable and distinct in their natural habitats. My suspicion is that many species have very narrow ecological ranges, and that if the limits of those ranges are transgressed, other organisms will displace them. In culture, we remove the competition factor and can gain a false impression of flexibility.

CONCLUSIONS

I interpret the foregoing to mean that in order to define many of our taxa properly, we must consider not simply the morphology of a sexual or an asexual state, nor morphology supplemented by development, nor even those augmented by knowledge of the entire life history, but all of the above, placed in an environmental context that specifies the ecological demands made by each species. This context may have been millions of years in the making, and will not easily reveal all of its secrets. Nevertheless, one shorthand form incorporating much of this ecological information already exists for many fungi, and is made use of in taxonomy. I refer to host or substrate range. It is significant that in M. B. Ellis's second book *More Dematiaceous Hyphomycetes* (*6*), the key to species of *Cladosporium* concerns itself almost entirely with host or substrate, rather than with other available characters. I am not promoting a return to the old days when 400 species of *Ramularia* were recognized on 400 different hosts; I am merely advocating a closer look at the factors that determine where, when, and on what a fungus will grow.

Fungal ecology and biogeography are in their infancy. We can say little about the ecological necessities of most fungi. Perhaps we need better communications. Much of the information we need could be supplied by applied mycologists. I will close by inviting these colleagues to include as much ecological information as possible with every fungus they submit for identification; in this way we could work together in building the twin, interconnected edifices of fungal ecology and taxonomy.

LITERATURE CITED

1. Arx, J. A. von. 1970. *The genera of fungi sporulating in pure culture.* Cramer, Lehre. 288 pp.
2. Arx, J. A. von. 1974. *The genera of fungi sporulating.in pure culture.* 2nd Ed. Cramer, Lehre. 315 pp.
3. Barron, G. L. 1968. *The genera of Hyphomycetes from soil.* Williams & Wilkins, Baltimore. 364 pp.
4. Bartnicki-Garcia, S. 1970. *Cell wall composition and other biochemical markers in fungal phylogeny.* Pages 81-103 *in* J. B. Harborne, ed. *Phytochemical phylogeny.* Academic Press, New York.
5. Ellis, M. B. 1971. *Dematiaceous Hyphomycetes.* Commonw. Mycol. Inst., Kew. 608 pp.
6. Ellis, M. B. 1976. *More dematiaceous Hyphomycetes.* Commonw. Mycol. Inst., Kew. 507 pp.
7. Hughes, S. J. 1953. *Conidiophores, conidia and classification.* Can. J. Bot. 31: 577-659.
8. Hütter, R., and J. A. DeMoos. 1967. *Organization of the trytophan pathway: a phylogenetic study of the fungi.* J. Bact. 94: 1896-1907.
9. Kendrick, W. B., ed. 1971. *Taxonomy of Fungi Imperfecti.* Univ. of Toronto Press. 309 pp.
10. Kendrick, W. B., and J. W. Carmichael. 1973. *Hyphomycetes.* Pages 323-509 *in* G. C. Ainsworth, R. K. Sparrow, and A. S. Sussman, eds. *The fungi: An advanced treatise. Vol. IVA.* Academic Press, New York.
11. Luttrell, E. S. 1951. *Taxonomy of the Pyrenomycetes.* Univ. Missouri Stud. 24: 1-120.
12. Luttrell, E. S. 1955. *The Ascostromatic Ascomycetes.* Mycologia 47: 511-532.
13. Luttrell, E. S. 1973. *Loculoascomycetes.* Pages 135-219 *in* G. C. Ainsworth, F. K. Sparrow, and A. S. Sussman, eds. *The Fungi: An advanced treatise. Vol. IVA.* Academic Press, New York.
14. Nag Raj, T. R., and W. B. Kendrick. 1975. *A monograph of Chalara and allied genera.* Wilfrid Laurier Univ. Press, Waterloo. 200 pp.
15. Skalicky, Y., and V. Jechova. 1964. *The importance of some microchemical colour reactions in mycology.* Biol. Plant. 6: 57-65.
16. Upadhyay, H. P., and W. B. Kendrick. 1975. *Prodromus for a revision of Ceratocystis (Microascales, Ascomycetes) and its conidial states.* Mycologia 67: 798-805.
17. Vuillemin, P. 1910. *Matériaux pour une classification rationelle des Fungi Imperfecti.* Compt. Rend. Acad. Sci. (Paris) 150: 882-884.
18. Vuillemin, P. 1910. *Les Conidiosporés.* Bull. Soc. Sci. Nancy, Sér. III, XI 2: 129-172.
19. Vuillemin, P. 1911. *Les Aleuriosporés.* Bull. Soc. Sci. Nancy, Sér. III, XI 3: 151-175.
20. Whittaker, E. H. 1969. *New concepts of the kingdoms of organisms.* Science 163: 150-160.

8] Hennigian Phylogenetics in Contemporary Systematics: Principles, Methods, and Uses

by DAVID H. KAVANAUGH*

ABSTRACT

When classifications are viewed as devices for storage and retrieval of information about groups of organisms, their formulation requires concern for both the quality and quantity of data stored and their intended purpose (whether as special or general reference classifications). An additional criterion by which the relative merits of different classifications can be assessed is that of predictive power—the degree to which the classification permits us to predict previously unobserved attributes of classified organisms or characters, and (even the existence) of organisms not yet discovered and classified. The methods and goals of classification as espoused by advocates of the four main contemporary "schools" of classification are briefly discussed and compared. Classification (ordering of organisms into classes) is compared with systematization (ordering of organisms according to relations within a system), and the role of each in systematics is reviewed. The primary goal of cladistics is viewed as systematization, with classification as a secondary goal. An overview of the principles, methods, and vocabulary of Hennig's phylogenetics (cladistics) is provided and simply illustrated. The relative usefulness and predictive power of strictly cladistic classifications in both theoretical and applied biology are considered.

INTRODUCTION

When I was asked to present a short paper on Hennigian phylogenetics (cladistics) for this symposium, I agreed, but questioned the usefulness of yet another reiteration of the principles and methods involved. In addition to Hennig's definitive treatment (*25*), a number of useful and detailed

*Department of Entomology, California Academy of Sciences, Golden Gate Park, San Francisco, California 94118.

CONTEMPORARY APPROACHES TO CLASSIFICATION

TYPE OF CLASSIFICATION	CRITERION(-A) FOR GROUPING
"Phenetic"	overall similarity
"Omnispective"	weighted similarity
"Cladistic"	strict monophyly
"Evolutionary"	common ancestry (monophyly or paraphyly) and weighted similarity

Figure 8.1. A comparison of contemporary approaches to classification. The four main types of classifications are compared with respect to criteria employed in grouping taxa above the species rank.

general discussions and summaries have appeared (*8, 9, 10, 14, 17, 22, 24, 29, 30, 32, 37, 42, 45*). A host of other papers (a small sample of which is cited below) have examined the merits of particular cladistic concepts and methods from various viewpoints. The literature of systematics seems, therefore, replete with discussions of this approach; and, as is readily apparent, repeated debate over specific methods and concepts has tended toward polemics rather than scientific inquiry. On the surface, the relative merits of different philosophies on (and types of) classification appear to be at issue. A review of the literature, however, indicates a general lack of perspective in relating classification to systematics in its broadest sense. I suggest that a clearer understanding of, and appreciation for, this relationship is prerequisite to valid comparisons of different approaches to classification. My comments here are directed toward this end.

In this paper, I briefly review the four main contemporary classificatory approaches and discuss the purposes of classification and the relationship between classification and systematics. From this perspective, a brief overview of cladistic principles and methods is provided. Finally, theoretical and practical uses of cladistics and cladistic classifications are considered.

CONTEMPORARY APPROACHES TO CLASSIFICATION

In the past two decades, systematic literature has served as a battleground for the conflicting views of advocates of the four main contemporary schools of classification. I avoid use of the expression "systematic philosophies" because, as should become apparent later, this controversy has been confined mainly to concepts and methods of classification, and has not extended to other areas of systematics (but see *22, 28, 29, 32, 33, 47*). These four approaches are known as the phenetic, omnispective, cladistic, and evolutionary schools (Fig. 8.1).

Pure pheneticists (*44*) work from the premise that evolutionary history (phylogeny in particular) is unknowable and, therefore, cannot be used as the basis for classification. The criterion both for grouping and ranking of taxa in a classification is overall similarity or difference. In general, all characters are weighted equally (each considered to be of equal significance); and the information content of a classification is proportional to the number of characters used in its formulation (up to a point of diminishing returns). This approach is usually considered synonymous with "numerical taxonomy" and gained impetus from introduction of the use of computers to systematics. Recently, however, several numericists have devised and used computer techniques in attempts to reconstruct phylogeny (*7, 11, 18, 21*); so not all numericists are strict pheneticists.

The so-called "omnispective" approach to classification as advanced mainly by Blackwelder (*6*), relies on intuitive and pragmatic techniques. Advocates agree with pheneticists that phylogeny cannot be faithfully reconstructed and used in classification, but also they contend that the great diversity of interests and objectives in biology precludes use of any one general reference classification. Taxonomists are therefore free to construct classification for their own use, using their own criteria. The criterion chosen for grouping and ranking is weighted similarity; I think of this approach as the "trust me, I know what I'm doing" school. Numerous taxonomists today, though fewer than in the past, still work from this perspective.

Cladists contend that testable hypotheses of phylogeny (the branching sequence of evolution) can be generated through appropriate handling of data on living organisms, and that, ultimately, these hypotheses should serve as the basis for classification. As discussed further below, the criterion for grouping is strict monophyly rather than degree of similarity. Ranking of taxa in a classification is dependent on their age rather than on degree of phenetic distinctiveness. The information content of the classification is proportional to the degree to which it reflects phylogeny; and cladists reason that their classifications contain the greatest potential predictive capability and are, therefore, most suitable as general reference classifications.

Advocates of evolutionary classifications as defined by Mayr (*31*) and Simpson (*43*) agree with cladists that hypothetical reconstructions of evolutionary history are possible and basic to classification. However, they attempt to incorporate both cladistic (branching sequence) and divergence data in their classifications as a more complete representation of evolutionary history and claim that this practice increases the information content of classifications. Criteria for grouping include less rigorous monophyly (*43*) and weighted similarity. [Cladists would recognize their concepts of monophyly, paraphyly, and even instances of polyphyly as embraced by Simpson's (*43*) definition of monophyly.] Ranking is ultimately based on divergence criteria.

In summary, the phenetic and omnispective schools deny the practicality of hypothetical reconstructions of evolutionary history, while the cladistic and evolutionary schools seek and depend on them. Cladists ignore data on degree of similarity and divergence in classification, while the other three approaches rely totally or in part on them for grouping and ranking. The second comparison illustrates legitimate differences in methods and concepts of classification; and following sections explore the bases of this controversy. The first comparison, however, is actually one between science and non-science. Science is the generation of hypotheses and subsequent testing of them, adjustment or replacement of hypotheses, further testing, and so on. If, as most contemporary biologists agree, evolutionary theory provides the unifying concept for all biology, then classification, above all other devices, should reflect evolutionary history. Rejection of hypothetical reconstructions of this history because their truth is unprovable is non-science; and classification without an evolutionary perspective is non-biology.

PURPOSES OF CLASSIFICATION

Many authors have discussed the purposes of biological classification (*6, 14, 16, 23, 37, 41, 43, 44*), and their works should be consulted for reviews of differing opinions. In summary, however, all agree that the basic function of classification is as an indexing device through which storage and retrieval on information about organic diversity is facilitated. In evaluating the relative merits of different classifications, both quantity and quality of information stored and ease of retrieval should be examined. Our only access to information about organisms classified is by reference to names. Names denote concepts, and properly devised concepts embody meaningful information. If the classification reflects organization intrinsic to the organisms classified (namely the natural hierarchy resulting from evolution [*23*]), its total information content is increased by the inclusion of information on these relationships.

A second general purpose of classification is as an aid in the generation of new hypotheses about organisms and relationships among them. Different classifications can be compared, therefore, by reference to the kinds of testable hypotheses they facilitate and by how successfully the latter survive appropriate tests. Predictiveness should, at least theoretically, be maximized in classifications which are based on and reflect the intrinsic order among organisms. I briefly discuss the predictiveness of cladistic classification below, but the paper by Batra, *et al.* (Chapter 16, this volume) greatly expands upon this theme.

Ultimately, a choice of one among several different classifications must depend on its intended use. Because nomenclature is our only access to information about organisms, and because interdisciplinary studies may require access to data from widely different sources, a single general reference classification (and therefore a single nomenclature) is highly desirable for biology. However, as discussed by Blackwelder (*6*), there are often needs for special purpose classifications to handle specific kinds of information and indicate particular relationships. These classifications should not be considered as competitors with a general reference classification because there are needs for both. In any given situation the classification of choice must be the one which best fulfills the need. Requirements of a general reference classification, however, include maximized information content, ease of information retrieval, predictiveness, and nomenclatural stability.

CLASSIFICATION AND SYSTEMATIZATION

Simpson (*43*) defined *systematics* as "the scientific study of the kinds and diversity of organisms and of any and all relationships among them," and *classification* as "the ordering of [organisms] into groups (or sets) [classes] on the basis of their relationships, that is, of associations by continuity, similarity or both." At the same time he defined *taxonomy* as "the theoretical study of classification, including its bases, principles, procedures, and rules." According to these clear definitions, classification is a subdiscipline of systematics and is the object of taxonomy. In what may prove to be the most important conceptual paper in systematics since Hennig's treatise (*25*), Griffiths (*22*) explored the distinction between classification and what he called *systematization*. His synthesis of philosophical concepts with those of traditional systematics is of direct significance for an understanding of the role and potential contributions of cladistics.

Systematization is defined by Griffiths (*22*) as "the ordering of organisms according to relations within a system." Simply stated, Griffiths depicts organic diversity as a hierarchy of element/system relations—an extension of the "levels of organization concept." Systems are unique wholes. Just as single cells are considered elements of tissues, tissues as elements of organs, organs as elements of organ systems, and so on, populations are viewed as elements of species and species as elements of a system at the next higher level of organization—namely, the system which has been generated through the speciational processes of evolution. Organization rather than randomness is intrinsic within the system, and relationships among elements correspond to the branching sequence of phylogeny.

Figure 8.2 compares two sets of objects, one (A) in which element/sys-

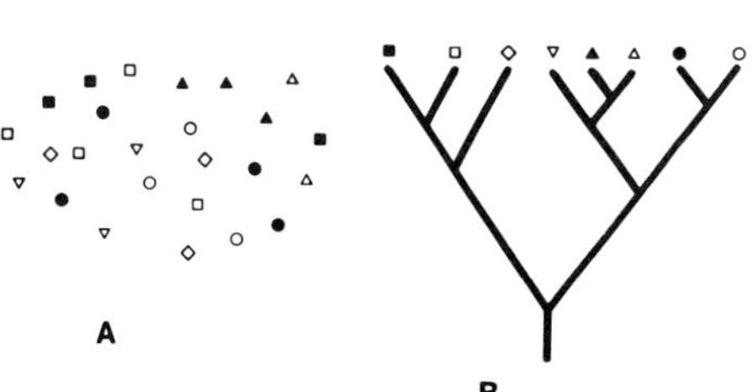

Figure 8.2. Two sets of objects, one (B) with and one (A) without intrinsic order. Set A is not a system (objects have no element/system relations). Set B is a system in which the objects are elements. Many classifications of both A and B are possible; but only one systematization of B reflects the order intrinsic to that system, and no systematization is possible for A.

tems relations do not exist and in which there is no intrinsic order, and the other (B) in which intrinsic order exists in the form of element/system relations. Species are elements of a system similar to B; and this system exists, independent of our perception of it, as an historical reality. The important consequences of adopting this perspective are: relationships among elements (taxa) in the system are historically determined by the branching pattern of phylogeny and are independent of any criterion of similarity; and, because systems are themselves wholes, elements are automatically and necessarily part of them, again without reference to any other criteria. Attempts to discover relationships within the system and to order taxa accordingly can be called systematization. Classes, on the other hand, are abstractions, and class membership is dependent on possession of certain essential or contingent attributes (or, for pheneticists, on overall similarity measures). Classification is, therefore, distinct from systematization. Again, in reference to Fig. 8.2, any number of classifications are possible for either A or B, depending on criteria for class membership. However, only one systematization mirrors the system intrinsic in B; and A is not a system and, therefore, cannot be "systematized."

HENNIGIAN PHYLOGENETICS (CLADISTICS)

Hennig (see Griffiths, *22*) stated that: "The task of a scientific systematics is not to introduce order into the manifold of particular phenomena, but to investigate and represent their intrinsic order." Such a view is in marked contrast to the traditional image of the taxonomist forcing order on a bewildering array of diversity; but it does indicate a clear link between Hennig's "systematics" and Griffiths' "systematization." I suggest that the primary objective of cladistics is, in fact, systematization rather than classification, with the latter as an important but subsequent and secondary objective. The following brief overview of cladistic concepts and methods is presented from this perspective.

Again, the primary goal of cladistics is systematization: the ordering of

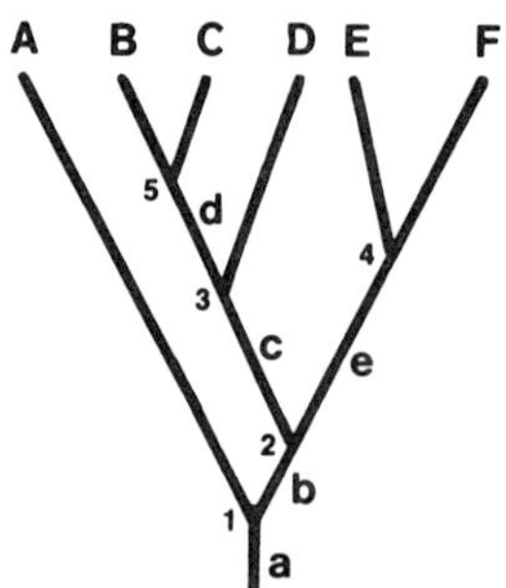

Figure 8.3. A cladogram. Species are represented by line segments, extant species denoted by upper case letters, stem species by lower case letters. Branching points correspond to speciation events, individually denoted by numbers.

taxa according to their phylogenetic relationships within the hierarchic system produced by speciation. Discussion of a few concepts is required at this point. Degree of phylogenetic relationship corresponds to recency of common ancestry and to no other criterion. In Fig. 8.3, E is more closely related to F than to D because E and F share the more recent common ancestor e, while the most recent common ancestor of E and D is the more remote b.

I agree with Brundin (*10*) that phylogenetic relationship exists only between distinct species or supraspecific taxa. Tuomikoski (*45*) and Whitehead (*46*) have tried to explore phylogenetic relationships among populations of a single species by discussing terms such as monophyly, paraphyly, and polyphyly in reference to these species. The resulting confusion inherent in their arguments reinforces the opinion that the species is the operational unit for cladistics.

In general, I accept Mayr's (*31*) "biological species" concept with the following addition: species are delimited temporally by successive speciation events. In Fig. 8.3, species are represented by line segments and speciation events by numbered branching points. By definition then, species e is delimited in time by events 2 and 4. The justification for and consequences of use of this modified definition have been discussed by Brundin (*10*).

Because the branching pattern inherent in the system under study is hierarchic, species are elements in a sequence of progressively more inclusive subsystems; and each subsystem is itself a unique whole. All subsystems necessarily fulfill criteria of strict monophyly (*1, 2, 3, 12, 24, 25, 27,*

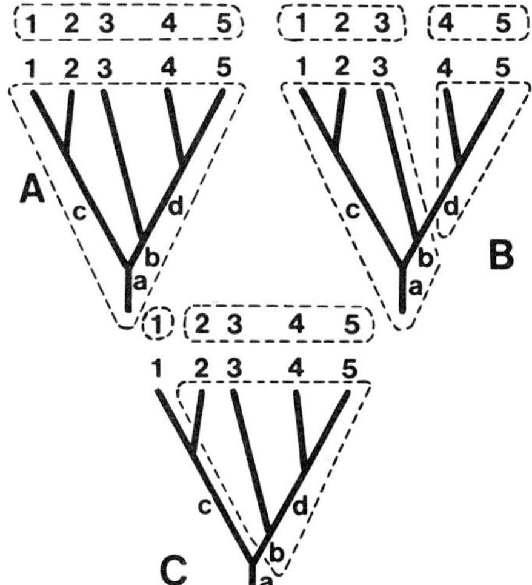

Figure 8.4. Illustration of concepts: monophyly (A), paraphyly (B), and polyphyly (C). Numbers refer to extant species, lower case letters refer to actual (not hypothetical) stem species; dotted lines circumscribe groups. Note that in B the group composed of d, 4, and 5 is monophyletic while that composed of a, b, c, 1, 2, and 3 is paraphyletic.

34, 35, 36, 45) (A in Fig. 8.4); namely, the subsystem includes the stem species and only and all its descendants. Paraphyly refers to a group including the stem species and only but not all its descendants. In B (Fig. 8.4), the group consisting of a, b, c, 1, 2, and 3 is paraphyletic while d, 4, and 5 form a monophyletic group. Polyphyly (C in Fig. 8.4) denotes a group from which the stem species for all group members is excluded. Paraphyly and polyphyly as defined here are inapplicable concepts in systematization, but their use in classification is hotly debated (*2, 3, 4, 7, 20, 24, 31, 46*).

When one uses these concepts, the procedure of cladistics (systematization) is as follows: Recognize one monophyletic group (subsystem) X. Search for the monophyletic group Y that demonstrates most recent common ancestry (closest phylogenetic relationship) with X. X and Y are called *sister groups*. Together they form a more inclusive monophyletic group (subsystem). Next, search for the sister group of X plus Y, and so on. The systematization resulting from repetition of this sequence should reflect the inherent order of the system. Hennig called this procedure *relative ranking*. In contrast, *absolute ranking* refers to the assignment of formal rank to the various monophyletic groups; therefore, it relates to classification rather than to systematization. The criterion for absolute ranking is age of origin, not degree of divergence. Crowson (*16*), Erwin (*19*), Hennig (*25*), Griffiths (*22, 23*), Mayr (*31*), and Whitehead (*46*) have discussed problems related to absolute ranking.

Up to this point, I have discussed only theoretical aspects of cladistics and have ignored Hennig's methods—although these always seem to be of greatest interest and controversy—because I considered it most important to provide a perspective and conceptual framework within which my few comments on methods and your own subsequent encounters with cladistic principles and methods might be assimilated and appreciated.

Cladistic methods can be briefly summarized as follows. Monophyly, sister group relationship, and the branching sequence of phylogeny can be inferred from analysis of the characters of organisms. An attempt is made to recognize homologous character states of each character and to order these in a sequence corresponding to the actual evolution of the character. This ordered sequence (Fig. 8.5, A) is called the *transformation series* of the character; and determination of the sequence and polarity of character transformations is the cornerstone of cladistic analysis. Hennig (*25*) discussed at length the problem of distinguishing homology from parallelism and convergence in different transformation series and suggested means for maximizing success in this effort. Platnick (*39*) has also discussed recognition of parallelism.

The initial character state in a series (for example, X in Fig. 8.5) is termed the pleisiotypic state; subsequent states (X′ through X‴) are apotypic states. Pleisiotypy and apotypy are relative terms. For example, X′ is apotypic with respect to X but pleisiotypic in relation to X″. B (in Fig. 8.5) illustrates Hennig's "scheme of argumentation." If two taxa (for example, 1 and 2) share a pleisiotypic character state, they are said to demonstrate

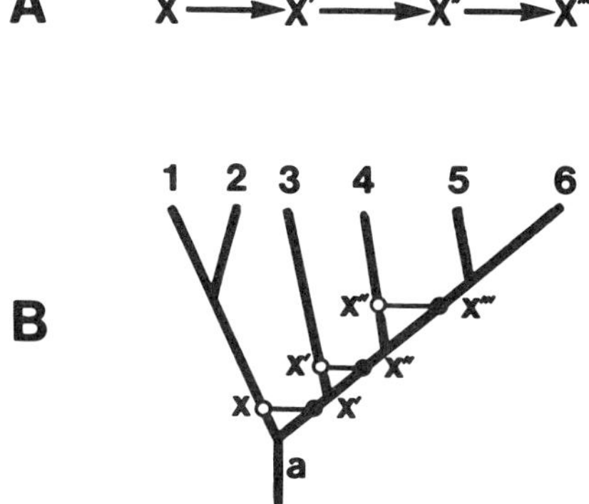

Figure 8.5. A illustrates the transformation of a character, with state X judged pleisiotypic and states X′ through X‴ being successively more apotypic. B illustrates Hennig's "scheme of argumentation". Open circles represent relatively pleisiotypic and solid circle relatively apotypic states of a character. Taxa 1 and 2 demonstrate sympleisiotypy, and their phylogenetic relationship cannot be established using this character.

sympleisiotypy. If two taxa (for example, 5 and 6) share an apotypic state, synapotypy exists. Hypotheses of close phylogenetic relationship and monophyly are supported only by synapotypy, not by sympleisiotypy. Therefore, the object of cladistic analysis is the recognition of synapotypy as the key to recognition of monophyly, sister group relationships, branching sequence and, hence, systematization. Hennig suggested several lines of evidence which can aid in the determination of sequence and polarity in transformation series. This included his so-called holomorphological, chorological, paleontological, and parasitological methods (*10, 24, 25, 30, 42*).

USES OF CLADISTICS

Ashlock (*3*) has provided an excellent discussion of the uses of cladistics. He emphasized the contributions of cladistic analyses and techniques to biogeography, chronistics, and studies of co-evolution. The use of cladistic concepts and methods in biogeography is now quite common (*5, 7, 13, 15, 38, 40, 46*), due mainly to success in such applications demonstrated by Hennig (*26*) and Brundin (*8*). Use of cladistic classifications has lagged behind, however, probably because most taxonomists still find the ranking criteria (especially the age criterion) unacceptable. However, once systematization is viewed as a prerequisite for classification and the role of cladistics in systematization becomes more widely recognized, more systematists may appreciate the benefits derived from precise representation of phylogeny in the classification itself.

CONCLUSIONS

My purpose in this paper has been to emphasize the conceptual distinction between systematization and classification, and between systems and classes, by restating and discussing ideas introduced by Griffiths (*22*). In this context, Hennigian phylogenetic (cladistic) principles and methods are viewed as directed primarily toward systematization, and secondarily toward classification. The usefulness of cladistic techniques in reconstructing systems (that is, in systematization) is generally recognized. The extent to which a formal classification should reflect the reconstructed system is, however, hotly debated and subject to differences of opinion based on differences in the intended use of the classification.

The use of cladistics in classification is, therefore, viewed as dependent on the purposes for which a given classification is proposed (that is, as a special or as a general reference classification). Relationships within the system generated by speciation are strictly phylogenetic and hierarchic. Consequently, classifications that most accurately reflect this system should exhibit the greatest predictiveness. Perhaps the ultimate stimulus for adoption of cladistic classifications will come from disciplines such as agriculture where predictiveness in and from classifications is of more than academic concern.

ACKNOWLEDGMENTS

I thank G. E. Ball, R. C. Drewes, and T. L. Erwin for their eager participation in discussions of topics presented here. Their insights and comments have contributed substantially to my appreciation of various concepts in systematics and thereby to this paper.

LITERATURE CITED

1. Ashlock, P. D. 1971. *Monophyly and associated terms.* Syst. Zool. 20: 63–69.

2. Ashlock, P. D. 1973. *Monophyly again.* Syst. Zool. 21: 430–438.

3. Ashlock, P. D. 1974. *The uses of cladistics.* Ann. Rev. Ecol. Syst. 5: 81–99.

4. Ball, G. E. 1975. *Pericaline Lebiini: notes on classification, a synopsis of the New World genera, and a revision of the genus Phloeoxena Chaudoir (Coleoptera: Carabidae).* Quaest. Entomol. 11: 143–242.

5. Ball, I. R. 1975. *Nature and formulation of biogeographical hypotheses.* Syst. Zool. 24: 407–430.

6. Blackwelder, R. E. 1967. *Taxonomy. A text and reference book.* Wiley, New York. 698 pp.

7. Brothers, D. J. 1975. *Phylogeny and classification of the aculeate Hymenoptera, with special reference to Mutillidae.* Univ. Kans. Sci. Bull. 50: 483–648.

8. Brundin, L. 1966. *Transantarctic relationships and their significance, as evidenced by chironomid midges, with a monograph of the subfamilies Podonominae and Aphroteniinae and the austral Heptagyiae.* Kungl. Svenska Vetenskapsakademiens Handl., Fjarde Ser. 11, 1: 1–472 (30 pls).

9. Brundin, L. 1968. *Application of phylogenetic principles in systematics and evolutionary theory.* Pages 473–495 *in* T. Orvig, ed. *Current problems of lower vertebrate phylogeny.* Interscience, New York.

10. Brundin, L. 1972. *Evolution, causal biology, and classification.* Zool. Scripta 1: 107–120.

11. Camin, J. H., and R. R. Sokal. 1965. *A method for deducing branching sequences in phylogeny.* Evolution 19: 311–326.

12. Colless, D. H. 1972. *A note on Ashlock's definition of "monophyly."* Syst. Zool. 21: 126–128.

13. Cracraft, J. 1974. *Continental drift and vertebrate distribution*. Ann. Rev. Ecol. Syst. 5: 215-261.

14. Cracraft, J. 1974. *Phylogenetic models and classification*. Syst. Zool. 23: 71-90.

15. Croizat, L., G. J. Nelson, and D. E. Rosen. 1974. *Centers of origin and related concepts*. Syst. Zool. 23: 265-287.

16. Crowson, R. A. 1970. *Classification and biology*. Heinemann Educ. Books, London. 350 pp.

17. Darlington, P. J., Jr. 1970. *A practical criticism of Hennig-Brundin "phylogenetic systematics" and Antarctic biogeography*. Syst. Zool. 19: 1-18.

18. Eades, D. C. 1970. *Theoretical and procedural aspects of numerical phyletics*. Syst. Zool. 19: 142-171.

19. Erwin, T. L. 1970. *A reclassification of bombardier beetles and a taxonomic revision of the North and Middle American species (Carabidae: Brachinida)*. Quaest. Entomol. 6: 4-215.

20. Farris, J. S. 1974. *Formal definitions of paraphyly and polyphyly*. Syst. Zool. 23: 548-554.

21. Farris, J. S., A. G. Kluge, and M. J. Eckardt. 1970. *A numerical approach to phylogenetic systematics*. Syst. Zool. 19: 172-191.

22. Griffiths, G. C. D. 1974. *On the foundations of biological systematics*. Acta Biotheoretica 23: 85-131.

23. Griffiths, G. C. D. 1974. *Some fundamental problems in biological classification*. Syst. Zool. 22: 338-343.

24. Hennig, W. 1965. *Phylogenetic systematics*. Ann. Rev. Entomol. 10: 97-116.

25. Hennig, W. 1966. *Phylogenetic systematics*. Univ. of Illinois Press, Urbana. 263 pp.

26. Hennig, W. 1966. *The diptera fauna of New Zealand as a problem in systematics and zoogeography*. Pacific Insects Monograph 9: 1-81. (Translated by P. Wygodzinsky from the German Original: Hennig, W. 1960. *Die Dipteren-Fauna von Neuseeland als Systematisches und Teirgeographisches Problem*. Beitr, Z. Entomol. 10(3/4): 221-329.)

27. Hennig, W. 1969. *Die Stammegeseschichte der Inseckten*. Senckenb. Naturforsch. Ges. (Frankfurt am Main). 436 pp.

28. Hull, D. L. 1967. *Certainty and circularity in evolutionary taxonomy*. Evolution 21: 174-189.

29. Hull, D. L. 1970. *Contemporary systematic philosophies*. Ann. Rev. Ecol. Syst. 1: 19-54.

30. Kavanaugh, D. H. 1972. *Hennig's principles and methods of phylogenetic systematics*. The Biologist 54: 115-127.

31. Mayr, E. 1969. *Principles of systematic zoology*. McGraw-Hill, New York. 428 pp.

32. Nelson, G. J. 1970. *Outline of a theory of comparative biology*. Syst. Zool. 19: 373-384.

33. Nelson, G. J. 1971. *"Cladism" as a philosophy of classification*. Syst. Zool. 20: 373-376.

34. Nelson, G. J 1971. *Paraphyly and polyphyly: redefinitions*. Syst. Zool. 20: 471-472.

35. Nelson, G. J. 1973. *Comments on Hennig's "phylogenetic systematics" and its influence on ichthyology*. Syst. Zool. 21: 364-374.

36. Nelson, G. J. 1973. *"Monophyly again?" A reply to P. D. Ashlock*. Syst. Zool. 22: 310-312.

37. Nelson, G. J. 1974. *Classification as an expression of phylogenetic relationship*. Syst. Zool. 23: 344-359.

38. Platnick, N. I. 1976. *Drifting spiders or continents?: Vicariance biogeography of the spider family Laroniinae (Araneae: Gnaphosidae)*. Syst. Zool. 25: 101-109.

39. Platnick, N. I. 1977. *Parallelism in phylogeny reconstruction*. Syst. Zool. 26: 93-96.

40. Rosen, D. E. 1975. *A vicariance model of Caribbean biogeography*. Syst. Zool. 24: 431-464.

41. Ross, H. H. 1974. *Biological systematics*. Addison-Wesley, Reading, Mass. 345 pp.

42. Schlee, D. 1971. *Die Rekonstruktion der Phylogenese mit Hennig's Prinzip*. W. Kramer, Frankfurt am Main. 62 pp.

43. Simpson, G. G. 1961. *Principles of animal taxonomy*. Columbia Univ. Press, New York. 247 pp.

44. Sokal, R. R., and P. H. A. Sneath. 1963. *Principles of numerical taxonomy*. Freeman, San Francisco. 359 pp.

45. Tuomikoski, R. 1967. *Notes on some principles of phylogenetic systematics*. Ann. Entomol. Fenn. 33: 137–147.

46. Whitehead, D. R. 1972. *Classification, phylogeny, and zoogeography of Schizogenius Putzeys (Coleoptera: Carabidae: Scaritini)*. Quaest. Entomol. 8: 131–348.

47. Wiley, E. O. 1974. *Karl R. Popper, systematics, and classification: a reply to Walter Bock and other evolutionary taxonomists*. Syst. Zool. 24: 233–243.

9] Biosystematics and Biogeography

BY BERYL B. SIMPSON*

ABSTRACT

Biosystematics, as opposed to systematics, is the study of taxonomy that encompasses experimental data at the population level. Such data include analyses of genetic variability, breeding systems, hybridization, competitive and colonizing ability and local adaptations. Only recently have these kinds of data been incorporated into biogeographical hypotheses. While unable to provide answers to questions concerning ancient lineages or phyletic changes, such studies, combined with reliable geological data, help to explain most of the present species distribution patterns. Many former biogeographical theories or dogmas such as land bridge connections, migrations from Old World centers of origin, age and area, and geological constancy of the tropics, have been supplemented or replaced by hypotheses such as continental drift, island biogeography, the taxon cycle, and tropical instability due to Pleistocene climatic changes. The value of these modern hypotheses, which incorporate new biosystematic and geologic data, is that they are predictive to some extent, more heuristic than descriptive "laws," and reveal the dynamic nature of biotic distributions. In addition, they focus attention on interactions at the population level as the ultimate determinants of the biogeographical patterns of species.

INTRODUCTION

Systematists have traditionally been interested in the study of biogeography, which we can define here as the elucidation of the geographical distributions of organisms and the historical and biological factors which have caused these distributions. Although Ball (*5*) credited Sclater (*58*) with providing the first major impetus to the study of biogeography, botanists such as von Humboldt (*38*), Meyen (*48*) and de Candolle (*14*) had previous-

*Department of Botany, U. S. National Museum of Natural History, Smithsonian Institution, Washington, D.C. 20560.

ly written extensively of the distributions of plants and remarked on the convergence of plant form of unrelated taxa in areas of similar climate. For the most part, however, initial biogeographical writings consisted of enumerations of mapped (or tabulated) distributions (the "descriptive phase" of Ball (5)). Later, as evolutionary processes became a dominant theme in biology, emphasis shifted to an interpretation of the directions of evolution and times of the origin of distribution patterns. At this point, notions of migration along with differentiation became increasingly important. Numerous "laws" or "theories" that best seemed to explain the patterns were advanced. Many of the authors of these hypotheses were, justifiably, ignorant of the geological history of the earth. Others appeared to be simply disdainful of the evidence that was available. Yet, as was emphasized by Rosen (55), cogent biogeographical hypotheses are those which are derived from an analysis of repetitive patterns and which are consistent with independently derived geological evidence. However, it has been within only the last 10 to 15 years that reliable geological data have been available to provide an independent body of comparative information.

Still another source of data, biosystematic data, can provide information that makes biogeographical interpretations dynamic rather than statically descriptive of distribution patterns and their possible geological causes. As in the case of adequate geological data, it has only been during the last ten years that sufficient biosystematic information has accrued to allow its incorporation into biogeographical models.

Since the term biosystematics has been used, misused, and abused *(69)* to the point where it has been suggested that the epithet be allowed to fade away *(41)*, I will define the way in which I will use it. The original application of biosystematics by Camp and Gilly [as biosystemy, *(13)*, recently reiterated by Raven *(51)*], was for taxonomy, primarily experimental, devoted to the study of evolution at the population level. Camp and Gilly specifically stated that the study of biosystematics should include such items as the genetic structure of populations, levels of genetic variation, breeding systems, hybridization, dispersal ability and local adaptations. I will use this application of the term for two reasons. First, as indicated above, it apparently agrees with the original use of the word. Second, I want to emphasize the impact on biogeography of studies dealing with the biological properties of populations and individuals. Biogeographical studies that incorporate such data are conceived at a very different level from those pursued by students of ancient phylogenies.

What I am attempting to do here, consequently, is to review some of the former dogmas of biogeography and to explain why, on the basis of the information at hand, they were often logical paradigms. I review these former hypotheses, often put forth as the only explanation for patterns, not to ridicule them, but to show how modern biosystematic and geologic studies have produced supplementary or alternative explanations for the same phenomena described by these former biogeographers. In particular, I shall

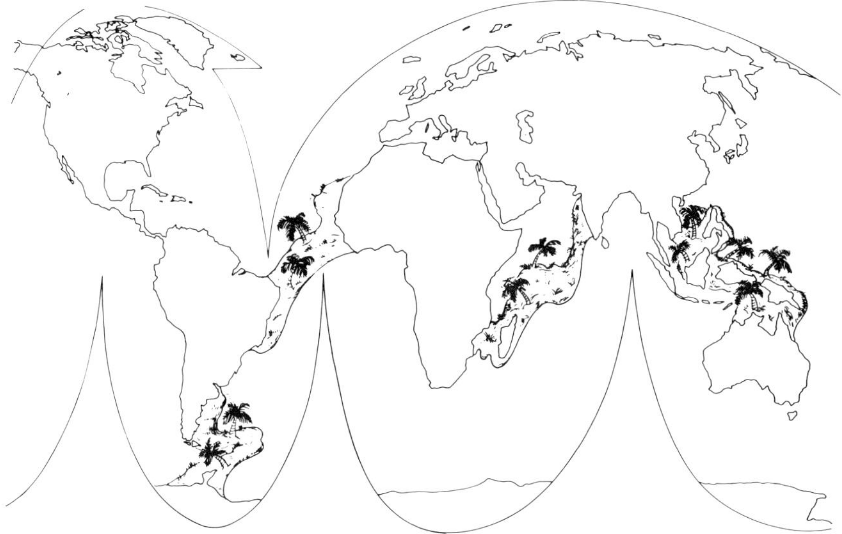

Figure 9.1. During the first third of this century, biogeographers often postulated the presence of direct land connections between continents in order to explain trans-oceanic disjunct distributions. The land bridges shown here (redrawn from Schuchert, ref. 56, Pl. 24) were hypothesized to have lasted from the Permian to the Cretaceous.

discuss the old theories of land bridges and migration routes from Old World centers or origin. I shall contrast these theories with the theory of island biogeography and assess the current state of this theory. Next I shall review the age and area hypothesis and contrast it with the modern concept of taxon cycles. Finally, I shall discuss the dogma which stated that the tropical areas of the earth were old and stable throughout geological time, and examine its implications for the process of speciation, and then enumerate the biosystematic and geological studies that have recently altered this idea. A major thought embodied in all of these modern interpretations is that biogeographical patterns are dynamic, not stable. Distributions are seen as the sum of present patterns of many populations. Such patterns are recent creations, constantly in a state of flux due to interactions between organisms and their continually changing biotic and physical environments.

LAND BRIDGES AND FIXED CONTINENTS

Once it was apparent that some taxa occur disjunctly in areas far apart, it became incumbent upon biogeographers to explain the disjunctions. Since the positions of the continents on the surface of the earth were considered fixed, disjunct distributions seemed to require that organisms either had superb powers of dispersal or that they had migrated via a land mass no longer present. Wegener had already proposed a theory of continental drift that could account for the separation of land masses, but the strength of belief in stable continents during the period between 1926 and 1966 by the geological community precluded its acceptance (see ref. *76* for an analysis of the mileu of the geological thinking during this period). Advocates for long distance dispersal (*54*) and for land bridge migrations (*56*) had their supporters, but for the first part of the century, the land bridge sympathizers led by Charles Schuchert (*56*) seemed to dominate the field of biogeography (Fig. 9.1). To show the extent to which various biogeographers had hypothesized intercontinental land connections, Handlirsch (*36*) provided a map of all of the land bridges proposed by the time of his article in 1913 (Fig. 9.2). Little room was left on the earth for the oceans.

Gradually, the lack of any geological support for such land bridges (that is, no sunken corridors), led to the dismissal of proposals with regard to the existence of such convenient migration routes. This does not mean that all concepts (*47*) of land bridge connections were denied. The Pliocene connection of North and South America (*34*), the Pleistocene low-sea-level

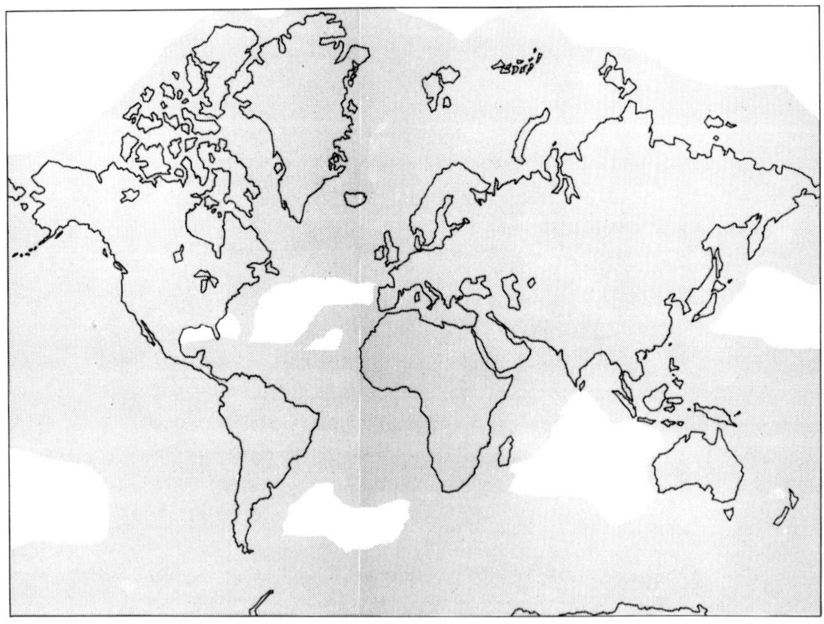

Figure 9.2. A superposition of all of the land bridges postulated by various biogeographers up to 1913 would have covered all of the area indicated in shading (redrawn from Handlirsch, ref. *36*, Map 5).

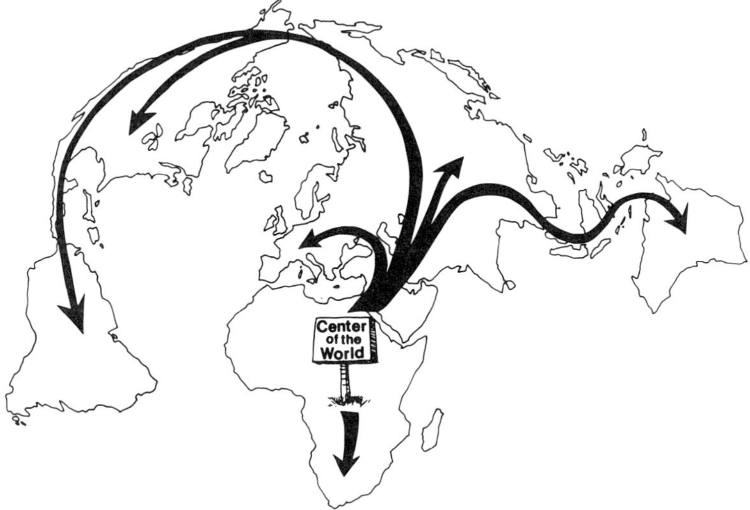

Figure 9.3. As an alternative to the construction of land bridges, but still denying the possibility of continental drift, Darlington (*20*) proposed that the dominant pattern of migration was from the Old World tropics to other regions of the earth (based on Darlington, ref. *20*, Figs. 27, 72, 73).

connections between Malaya, Borneo, and Sumatra (*3*) and between New Guinea and its surrounding islands (*22*) are all examples of land bridges that were recognized as having effectively served as migration routes. However, biogeographers still clung to the belief that distribution patterns reflected some sort of ancient overland migration from regions of origin to other regions apparently never linked to the original areas by direct land connections (for example, South America and Africa, or Africa and Australia). The most comprehensive theory of biogeography that tried to explain distributions in terms of more or less direct migration routes and fixed continents was elaborated by Darlington (*20, 21*), who amassed a wealth of data that he believed indicated major dispersal patterns from Old World tropical centers of origin (Fig. 9.3). It should be noted that at the time of the publication of Darlington's *Zoogeography* in 1957 no other hypothesis would have been accepted by either biologists or geologists. The dogma of the fixity of continents was still firmly entrenched in geological doctrine, but paleontological studies had shown relationships of taxa, primarily vertebrates, on various continents. In essence, however, Darlington did little more than describe a staggering number of carefully constructed modern and fossil distributions.

After 1960, as was clearly described by Marvin (*46*), the evidence for mobility of continents and for the process of continental drift rapidly and

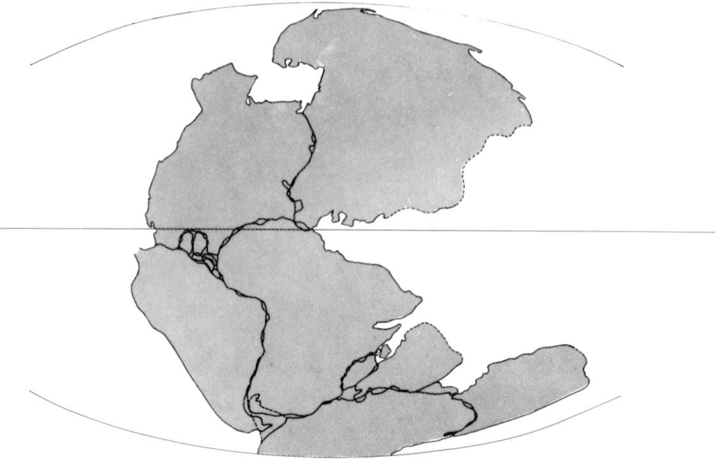

Figure 9.4. The acceptance of continental drift after 1960 gave to biogeographers a simple explanation for many trans-continental distributions. Such disjunctions, in general, involve ancient groups or involve only high order categories since the continents were all separated by the end of the Cretaceous. The Permian continent, Panagea, is shown here (redrawn from Dietz and Holden, ref. *25*, p. 35).

completely replaced the former idea of fixed continents. Correspondingly, numerous articles were written to show how the theory of drifting continents best explained many distribution patterns of both plants and animals (*17*, *52*, *57*). Even though the fact that the continents were united at an earlier period (Fig. 9.4) does explain the continental distributions of some higher order categories, it cannot account for transoceanic disjunctions at the specific (or usually generic) level. For example, Raven and Axelrod (*52*) indicated that about 24% of the angiosperm families have had representatives that at one time were shared between South America and Africa. The percentage shared because of continental connections between Africa and Australia was lower, and that between North and South America was lower still. All cases of plant disjunctions at the specific level were considered to have resulted from recent, long distance dispersal or human introduction. Just as continental drift cannot explain continental disjunctions at low taxonomic levels, neither can it provide explanations for intracontinental and intra-archipelago radiations. For many ancient (pre-Cretaceous) groups, continental drift is regarded as the underlying cause of their present intercontinental, and to some extent their intracontinental, distributions. Nevertheless, it seems that the present continental and islandic patterns for most groups are the result of recent radiations and dispersals. It is in the elucidation of these patterns that biosystematic studies play a role.

PATCHY ENVIRONMENTS, DISPERSAL, AND ISLAND BIOGEOGRAPHY

As ecological and biosystematic studies of populations progressed, it became apparent that propagules could, and did, disperse over areas of inhospitable habitat. All that was necessary to insure eventual colonization, irrespective of the distance traversed, was a probability of reaching a favorable patch of habitat. Using the concept of a fixed rate of dispersal, the assumption of a suitable habitat for establishment, the stochastic probabilities of extinction, and the principle of interspecific competition, MacArthur and Wilson (*44, 45*) proposed one of the dominant theories of modern biogeography—the theory of island biogeography. This theory is based on a model that incorporates relationships that had been documented in the field and/or laboratory, namely, that patch size shows a positive correlation with the diversity of taxa in it, that small populations have a higher probability of becoming extinct than large populations and that the probability of inter- and intraspecific competition increases as population sizes and the number of species increases. According to this model, the number of species of a taxonomic or functionally similar (in an ecological sense) group (very dissimilar organisms such as plants and birds cannot be indiscriminately compared) will reach a constant, or equilibrium value, on a given island. The equilibrium value on an island is determined by the size of the island and the distance of the island from the source of colonizing propagules. The model assumes both a constant and equal rate of immigration for all members of the group and fixed island sizes and distances.

It is easy to visualize (Fig. 9.5) that for two islands equidistant from a source of propagules, a large island will have a greater probability of receiving colonists. Likewise, given two islands of equal size, an island close to the source will have a higher probability of receiving propagules than one far from the source. The model thus states that near and large islands have a higher rate of immigration than far and small islands. Immigration is positively correlated with island size and negatively correlated with island distance. The second part of the theory states that extinction rates will be higher on small than on large islands because the small population sizes of small islands will have higher probabilities of chance extinction. Extinction rates are thus negatively correlated with island size.

The theory of island biogeography met the criterion of falsification or testability in a series of experiments conducted off the coast of Florida on man-made mangrove islands that had been divested of their previous arthropod fauna by fumigation. In the course of these experiments Simberloff and Wilson (*63, 64, 85*) measured the rates of immigration and the amount

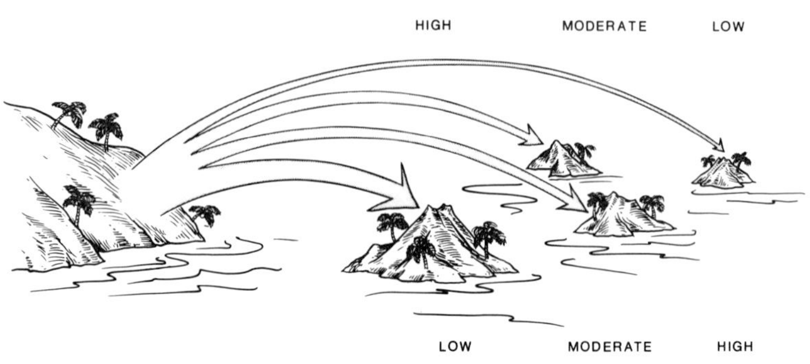

Figure 9.5. The theory of island biogeography predicts that an equilibrium number of species will eventually be achieved on an island when the rate of immigration equals the rate of extinction of species on that island. Both the rate of immigration to, and the rate of extinction on, an island vary with the size of an island and the distance of the island from the source of colonizing propagules (based on MacArthur and Wilson, ref. *45*).

of faunal turnover, and monitored the arthropod fauna for several years to see if an equilibrium number of arthropod species was reached on islands of various sizes and distances from the mainland. The results of these studies were consistent with the model, and the data from a three-year period indicated that equilibrium numbers of taxa were, indeed, established.

The importance of this theory is twofold. First, it allows predictions, rather than simple descriptions, to be made. Second, since habitats are patchy in nature, it transcends the study of simple islands and can be applied to almost any type of habitat. Good correlations between the sizes of habitat patches and distances from sources of colonization were found between isolated areas of paramo vegetation and numbers of species of resident birds (*77*), caves and the number of invertebrates inhabiting them (*19*), disjunct bodies of fresh water and the diversity of protozoans found in them (*12*), the areal distribution of given species of oaks and the numbers of specialized insect herbivores feeding on them (*50*), extent of Mediterranean habitat types of bird species diversity (*16*), and between the areas of cultivated, introduced crops and the numbers of insect pests infesting them (*70*).

In addition to its extension to habitat islands, the theory of island biogeography provided a general framework within which studies could be made that relaxed one or more of the original assumptions. Some of these studies took into account historical changes in island sizes and distances, while others utilized biosystematic information on the different properties of individual species.

The first extension of the theory of island biogeography to historical biogeography was by Diamond (*22*), who found that some islands off the coast of New Guinea were inhabited by a higher number of species than would have been predicted on the basis of their present sizes and distances from New Guinea. A straightforward application of the model of island biogeography assumes constant island shape and distance, but geological data showed that, during the Pleistocene, the lowering of sea level had greatly altered the sizes of various islands and their distances from a source of propagules. In fact, some of the islands would have been connected to the mainland during such periods. By assigning the number of bird species to such islands equal to the number now found on the mainland and allowing an elapsed time of 10,000 years (the time since the last raising of sea level resulted in restoration of the islands), Diamond calculated the time it would have taken to reach 37% of the predicted equilibrium number of species. The calculated time (the "relaxation" time) made sense in terms of the current numbers of species on the islands, especially when the accelerating or inhibiting effects of island size on extinction were considered. Historical events were also taken into consideration in a study of the modern configurations and numbers of plant taxa on the Galapagos Islands and the paramos of Venezuela and Colombia compared to the glacial configurations and modern species numbers (*65, 66*). The glacial shapes and distances of these islands were better predictors of modern species diversities than the modern areas and distances. In all of these cases, the apparent supersaturation of the "islands" was due to recent geological events that had altered the sizes and distance parameters. The theory of island biogeography was applicable if these changes were taken into account. These studies further showed that, even more than had been originally proposed, the number of species on islands—both true islands and habitat islands—was in a dynamic state. Both biological and historical factors could be involved. The latter could produce a number of species that might or might not eventually lead to an equilibrium value calculated on the basis of the modern island parameters.

Refinements in the theory of island biogeography have been made by researchers who have carried out more detailed biosystematic studies of the actual taxa found on islands and/or reanalyzed the assertion of rapid turnover rates on islands. In an additional study on the avifauna of the islands around New Guinea, Diamond (*23*) carefully assessed the actual taxonomic composition of the bird faunas of the different islands. The original model assumed that all taxa had equivalent powers of dispersal and establishment and equal probabilities of extinction. Diamond, in contrast, found that some taxa occur only on large islands; some species are widespread while others are narrowly restricted to one or a few islands. He also found that some groups of species occur together with a frequency higher than that which would be predicted on the basis of random association alone. These biosystematic studies have several important implications. They imply that area per se is not the factor determining the number of species on an island, but rather the number of available habitats. In general, since the number of habitats is usually correlated with area, area could be used in place of habitat diversity as a predictor of the number of species. The past failure of attempts to use indices of habitat complexity as a predictor of diversity (*59*) can probably be attributed to the fact that indices were not adequate. Diamond's analysis also showed that the colonizing ability of some species was greater than that of others, a fact well appreciated by students of weed biology (*4*). Finally, Diamond's results indicated that the diversity and composition of an island's fauna is determined, in part, by the order in which taxa are established. The ability to colonize successfully depends upon the competitors already present.

Another refinement of the original theory of island biogeography was recently pointed out by Simberloff (*60*). He reanalyzed the resident faunas of the mangrove islands to ascertain if turnover rates had been as high as was formerly postulated. His careful taxonomic reassessment indicated that turnover rates were much lower than had originally been ascertained due to the presence of "waif" or transient species that were not true colonists.

All these studies show that the theory of island biogeography provides a valuable general framework to explain the causes leading to many modern distribution patterns, both islandic and continental. It has made biogeographers aware of the dynamic nature of geographical distributions and the patchiness of the environment for almost all organisms. Extinction is viewed as a population event, not a global event, and large scale distribution patterns are seen as the sum, at a given moment, of a fluctuating array of population distributions. It remains to be seen if the theory will stand as originally stated or whether it can be modified or replaced by a theory that incorporates more biological information and still retains general properties of predictability.

AGE AND AREA

Just as the denial of continental drift led to the postulation of land bridges, the absence of fossils and the lack of data indicating degrees of relatedness allowed the formulation of laws based on age and area. The resulting hypothesis, put forth and vigorously defended by Willis (*82*), had a wide acceptance among many systematists, including Guppy (*31*). Willis, arguing against the relic nature of restricted endemic species, postulated that the area occupied by a taxon was a direct reflection of its age. By inspection, the "wheel within a wheel" (*81*) pattern of many distribution patterns (Fig. 9.6) gave evidence of the radiating nature of the evolutionary history. Endemic species were most numerous at points at which the ancestral stock was presumed to have entered the island because, as this original intro-

Figure 9.6. In 1922 Willis (*82*) proposed his "Age and Area" hypothesis which stated the age of a taxon to be directly proportional to the size of its geographical range. The more ancient members of a group had broad distributions whereas the recently evolved members had restricted distributions. The distributional limits of several species of *Ranunculus* species shown here demonstrate the "wheel within a wheel" pattern that Willis used as evidence for his theory. The distributions are from Willis (*82*, Diagram 3). A biosystematic study of the alpine species of *Ranunculus* in New Zealand by Fisher (*29*) showed Willis' interpretation to be incorrect and provided an explanation for the adaptive radiation of buttercups in the alpine habitats. (The drawings of the two species of *Ranunculus* are based on Fisher, ref. *29*, Figs. 102 and 115.)

duction spread, it gave rise to a new taxon at the center of the range. This second species then began to spread outward and subsequently gave rise to another species. As the procedure continued, it resulted in a pattern of concentric circles, with the centers of distribution of the endemic taxa coincident with the point from which the original introduction had spread (the point of introduction).

The distribution pattern of the various species of *Ranunculus* in New Zealand (Fig. 9.6) was used by Willis several times (*81, 82*) to illustrate his law. In discussing this and other examples from New Zealand, Willis (*82*) specifically dismissed the influence of historical processes such as glaciations (New Zealand is, after all, under maritime influences) even though he admitted their effect in high latitude areas of the northern hemisphere. Willis's ideas might have been more widely accepted and longer lasting had he not denied the process of natural selection. In a century in which the concept of Darwinian evolution was to become the greatest underlying principle of biology, Willis's idea was doomed to failure. Nevertheless, his conclusion about the age of some of the New Zealand endemics was correct—albeit the result of erroneous reasoning. Careful biosystematic studies of the same group of Ranunculaceae from the same area demonstrated the recent origin of many of the endemic taxa. Combined with modern geological data, however, a casual explanation for their origin was provided.

In a careful study of the geographical variation, distribution patterns and natural hybridization of the alpine *Ranunculus* species of New Zealand, Fisher (*29*) assessed the relationships of the various taxa and concluded that the genus had undergone a recent, extensive radiation within the alpine habitats. Modern geological studies have shown that the high peaks of the New Zealand mountains had been glaciated and that the lower elevational limits of the alpine zone were significantly lowered during glacial periods (*79*). The radiation of species was apparently caused by repeated expansions and contractions of the alpine habitat which led to migrations followed by periods of isolation of alpine plant species such as those of the genus *Ranunculus.*

On the other hand, not all of the endemic taxa of New Zealand can be considered recently derived, or caused by Pleistocene events. Many of the lowland endemic taxa, as pointed out by Raven and Axelrod (*52*), are relics of an ancient Gondwana distribution. According to these authors, several ancient taxa have survived in New Zealand, whereas they have become extinct in other members of the ancient Gondwana land mass because of competition from more recent colonists. In contrast to the other Gondwana lands (South America, Africa, and Australia), New Zealand has remained isolated and relatively free from the effects of recent colonizers. In addition, New Zealand has been drifting northward and has retained a mild climate conducive to the continued existence of these taxa.

TAXON CYCLES AND INTRAISLAND DISTRIBUTION PATTERNS

Within the last ten years, an hypothesis for the direction of island evolution has been proposed that essentially espouses an idea opposite to that of Willis. This theory, which Wilson (*84*) (Fig. 9.7) labeled as the taxon cycle, states that, in many situations, the widespread low elevation taxa on an island are the most recent colonists and that the taxa restricted to montane rainforest habitats are the older taxa on that island. The present restricted distribution of these latter taxa is presumably caused by competitive pressure exerted by the new invaders.

The cyclical evolutionary process that produces these patterns was divided into three stages. Stage I is the colonization and rapid spread of genetically variable aggressive taxa. Stage II is the initial retreat of this taxon in the face of a newer, more aggressive arrival. Stage III is the most habitat-restricted phase with reduced population sizes and an increased predation and parasitism load due to counteradaptation by the resident biota. Extinction of this last group was regarded as likely in view of the fact that

Figure 9.7. The taxon cycle proposed by Wilson (*83*) postulates the opposite pattern from the "Age and Area" hypothesis of Willis (*81*). Widespread species of marginal habitats are the young members in stage 1 of the cycle. In the last stage of the cycle, a taxon has a restricted distribution and occupies a narrow range of habitats. Unless a taxon in the last stage is able to again invade marginal habitats, it will eventually become extinct. The taxon cycle has been applied only to islands and fits the patterns shown by ants of the Melanesian Islands (*83*) and by birds of the West Indies (*53*).

population sizes are reduced, predation and parasitism are increased, and extreme specializations are disadvantageous if environmental permutations occur. This model was originally conceived to explain the ecological relationships and behavioral characteristics of the Melanesian ant fauna. Studies by Wilson (*83*) and Brown (*10*) had shown that the taxa of marginal, widespread habitats were aggressive in encounters with resident taxa, presumably forcing them into habitats in which they could maintain a competitive superiority.

Williams found a somewhat similar (*80*) situation in the *Anolis* faunas of the West Indian islands. He found that on islands with a single species, that species ranged over a wide variety of habitats. However, where two species were present on an island, one occupied the marginal habitats and the other had a reduced range of habitats, usually in the highlands of the island. In some cases, the same species would have a different range of habitats on different islands, depending upon the number of competitors. Ricklefs and Cox (*53*) later applied the concept of taxon cycles to the avifauna of the West Indies (Fig. 9.7) but suggested that a fourth stage, restriction and substantial differention, be added. Thus, the taxon cycle can often set the stage for development of an endemic species.

It is obvious that the concept of taxon cycles embodies several of the principles of island biogeography such as continual immigration and species turnover. If the theory is examined in the light of Diamond's (*23*) analysis of the avifauna of New Guinea and surrounding islands, additional parallels are evident. The probability of colonization is not equal for all species. Some are better colonizers ("tramps" or "weeds") of marginal habitats than others. Moreover, the outcome of colonizations will depend on the numbers and types of taxa already present on the island. The concept of taxon cycles, like the theory of island biogeography, thus focuses on the fact that species distributions and the kinds and numbers of taxa on islands are in a dynamic state and continuously subject to change.

GEOLOGICAL STABILITY OF THE TROPICS AND THE NEED FOR SPECIALIZED SPECIATION MECHANISMS

One of the former central dogmas of biogeography (embodied, for example, in Darlington's hypothesis, Fig. 9.3), was that the lowland tropical areas of the world constitute uniform (*67*), geologically ancient and historically stable parts of the earth. Habitats might seem bewilderingly complex, but, at a gross level, they appear uniform over large areas. Fossil plant and animal associations from other parts of the world not now covered by tropical rainforest indicate that at the beginning of the Tertiary much of the terrestrial surface of the earth was covered by tropical or subtropical associations. The gradual development of modern temperate and Arctic biotas during the Tertiary and the effects of the Pleistocene glacial advances (*15, 30, 86*) in many areas has been known for decades. Similar historical changes appeared to have been absent from tropical lowland areas. Consequently, in view of the theory of geographical isolation as a prerequisite for speciation, the origin of the wealth of tropical taxa presented an enigma. Geological data were so scanty in tropical areas that geology provided no solution to the problem. In addition to the absence of geological data, little was known about the biology or even the systematics of most lowland tropical taxa.

Within a state of such ignorance, almost any explanation for the origin or maintenance of tropical species diversity could be proposed. Consequently, Dobzhansky (*26*) in his commentary on speciation in the tropics simply speculated that natural selection via physical factors played a minor role in these habitats and that "biotic complexity" somehow permitted the accumulation of many closely related taxa, perhaps by reducing extinction. Federov (*27*) went so far as to propose that speciation of tropical trees was due to the random genetic fixation of alleles following extensive inbreeding. Using the suggestion of Corner (*18*) that most trees were self-fertilizing, Federov theorized that tropical trees were self-compatible and normally self-fertilizing in nature. The breeding unit was essentially one tree because of the hyperdispersed distribution of trees in the tropical forest.

Later, Bawa (*6*) undertook a study to determine whether numerous tree species of a seasonal forest in Costa Rica were, in fact, self-pollinating. Contrary to previous statements, he found that almost all taxa were obligate outbreeders. Subsequent studies by Ashton (*2*) on canopy trees of the Malayan tropical forest showed that these species were also obligate outbreeders. Ashton has most recently suggested (*2*) that there is very little pollen and seed movement within the tropical rainforest and that a few clumped, related individuals represent the interbreeding population. These isolated clumps would be effectively isolated from other such clumps and would rapidly give rise to new taxa. Studies of the leaf extract isozymes of neighboring trees of the same species indicated that trees clumped together did tend to be more similar in their isozymes than trees distant from one another (*88*). Yet, more studies must be done to prove that the similarity of neighboring trees is not due simply to selection at the microhabitat level.

In fact, recent systematic and geological studies in many areas of lowland tropics have indicated that numerous possibilities for geographical isolation of populations have occurred within the last few million years. Such studies alleviate, to some extent, the need to postulate specialized speciation mechanisms for tropical taxa.

PLEISTOCENE CLIMATIC CHANGES IN THE TROPICS AND EVIDENCE FOR GEOGRAPHICAL SPECIATION

Because of the comparatively advanced state of basic systematics of avian groups, it is not surprising that birds were among the first animals of the Amazon Basin to be studied biosystematically (*32*). Over the course of several years, Jürgen Haffer studied intensively 33 species of toucans and toucanets (Ramphastidae) and 17 species of jacamars (Galbulidae). These investigations included sampling throughout the ranges of the species of these two families and the location and description of hybrid zones and areas of secondary contact between subspecies or semispecies of a superspecies. A superposition of the areas of contact between differentiated forms showed that there was a coincidence of hybrid zones and areas of narrow overlap among many of these forest-restricted birds. A more cursory analysis of the distributions of many other forest birds (*33*) showed that distributional limits of these taxa also occurred in the same regions. Although the forest appears to present a uniform habitat in these areas, the presence of large numbers of coincident contact zones indicated that forest habitats had formerly been fragmented and recently reexpanded. The areas of secondary contact were presumably along zones where forest patches had merged. The isolated pockets formed during times of fragmentation were labeled "refuges" (*32*).

At almost the same time that Haffer was carrying out his studies, P. E. Vanzolini and E. E. Williams (*75*) made a careful biosystematic study of the species group represented by *Anolis chrysolepis* Duméril and Bibron in northern South America. In this study, they measured numerous characters along a series of transects across and around the Amazon Basin and subjected the measurements to statistical analysis. Their results pointed out not only areas of apparent uniform habitat in which complexes of characters were constant (i.e. the characters showed high correlation with one another) but also areas in which characters were very variable and not correlated with one another. They concluded that the areas of little character variation were "core" areas in which long-term evolution in isolation had resulted in a relatively fixed complex of characters. The areas of variability represented areas of integration in which populations that had formerly been isolated in the core areas had come into contact and interbred. As in the case of the forest birds, the biosystematic evidence from this group of forest-dwelling lizards pointed to recent episodes of restrictions and expansions of populations.

Additional evidence comes from biosystematic studies of two groups of insects: *Drosophila* fruit flies and *Heliconius* butterflies. An initial study of a group of *Drosophila* revealed the presence of six broadly sympatric (*11*) sibling species labeled the *willistoni* species group. Little could be deciphered about their history since their geographical overlap was so extensive. Later studies (*68*), however, showed that several of the species contained semispecies that were all sympatric over much of their ranges and which would hybridize to varying degrees. Spassky and coworkers (*68*) interpreted their results as indicating that the semispecies of a taxon (that is, *D. paulistorum* Dobzhansky and Pavan) had differentiated very recently in isolation and that the present zones of secondary contact were roughly located along the areas in which the populations had formerly been isolated. This distribution pattern matched those found in avian and lizard groups and were thus also ascribed to Pleistocene periods of climatic change. The broadly overlapping ranges of the sibling species seemed to reflect an earlier radiation of the same type but far enough in the past for the originally allopatric semispecies to have reinforced reproductive barriers and expanded over larger geographical regions.

In addition to the genetic work with *Drosophila,* entomologists in Brazil carried out biosystematic studies of numerous species of heliconid butterflies (*8, 72*). Many of these species have wide ranging color morphs across northern South America and the West Indies. Throughout their ranges, many species parallel one another in patterns of wing coloration (*73, 74*). For biogeographical purposes, the heliconids are interesting because the subspecies (or color races) often hybridize where they meet and because the boundaries of the races correspond to contact zones found in other taxa of the lowland tropical rainforests of South America (*9, 72*).

Documentation of paleoecological changes that help to substantiate the biosystematic evidence for recent climate changes come from two sources: datings and analyses of lacustrine cores, and geomorphological studies. Cores taken from lake sediments show changing pollen (and presumably vegetation) composition within the last hundred thousand years at several places in lowland tropical South America. One core from western Brazil (*35*) shows previous dominance of savanna plants over forest plants in an area which is now covered by tropical forest. A short pollen sequence from the western interior of Guyana (*87*) indicates a former coverage in this area by more closed forests than are currently found. Finally, samples from the coastal areas of Surinam (*35*) show that during times of low sea level (maximum glaciations) savanna vegetation had a more extensive range than at the present time.

Geomorphological studies and aerial surveys throughout the Amazonian lowlands indicate that regions now covered by tropical forest show landforms that were produced under a different climatic and vegetative regime (*7, 42, 49*). As a result of these studies, Journaux (*42*) concluded that there was a phase more humid than that of the present between 90,000 and 21,000 years ago and a dry period between 21,000 and 13,000 years before the present. Another dry phase may have occurred between 4000 and 2000 years ago.

The fluctuations in climate, demonstrated by independent geological evidence, provide a mechanism to explain the patterns revealed by biosystematic studies of lowland forest taxa. The biogeographical explanation for the modern patterns in tropical South America derived from a combination of biosystematic and geological data is that populations of forest-restricted species were fragmented during Pleistocene dry periods and re-united during humid periods. Where expanding populations came together, hybrid zones or narrow zones of overlap are often visible. Presumably, this same process occurred several times during the Pleistocene and the more ancient episodes are now visible only as broadly sympatric species such as seen in the sibling species of the *Drosophila willistoni* group.

Climatic and vegetational changes in lowland tropical areas during the Pleistocene were by no means restricted to South America. Livingstone (*43*, p. 275) states that "the African environment is capricious, not stable, and apparently has been for at least several million years." New Guinea (*37*), Australia (*39, 40*) and even Melanesia to some extent (*1*) seem to have experienced similar kinds of biotic changes within the Pleistocene.

Not only did such changes occur in the past, but there is ample evidence that the lowland biotas have not "smoothed out" the influences of the most recent changes. The biosystematic studies of lowland tropical organisms thus show, when combined with geological data, a biogeographical picture of recently determined distribution patterns that, even apart from effects caused by man, are still in a dynamic state. The tropics do not represent a museum of species that have been accumulating since the Cretaceous. Superimposed upon ancient distribution patterns, often attributable to continental drift, are complexes of specific patterns caused by climatic changes occurring within the last three million years. There has certainly been adequate opportunity within the Pleistocene for geographical isolation and conventional processes of speciation for tropical organisms.

SUMMARY AND APPLICATIONS

Within the last fifteen years, biogeographical studies have expanded in a new direction. In general, this shift has been caused by the inclusion of biosystematic, or population level, data into hypotheses. These new concepts neither exclude nor deny the importance of studies dealing with ancient branchings, phylogenetic evolution, or traditional biogeographical principles (*47*). Rather, they supplement many former theories with data on the dynamics of population interactions. Theories derived from such biosystematic data allow predictive statements to be made about the abundance, kinds, and distributions of taxa in particular areas or habitats. In addition, recent biosystematic studies combined with reliable geological data have allowed new insights into processes of speciation of many lowland tropical organisms.

One of these modern theories, the theory of island biogeography, makes statements about the diversity of organisms that can exist on islands of various sizes and at various distances from a source of propagules. Although originally applied to true islands, the theory has been extended to a number of continental situations in which habitats occur as patches in space. In addition, the theory has applications for spacing and plot sizes of agricultural crops (*28, 70*). By reducing the continuous area of a crop by plot- or strip-cropping in a field, a practice long recognized by primitive agriculturalists, pest damage can be lessened. In terms of island biogeography, such a practice reduces the "island" size and increases the distance a potential pest must travel to find more of the same food. Another application of the theory of island biogeography, although debated to some extent (*24, 61, 62, 71, 78*), is in the planning of preserves for conservation purposes. The theory of island biogeography relies in part on the fact that, in reduced areas, population sizes are small and the probability of extinction is high. This is true particularly when the individualistic properties of species are also taken into consideration. This phenomenon should serve as a warning to conservationists.

Another modern concept in biogeography is that of the taxon cycle, which describes the cyclical process of invasion by highly vagile, variable taxa and their subsequent restriction to more specialized habitats. Part of the success of taxa in the first stage of this cycle is due to the fact that recent colonizers have left behind their normal parasites and predators. A parallel is evident between this initial stage and the success of introduced weeds in marginal habitats (*4*).

Finally, recent biosystematic and geological studies have shown that it is unnecessary to postulate special speciation mechanisms for organisms of the lowland, humid tropics. Quaternary climatic changes which occurred in the tropics as well as elsewhere in the world have provided opportunity for geographical isolation and subsequent differentiation. Present-day patterns are not the result of ancient, unchanged distributions.

All of these modern concepts of biogeography stress the dynamic component of the distributions of organisms. Especially in a world in which man is causing rapid environmental changes, principles of biogeography that incorporate such phenomena as dispersal, establishment, extinction, and cyclical perturbations are of particular importance for the potential management and conservation of species.

ACKNOWLEDGMENTS

I acknowledge the invaluable help of F. C. Thompson and J. L. Neff in the preparation of the manuscript and also that of A. R. Tangerini in the execution of the illustrations.

LITERATURE CITED

1. Ashton, P. S. 1972. *The Quaternary geomorphological history of western Malesia and lowland forest phytogeography*. Pages 35–49 *in* P. S. and M. Ashton, eds. *The Quaternary Era in Malesia*. Trans. Second Aberdeen-Hull Symposium on Malesian ecology (Aberdeen, 1971). Dept. Geogr., Univ. Hull. 122 pp.

2. Ashton, P. S. 1977. *The contrasting breeding system of trees that flower above and within the forest canopy*. Paper presented at the IV Simposio Int. Ecol. Tropical. Panama City, 7–11 Mar., 1977. Abstr. pp. 9–10.

3. Ashton, P. S., and M. Ashton, eds. 1972. *The Quaternary Era in Malesia*. Trans. Second Aberdeen-Hull Symposium on Malesian Ecology (Aberdeen, 1971). Dept. Geogr., Univ. Hull. 122 pp.

4. Baker, H. G., and G. L. Stebbins, eds. 1965. *The genetics of colonizing species*. Academic Press, New York. 588 pp.

5. Ball, I. R. 1975. *Nature and formulation of biogeographical hypotheses*. Syst. Zool. 24. 407–430.

6. Bawa, K. S. 1974. *Breeding systems of tree species of a lowland tropical community*. Evolution 28: 85–92.

7. Bigarella, J. J., and G. O. de Andrade. 1965. *Contribution to the study of the Brazilian Quaternary*. Geol. Soc. Am. Spec. Paper 84: 433–451.

8. Brown, K. S., Jr. 1975. *Geographical patterns of evolution in Neotropical Lepidoptera. Systematics and derivation of known and new Heliconiini (Nymphalidae: Nymphalinae)*. J. Entomol. (B) 44: 201–242.

9. Brown, K. S., and O. H. H. Mielke. 1972. *The heliconians of Brazil (Lepidoptera: Nymphalidae). Part II. Introduction and general comments with a supplementary revision of the tribe*. Zoologica (New York) 57: 1–40.

10. Brown, W. L. 1958. *Contributions toward a reclassification of the Formicidae. II. Tribe Ectatommini*. Bull. Mus. Comp. Zool., Harvard 118: 175–362.

11. Burla, H., A. B. da Cunha, A. R. Cordeiro, Th. Dobzhansky, C. Malogolowkin, and C. Pavan. 1949. *The willistoni group of sibling species of Drosophila*. Evolution 3: 300–314.

12. Cairns, J., M. L. Dahlberg, K. L. Dickson, N. Smith, and W. T. Waller. 1969. *The relationship of fresh-water protozoan communities to the MacArthur-Wilson equilibrium model*. Amer. Nat. 103: 439–534.

13. Camp, W. H., and C. L. Gilly. 1943. *The structure and origin of species*. Brittonia 4: 323–385.

14. Candolle, A. L. P. P. de. 1855. *Géographie botanique raisonée; ou exposition des faits principaux et des lois concernant la distribution geographique des plants de l'époque actuelle*. Masson, Paris. 2 vols. 1365 pp.

15. Charlesworth, J. K. 1957. *The Quaternary Era, with special reference to its glaciations*. Edward Arnold, London. 2 vols. 1700 pp.

16. Cody, M. L. 1975. *Toward a theory of continental species diversity: bird distributions over Mediterranean habitat gradients*. Pages 214–257 *in* M. L. Cody, and J. M. Diamond, eds. *Ecology and evolution of communities*. Harvard Univ. Press (Belnap), Cambridge. 545 pp.

17. Colbert, E. H. 1973. *Wandering lands and animals*. Dutton, New York. 325 pp.

18. Corner, D. J. H. 1954. *The evolution of tropical forest*. Pages 34–46 *in* J. S. Huxley, A. C. Hardy, and E. B. Ford, eds. *Evolution as a process*. Allen and Unwin, London.

19. Culver, D. C. 1970. *Analysis of simple cave communities. I. Caves as islands.* Evolution 29: 463–474.

20. Darlington, P. J., Jr. 1957. *Zoogeography: the geographical distribution of animals.* Wiley, New York. 675 pp.

21. Darlington, P. J., Jr. 1965. *Biogeography of the southern end of the world; distribution and history of far-southern life and land, with an assessment of continental drift.* Harvard Univ. Press, Cambridge. 236 pp.

22. Diamond, J. M. 1973. *Distributional ecology of New Guinea birds.* Science 179: 759–769.

23. Diamond, J. M. 1975. *Assembly of species communities.* Pages 342–444 *in* M. L. Cody and J. M. Diamond, eds. *Ecology and evolution of communities.* Harvard Univ. Press (Belnap), Cambridge.

24. Diamond, J. M. 1976. *Island biogeography and conservation: strategy and limitations.* Science 193: 1027–1029.

25. Dietz, R. S., and J. C. Holden. 1970. *The breakup of Pangaea.* Sci. Amer. 223(4): 30–41.

26. Dobzhansky, Th. 1950. *Evolution in the tropics.* Amer. Sci. 38: 209–221.

27. Fedorov, An. A. 1966. *The structure of the tropical rainforest and speciation in the humid tropics.* J. Ecol. 54: 1–11.

28. Feeney, P. 1976. *Plant apparency and chemical defense.* Pages 1–92 *in* J. W. Wallace and R. L. Mansel, eds. *Recent advances in phytochemistry* (Vol. 10). Plenum Press, New York.

29. Fisher, F. J. F. 1965. *The alpine Ranunculi of New Zealand.* New Zealand Dept. Sci., Indust. Res., Bull. 165. 192 pp.

30. Flint, R. F. 1971. *Glacial and Quaternary geology.* Wiley, New York. 892 pp.

31. Guppy, H. B. 1921. *The testimony of the endemic species of the Canary Islands in favor of the age and area theory of Dr. Willis.* Ann. Bot. 35: 513–521.

32. Haffer, J. 1969. *Speciation in Amazonian forest birds.* Science 165: 131–137.

33. Haffer, J. 1974. *Avian speciation in tropical South America.* Publ. Nuttall Ornithological Club 14: 1–390. Nuttall Club, Cambridge.

34. Hallam, A. 1972. *Continental drift and the fossil record.* Sci. Amer. 227(5): 56–66.

35. Hammen, T. van der. 1972. *Changes in vegetation and climate in the Amazon Basin and surrounding areas during the Pleistocene.* Geol. en Mijnbouw 51: 641–643.

36. Handlirsch, A. 1913. *Beiträge zur exacten Biologie.* Sitzungsber. Kaiser. Akad. Wissenschaft., Math-Natur. Kl. 122(3): 361–481.

37. Hope, G. S. 1976. *The vegetational history of Mt. Wilhelmy, Papua, New Guinea.* J. Ecol. 64: 627–664.

38. Humboldt, A. F. von. 1807. *Essai sur la geographie des plantes.* Levrault, Schoell and Co., Paris. 35 pp.

39. Jersey, W. F. de. 1968. *Palaeobotany and palynology in Australia: a historical review.* Rev. Palaeobot. Palynol. 6: 111–136.

40. Jessup, R. W. 1961. *A Tertiary-Quaternary pedological chronology for the southeastern portion of the Australian zone.* J. Soil Sci. (Britain) 12: 199–213.

41. Johnson, L. A. S. 1970. *Biosystematics alive?—A discussion.* Taxon 19: 152–153.

42. Journaux, A. 1975. *Recherches géomorphologiques en Amazonie brésilienne.* Centre Nat. de la Recherche Scientifique. Centre de Geomorphologie de Caen, Bull. Semestriel. 20: 1–68.

43. Livingstone, D. A. 1975. *Late Quaternary climatic change in Africa.* Ann. Rev. Ecol. Syst. 6: 249–280.

44. MacArthur, R. H., and E. O. Wilson. 1963. *An equilibrium theory of insular zoogeography.* Evolution 17: 373–387.

45. MacArthur, R. H., and E. O. Wilson. 1967. *The theory of island biogeography.* Monographs in population biology. I. Princeton Univ. Press, Princeton. 203 pp.

46. Marvin, U. B. 1973. *Continental drift: the evolution of a concept.* Smithsonian Institution Press, Washington. 239 pp.

47. McKenna, M. C. 1973. *Sweepstakes, filters, corridors, Noah's arks, and beached Viking funeral ships in palaeogeography.* Pages 295–308 *in* D. H. Tarling and S. K. Runcorn, eds.

Implications of continental drift to the earth sciences. Vol. 1. Academic Press, New York. 622 pp.

48. Meyer, F. J. F. 1836. *Grundriss der Pflanzengeographie*. Translated by M. Johnston (1946) as *Outline of the geography of plants*. Ray Society, London, 422 pp.

49. Moushino, M. R. 1971. *Quaternary process changes of the middle Amazon area*. Bull. Geol. Soc. Amer. 82: 1073-1078.

50. Opler, P. A. 1974. *Oaks as evolutionary islands for leaf-mining insects*. Amer. Sci. 62: 67-73.

51. Raven, P. H. 1977. *Systematics and plant population biology*. Syst. Bot. 1: 284-316.

52. Raven, P. H., and D. I. Axelrod. 1974. *Angiosperm biogeography and past continental movements*. Ann. Missouri Bot. Gard. 61: 539-673.

53. Ricklefs, R. E., and G. W. Cox. 1972. *Taxon cycles in the West Indian avifauna*. Amer. Nat. 106: 195-219.

54. Ridley, H. N. 1930. *The dispersal of plants throughout the world*. Reeve and Co., Ashford, Kent. 744 pp.

55. Rosen, D. E. 1975. *A vicariance model of Caribbean biogeography*. Syst. Zool. 24: 431-464.

56. Schuchert, C. 1932. *Gondwana land bridges*. Bull. Geol. Soc. Amer. 43: 875-916.

57. Schuster, R. M. 1976. *Plate tectonics and its bearing on the geographical origin and dispersal of angiosperms*. Pages 48-138 *in* C. B. Beck, ed. *Origin and early evolution of angiosperms*. Columbia Univ. Press, New York.

58. Sclater, P. L. 1958. *On the general geographical distribution of the members of the class Aves*. J. Linn. Soc. (Zool.) 2: 130-145.

59. Simberloff, D. S. 1974. *Equilibrium theory of island biogeography and ecology*. Ann. Rev. Ecol. Syst. 5: 161-182.

60. Simberloff, D. S. 1976. *Species turnover and equilibrium island biogeography*. Science 194: 572-578.

61. Simberloff, D. S., and L. G. Abele. 1976. *Island biogeography theory and conservation practice*. Science 191: 285-286.

62. Simberloff, D. S., and L. G. Abele. 1976. *Island biogeography and conservation: strategy and limitations*. Science 193: 1032.

63. Simberloff, D. S., and E. O. Wilson. 1969. *Experimental zoogeography of islands. The colonization of empty islands*. Ecology 50: 278-296.

64. Simberloff, D. S., and E. O. Wilson. 1970. *Experimental zoogeography of islands. A two-year record of colonization*. Ecology 51: 934-937.

65. Simpson, B. B. 1974. *Glacial migrations of plants: Island biogeographical evidence*. Science 185: 689-700.

66. Simpson, B. B. 1975. *Pleistocene changes in the flora of the high tropical Andes*. Paleobiology 1: 273-294.

67. Simpson, B. B. 197-. *Stability and the high central Andes*. IV Int. Symposium Tropical Ecol., 7-11 Mar. 1977. Panama City, Panama (in press).

68. Spassky, B., R. C. Richmond, S. Perez-Salas, O. Pavlovski, C. A. Mourao, A. S. Hunter, H. Hoenigsberg, Th. Dobzhansky, and F. J. Ayala. 1971. *Geography of the sibling species related to Drosophila willistoni and of the semispecies of the Drosophila paulistorum complex*. Evolution 25: 129-143.

69. Stafleu, F. 1969. *Biosystematic pathways anno 1969*. Taxon 18: 485-500.

70. Strong, D. R. 1974. *Nonasymptotic species richness models and the insects of British trees*. U. S. Natl. Acad. Sci., Proc. 71: 2766-2769.

71. Terborgh, J. 1976. *Island biogeography and conservation: Strategy and limitations*. Science 193: 1029-1030.

72. Turner, J. R. G. 1971. *Two thousand generations of hybridization in a Heliconius butterfly*. Evolution 25: 471-482.

73. Turner, J. R. G. 1971. *Studies in Mullerian mimicry and its evolution in Burnet moths*

and heliconid butterflies. Pages 224-260 *in* R. Creed, ed. *Ecological genetics and evolution.* Blackwell, Oxford. 319 pp.

74. Turner, J. R. G. 1973. *Passionflower butterflies.* Animals 15 (Jan): 15-21.

75. Vanzolini, P. E., and E. E. Williams. 1970. *South American anoles: The geographic differentiation and evolution of the Anolis chrysolepis species group (Sauria, Iguanidae).* Arq. Zool. (São Paulo) 19: 1-298.

76. Vine, F. J. 1977. *The continental drift debate.* Nature 266: 19-22.

77. Vuilleumier, F. 1970. *Insular biogeography in continental regions: The northern Andes of South America.* Amer. Nat. 104: 373-388.

78. Whitcomb, R. F., J. F. Lynch, P. A. Opler, and C. S. Robbins. 1976. *Island biogeography and conservation: Strategy and limitations.* Science 193: 1030-1032.

79. Willett, R. W. 1950. *The New Zealand Pleistocene snowline, climatic conditions and suggested ecological effects.* New Zealand J. Sci., Tech. Bull. 32: 18-48.

80. Williams, E. E. 1969. *The ecology of colonization as seen in zoogeography of anoline lizards on small islands.* Quart. Rev. Biol. 44: 345-389.

81. Willis, J. 1921. *Endemic genera of plants in their relation to others.* Ann. Bot. 35: 493-510.

82. Willis, J. C. 1922. *Age and area. A study in geographical distribution and origin of species.* Cambridge Univ. Press. 259 pp.

83. Wilson, E. O. 1958. *Studies on the ant fauna of Melanesia, I-IV.* Bull. Mus. Comp. Zool., Harvard, 118: 101-153, 303-371.

84. Wilson, E. O. 1961. *The nature of the taxon cycle in the Melanesian ant fauna.* Am. Nat. 95: 169-193.

85. Wilson, E. O., and D. S. Simberloff. 1969. *Experimental zoogeography of islands. Defaunation and monitoring techniques.* Ecology 50: 267-278.

86. Wright, H. E. Jr., and D. G. Frey. 1965. *The Quaternary of the United States.* Princeton Univ. Press, Princeton. 922 pp.

87. Wymstra, T. A., and T. van der Hammen. 1966. *Palynological data on the history of tropical savannas in northern South America.* Leidse Geol. Mededel. 38: 71-90.

88. Yap. Y. 1977. *Genetic variation within and between populations of Malaysian rain forest trees.* Pap. presented at the IV Simposio Int. Ecol. Tropical. Panama City, 7-11 Mar. 1977. Abstr. pp. 9-10.

10] Influence of Morphology, Biology, and Ecology on Evolution of Parasitism in Nematoda

BY ARMAND R. MAGGENTI*

ABSTRACT

Two major lines for the development of parasitism in Nematoda are discussed. The primary predecessor group of nematode parasites of vertebrates are the free-living bacterial feeders of the order Rhabditida. Insect and plant parasitic nematodes are derived from free-living fungus feeders found in soil and as insect associates. The inherent preadaptive features of both groups are discussed. The adaptive steps from free-living bacteria and fungus-feeding nematodes to parasites of vertebrates, insects and plants are outlined. The hypothetical transition draws on biological (life cycle), ecological (host-parasite relationships) and morphological data to reconstruct the phylogeny of parasitism in Nematoda. Contemporary forms exhibiting preadaptive features are utilized as models in this reconstruction. A change in inclination and deceleration of the earth's angular velocity are postulated as one mechanism mediating the Cretaceous disaster and its coincident impact on the development of nematode parasitism in plants and animals.

INTRODUCTION

Nematodes are the largest group of lower metazoan parasites of plants and animals with over twenty thousand nominal species currently recognized in the phylum. Among the Invertebrata they are third in speciation, being exceeded by Mollusca and Arthropoda. There are over two thousand plant parasitic nematodes, five thousand animal parasites, and some thirteen thousand marine, soil, and freshwater species. Even so this represents fewer than two percent of the nominal species of insects. Paramanov (*25*) speculated that if all the species of nematodes were known the number would exceed 500,000. This estimate is not unreasonable; the number of genera

*Division of Nematology, University of California, Davis, California 95616.

and species of plant parasitic nematodes has increased tenfold in the last ten to fifteen years. Similar increases have occurred among the animal parasites. There has been a decline in taxonomic effort with the soil, freshwater, and marine nematodes because of a lack of incentive to study these forms.

Nematodes have had a long and unsettled taxonomic history. Lack of agreement as to their placement in the animal kingdom as an independent phylum persists today. This presentation will conform to the concept that they are an independent phylum, as set out by Cobb (*3*), Pearse (*26*), and Chitwood (*4*), and to an internal classification as I proposed (*19*), which is a modification of that found in P.-P. Grasse's treatise (*6*). As such, Nematoda is recognized as a phylum subdivided into two classes, Adenophorea and Secernentea. The Adenophorea are further subdivided into two subclasses, Enoplia and Chromadoria.

Within Nematoda, animal parasitism is distributed throughout the class Secernentea among the five currently recognized orders: Rhabditida, Tylenchida, Strongylida, Ascarida, and Spirurida. Plant parasitism in this subclass is limited to the order Tylenchida. Among the Adenophorea parasitism is limited to the subclass Enoplia and plant parasitism to the order Dorylaimida. Animal parasitism in Adenophorea is found within the orders Trichosyringida, and Trichocephalida. Parasitism is unknown in the subclass Chromadoria. Accepting the premise of a relatively stable classification, as established above, one can proceed to the development of parasitism within the phylum. For the most part emphasis will be on the secernentean parasites of plants and animals. All of the secernentean parasites of vertebrates appear to have arisen from bacterial feeders of the order Rhabditida and the related orders Strongylida, Ascaridida, and Spirurida. Whereas plant parasitism, and for the most part insect parasitism, arose from fungus-feeding ancestors in the orders Diplogasterida and Tylenchida.

APPROACHES TO NEMATODE PHYLOGENY AND THE DEVELOPMENT OF PARASITISM

The science of nematology is fortunate because of the tremendous array of modern species both free-living and parasitic. The habitats, niches, and behavioral activity of these nematodes appear almost limitless. Consequently there exists among contemporary species a logical sequence from free-living to the parasitic mode of existence. The known nominal species of nematodes and their variation does allow for cautious interpretation of relationships and the phylogenetic evaluation of characters as primitive or advanced. Such an approach is not unusual for taxonomists or unique to nematologists. It is the everyday concern of any taxonomist involved in more than species descriptions.

The available fossil record in botany and zoology probably represents less than one percent of the plant and animal life that has occurred on this planet. The lack of fossils has not prevented paleobotanists or paleontologists from arriving at conclusions regarding the evolution of life forms. This has been done through knowledge and understanding of the embryology and comparative morphology of contemporary species which may provide convincing circumstantial evidence, but the fossil record remains the only source of documentary evidence.

Nematological approaches to phylogeny without fossil evidence. One approach used by nematologists to study nematode phylogeny is the study of individual organ systems in a manner that allows reasonable speculation as to the ancestral condition among modern forms. Morphological characters used in this interpretation should be selected and evaluated objectively and independent of a classification. Wherever possible characters should be utilized that are variable and that have been accepted as taxonomically profound throughout the phylum. No attempt is made here to propose any species as an ancestral taxon. Therefore, throughout this discussion I am accepting the concept of ancestor-descendant relationships as outlined by Engelmann and Wiley (*8*).

Figure 10.1. Superimposition of the analyses of nemic esophagi, male reproductive system and system of excretion. Illustrates areas of correlation with accepted higher classification and phylogeny of nematodes as well as areas of disagreement.

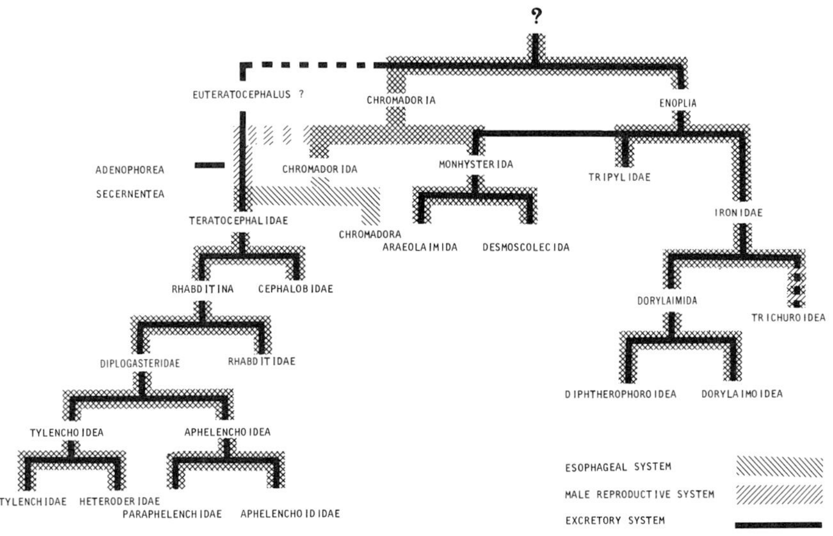

Previously (*17*) I tried to show the potential of organ system analyses as a means of studying the stability of higher classifications of nematodes. At that time three organ systems were employed to illustrate the feasibility of individual system analysis followed by intersystem correlation. The systems selected were the esophagus, the excretory system, and the male reproductive system. When the independent analyses were superimposed, various degrees of correlation were noted and compared to the current higher classification of nematodes (Fig. 10.1). Obvious discrepancy of correlation reflected the differential between a phenetic and cladistic approach to nematode classification. By the use of this method it was apparent that confidence could be given to the basic classification currently followed and that specific areas in the classification, that is, Chromadoria, needed more study. In any event it proved to be a plausible approach in the absence of fossils.

Diplogasterida and the development of insect and plant parasitism. The available evidence, morphological and biological as put forth by Paramanov *(24)*, justifies the recognition, as here proposed, of the Diplogasterida as an order separate from Rhabditida. There are representatives among the Diplogasterida that exhibit both the bacterial and fungal-feeding habit. Hirschmann (*10*) demonstrated the unusual genetic variability within the species *Mesodiplogaster lheritieri* (Maupas, 1919) Goodey, 1963. In this species two distinct stoma types are manifested; the extent of difference would, without prior knowledge, lead to the placement of the two forms into separate species, if not genera. In addition, the dauerlarva (= ölhullenlarven) has esophageal and stomal characteristics reminiscent of Tylenchida (Fig. 10.2). Other morphological developments in the Diplogasterida, which seem to be logical steps toward Tylenchida, occur in the genera *Neodiplogaster* and *Tylopharynx* (Fig. 10.2). These genera show evidence of a primitive stomatal stylet as well as the tylenchid-like esophagus, and we suspect that they are fungal-feeders (*18*). This line of development is important to the development of both plant and insect parasitism. Most of the Diplogasterids are insect associates or are found associated with decomposing plant or animal matter rich in bacteria, fungi, and other microorganisms. Such an environment, coupled with the precursor spear for feeding on discrete cells, is an important step directly toward plant parasitism and indirectly toward insect parasitism. The closely related development of these two forms of parasitism is not surprising when one reflects on the closely timed evolutionary radiations of the species of angiosperm plants and insects. Poinar (*27*) reviewed the relationship of Diplogasterids to insect disease. Their role in plant parasitism is primarily surmised from morphologic similarities to Tylenchida.

Tylenchida are distinguished among Secernentea by the conversion of the stoma and its armature to a protrusible hollow stylet. Food is drawn in through the hollow stylet by action of the muscular esophagus. This

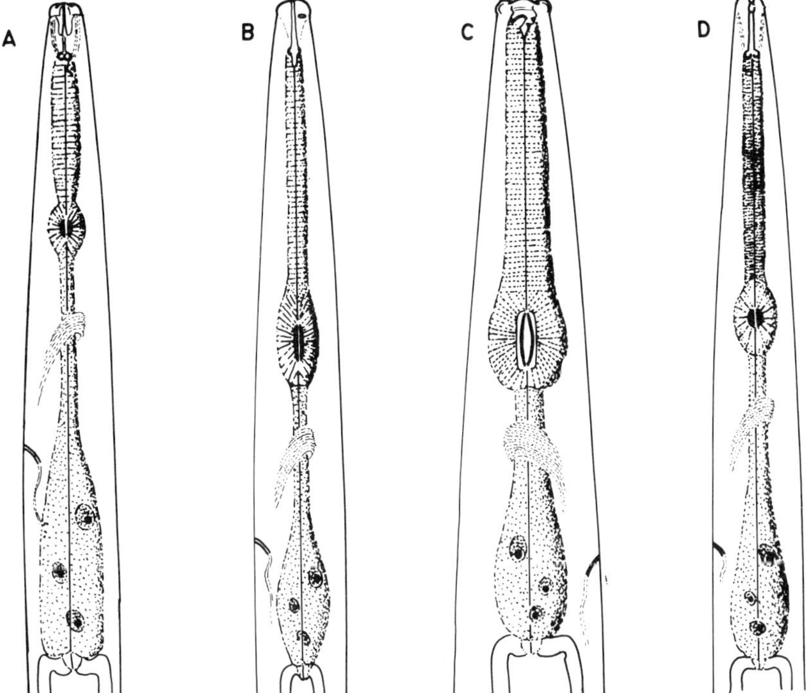

Figure 10.2. Comparison of Diplogasterid esophagi and Tylenchid esophagus.
A. Fourth-stage larva, *Ditylenchus dipsaci* (Tylenchida).
B. Third-stage dauerlarva, *Mesodiplogaster lheritieri.*
C. Adult female, *Mesodiplogaster lheritieri.*
D. Esophageal region, adult female *Tylopharynx* sp. (Diplogasterida) perhaps related to the ancestral form that gave rise to Tylenchida.

stomatal conversion, a preadaptation evident in Diplogasterida, allows tylenchid nematodes to feed on simple forms of cellular life such as fungi. Another preadaptive feature evident in Diplogasterids and in Tylenchida is the reduction of posterior esophageal musculature and accentuation of the esophageal glands. The secretions from these glands play an important part in plant and insect parasitism. In plants they are utilized to dissolve cell walls, break down intercellular cement, and to effect cellular changes. In insects they have an important role in intestinal penetration, cellular dissolution, and predigestion. These fungus-feeding nematodes must have followed two lines of development. One line, which involves fungus-feeders in the soil environment or associated with fungi in insect galleries, retained the small, slender spear and is the likely precursor of aboveground plant parasitism and insect parasitism. The other line developed stout spears which became too large to feed on fungi but were well adapted to feed on the subterranean parts of plants.

Phylogenetic steps in the development of nematodes as parasites of the subterranean parts of plants. It is important to emphasize that the modes of parasitism being proposed here represent a line of development of plant parasitism by nematodes and should not be confused with the phylogeny of the nematodes discussed. Paramanov (*25*) proposed that plant parasitism evolved through the stages of ectoparasitism and semiendoparasitism to endoparasitism. An emended sequence would better reflect the developmental processes. A more complete series would be facultative-ectoparasites, ectoparasites, ecto-endoparasites, migratory endoparasites, culminating in sedentary endoparasites. This line of development can be followed through the families Tylenchidae, Hoplolaimidae, Pratylenchidae, and Tylenchulidae to Heteroderidae. This biological sequence is accompanied by morphologic changes such as an increasing stoutness of the spear, increasing size and overlap of the intestine by the esophageal glands, the development toward sedentary swollen females, and the reduction in larval stages. To some degree each family listed above recapitulates in part the modes of parasitism evolved in Tylenchida.

Phylogenetically the most fascinating members of Tylenchida are found in the family Tylenchidae. This family, in particular its slender-speared members, apparently represents either an ancestral prologue of plant and insect parasitism or a contemporary epilogue of the development of parasitism. The family contains parasites of the subterranean and aboveground parts of plants, including fungal feeders *Tylenchus davainei* Bastian, 1865; facultative-ecotoparasites of fungi and higher plants *Tylenchus costatus* de Man, 1921; ectoparasites of higher plants *Tylenchorhynchus claytoni* Steiner, 1933; and migratory endoparasites of higher plants *Cynipanquina danthonia* Maggenti, Hart, and Paxman, 1974. In addition, many members of the family are recorded to have phoretic relations with wood-boring beetles (*30*).

Ectoparasites and the beginnings of migratory endoparasitism. The Hoplolaimidae are principally root ectoparasites. Their life cycle is straightforward from egg to adult with four molts, one of which occurs within the egg. The eggs, with few exceptions, are laid singly in the soil. Some members of the family, in particular *Helicotylenchus*, are able to cross over from ectoparasitism to endoparasitism. Adults, which normally are in the soil environment feeding as obligate ectoparasites, are often found completely within root tissue laying eggs. The latter is very important for it means introducing various life stages to endoparasitism and to a separation from the soil for egg laying. The hatched larvae often remain in the root, feeding and developing in the same manner as those that migrate out and become ectoparasitic. This is a step toward migratory endoparasitism, characteristic of the family Pratylenchidae in which endoparasitism is obligatory. In Pratylenchidae all larval stages and the adults move freely in and out of the roots; however, feeding is done only within root tissue. Eggs are laid randomly in root tissue as the females wander in search of new feeding sites in advance of tissue necrosis.

The development of sedentary endoparasitism. Among the Pratylenchidae, *Nacobbus batatiformis* Thorne and Schuster, 1956, illustrates the beginning of sedentary endoparasitism and the life cycle modifications that must be incorporated to make such a life style successful (*32*). The sedentary endoparasites show the most striking modifications in the life cycle. *Nacobbus* in the early larval stages not only migrates in advance of tissue necrosis but limits necrotic plant reaction to a critical time, thus forcing migration. Upon hatching from the egg, which is the overwintering stage in the soil, the larvae migrate in the soil and seek new roots. At the time of becoming preadults the larvae leave the roots in which they have developed because their feeding has caused necrosis or complete death of the root. The immature females enter new and larger roots where galls are formed. The posterior portion of the females protrudes from the root; at this stage the females are fertilized by the male. Eggs are laid in a gelatinous matrix outside the body in the soil. Gall formation and egg laying is usually followed by root necrosis. Therefore this cycle is distinguished by larvae moving away from necrotic tissue and by the eggs being deposited out of the necrotic environment created by the adult female.

In Tylenchidae we can see the evolution of sedentary endoparasitism by essentially ectoparasites, such as citrus nematode *Tylenchulus semipenetrans* Cobb, 1913, and the reniform nematode *Rotylenchulus reniformis* Linford and Oliveira, 1940. The first three active larval stages of citrus nematode feed ectoparasitically whereas the adult female feeds as a sedentary endoparasite. The male does not feed at all (*32*). The swollen female body is exposed on the outside of the root while the head and long misshapen neck are deeply embedded in the root. Eggs are protected from the soil environment by a gelatinous matrix, which is the product of the excretory cell (adult females lack a functional anus). Since their diet consists of liquids taken from plant cells and the amounts of nondigestible solid particles ingested are minimal, these animals, with their short life cycle, do not require an anus to carry out normal body functions (*15*).

In *Rotylenchulus reniformis* Linford and Oliveira, 1940, neither male nor larval stages feed (*32*). The male and female larvae emerge from the egg as second instars; after three superimposed molts, without change in body size, they become young infective females and adult males (*13*). From this point on, their life cycle closely resembles that of citrus nematode. The life cycle of the reniform nematode represents, among the plant parasitic nematodes, the ultimate in larval stage reduction and to some extent the potential elimination of males; this trend is not unusual among nematode parasites of plants and animals.

Root-knot nematodes and cyst nematodes are the most highly developed in terms of host-parasite relationships among the sedentary endoparasites of plants. Both are obligatory endoparasites in all larval stages. The principal difference between the two is that the root-knot type lays its eggs in a gelatinous matrix outside the body and the cyst nematode retains all or some of the eggs inside the body of the dead female, which becomes tanned and hardened. The so-called cyst is more functional in keeping the eggs together in a "nest" than it is in any other form of protection, since it offers no resistance to moisture loss or gas diffusion.

It is necessary to outline the life cycle of the root-knot nematode in order to illustrate modifications essential to successful obligatory sedentary endoparasitism without the larval migration of *Nacobbus* or the larval ectoparasitism of the citrus nematode. The second-stage larva of the root-knot nematode is the dispersal stage and also the infective stage. Soon after root penetration the larva settles near the stele of the root and initiates cell proliferation and giant cell production (*11*). These are special food cells that avoid necrosis in the vicinity of the developing larvae and mature females. The total initiation of plant response by the second-stage larva has lessened the necessity of the third and fourth stages, which are reduced to two rapid molts without feeding. In many species the male is no longer necessary for reproduction. However, under conditions of stress, larvae that were to become females are capable of reversing to the male sex. Again, the female has an undefined intestine and nonfunctional anus. The ovaries are bathed in a nutritional soup as in a culture medium and the rectal glands, which open through the nonfunctional anus, produce the gelatinous matrix (*14*).

Judging from the representatives I have observed, the lack of a defined intestine seems to be an inherent characteristic of the tylenchoidean sedentary endoparasites of plants. The same ability for ovarial development in a nutritional bath is also evident among insect parasites originating from fungus-feeding Tylenchida.

Nematode parasites of aerial plant parts and insects. In order to follow the hypothetical evolution of insect parasitism it is necessary to return the discussion to the slender-speared fungus-feeders in the suborders Tylenchina and Aphelenchina. Both suborders contain taxa that withstand desiccation, are insect associates, insect parasites, or parasites of the aboveground parts of plants. It would appear that these suborders exemplify convergent evolution. Evidence indicates that the greatest development of Tylenchina as plant parasites coincided with the radiation of the Angiospermae (monocotyledonous plants), while Aphelenchina first parasitized Filicineae (ferns) (*18*). Indications are that within both suborders plant and insect parasitism developed from fungus-feeding insect associates. As plant parasites they attack the leaves, stems, and seed heads. Paramanov's (*25*) conclusion, that the top-feeding plant parasites evolved from root parasites that sought the tops of plants because root niches were already occupied, does not seem as plausible as an origin from insect associates. The phoretic association with insects established for so many forms among these suborders (*30*) seems a more likely method for the evolution of both insect and aboveground plant parasitism. Their ability to withstand desiccation is an ideal preadaptation to becoming successful insect parasites or plant seed head parasites. Several examples exist among contemporary forms that illustrate this bridge from fungus-feeding insect associates to either insect or plant parasitism.

As contemporary illustrations of the development towards insect and plant parasitism there are three fine examples from Tylenchina and Aphelenchina. The first illustrates an external phoretic relationship between an insect and a plant parasitic nematode; the second is an internal phoretic relation between an insect and a plant parasitic nematode; and the third involves parasitism of both insect and plant by the nematode.

An external phoretic relationship between an insect and a plant parasitic nematode. Members of the genus *Bursaphelenchus* have been known for many years as insect associates, but until Mamiya and Kiyohara (*22*) described *B. lignicola* nothing was known about their life history, feeding habits, or how they were carried from tree to tree by beetles. Since 1948 severe damage to pine forests in central and southwestern Japan have been reported. Tokushige and Kiyohara (*33*) were the first to observe the then undescribed nematode in the wood of dead pines. Mamiya (*20*) recognized *B. lignicola* as the causal agent of pine wilt disease in Japan and *Monochamus alternatus* (Cerambycidae) as its main transport host. This must be one of the most serious plant nematode diseases known, because trees are killed in the span of a single summer. The nematodes are carried from tree to tree as third-stage dauerlarvae under the beetles' elytra or in their tracheae. The number of larvae carried in this manner by a single beetle varies from 15,000 to 175,000. Mamiya and Enda (*21*) showed that newly emerging adult beetles from nematode-infested trees carry the dauerlarvae to fresh living pine tissues during maturation feeding. Inoculation of healthy trees is accomplished in this manner. Maturation feeding by the beetle spans about 30 days. Of special interest is the fact that the adult beetle lays eggs that develop successfully only on the weakened, suppressed, and dead trees. Trees weakened by the nematodes inoculated during beetle maturation feeding become the target of beetle oviposition the next year.

An internal phoretic relationship between an insect and a plant parasitic nematode. Rhadinaphelenchus cocophilus (Cobb, 1919) Goodey, 1960, illustrates internal transport of a third-stage plant parasite from tree to tree. The nematode is a parasite of coconut palm and oil palm and is the causal agent of a disease known as red-ring. A cross section of the trunk of diseased trees between ground level and two meters height reveals a dull red internal ring 2 to 5 cm wide and 5 cm in from the periphery (*32*). The red ring is the result of cell necrosis associated with nematode feeding. Within this tissue are found resting third-stage larvae of the nematode. The adults and larvae of the nematode are found in the nonnecrotic tissue above the red-ring area. The coconut weevil, *Rhynchophorus palmarum*, preferentially seeks the necrotic red-ring area for its development. During its development each beetle larva consumes about 500 grams of plant tissue, thus ingesting approximately 500,000 third-stage resting larvae of the nematode. The larvae penetrate the gut wall and move into the hemocoel of the insect where they survive through insect molts and pupation. No known damage is evident in the weevil, though some measurements indicate that infested weevils are smaller. When the adult weevil seeks healthy palms for oviposition, nematodes are deposited into the tree along with the weevil eggs. The nematodes quickly attack the plant tissue causing the plant to produce the red ring of necrotic tissue within 14 to 21 days.

Insect and plant parasitism by the same species of nematode. The final example in this series from the Tylenchida is *Fergusobia curriei* (Johnston, 1938) Christei, 1941, which is apparently a parasite of the eucalyptus gall fly *Fergusonina tillyardi* Tonn. and a plant parasite of *Eucalyptus camaldulensis* Dehn. in Australia (*9*). A gametogenetic generation of the parasite occurs in the fly and a parthenogenetic generation in the eucalyptus. Gravid females of the nematode are found in the haemocoel of the gall fly, where eggs are laid. When the eggs hatch, the larvae make their way into the oviduct of the female fly. When the fly deposits eggs in eucalyptus tissue, larvae of the nematode are also deposited. Since gall formation begins before the insect eggs hatch, it is unknown what role the nematode versus the insect plays in gall production. The nematodes proceed to molt and develop into parthenogenetic females, apparently taking nutrition from the plant. Initial eggs produce males; later eggs produce females. At this time the insect has developed into a third instar maggot. The nematodes, males and females of the gametogenetic generation, mate in the gall and the female nematodes penetrate the maggot just before pupation. During pupation the parasites grow rapidly and are associated with the fat bodies of the pupal insect. Shortly after the fly emerges from the pupa the parasitic females begin to lay eggs, beginning the cycle again.

From the foregoing series it is clear that intimate phoretic relationships between insects and nematodes still exist among contemporary taxa and these can give clues to ancestral development of parasitism. I do not wish to imply that other mechanisms were not operative in the development of both nematode parasites of insects and nematode parasitism of the aboveground parts of plants. Even though, in all probability, insect parasitism developed directly from facultative parasites, the intimacy with plant parasitism cannot be overlooked as one operative mechanism. Nor should the role insects played in introducing nematodes to the tops of plants be ignored. It still remains likely that insect parasitism per se developed from fungus-feeding associates completely independent of any plant parasitism as illustrated by *Deladenus wilsoni* Bedding, 1968, and *Deladenus siricidicola* Bedding, 1968. Both of these species are parasitic in siricid wood wasps but are capable of maintaining a facultative existence on the symbiotic fungi of siricids (*1*).

Mechanisms whereby insect parasites obtain nourishment. The question of what mechanism a spear-bearing fungus-feeder utilizes to obtain nourishment as a parasite is still not answered. Obviously the nematode does not obtain food by inserting the spear into insect tissue as it does with plants. If this is accepted, then the nutritional requirements for development must be obtained in some other fashion and must be correlated with unusual morphologic changes. Among the obligate parasites under discussion there is evidence for transcuticular feeding or direct organ absorption from the haemocoelomic fluids. There is evidence from parasites of plants, and animals other than insects, that structural changes in the cuticle occur. Bird (*2*) reports that the loss of the median layer and/or internal cortical layer occurs in the root-knot nematodes, *Meloidogyne* spp and *Nippostrongylus* spp. Such a loss of certain layers in adult parasites may well increase the pereability of the cuticle. Microvilli have been reported on the outside of the cuticle of Tylenchida parasites of insects. Riding (*29*) reports that microvilli cover the entire surface of *Bradynema* spp, parasites of the fly pest *Megaselia halterata* of mushrooms, and also reports partial external coverage by microvilli in the adult female parasite *Deladenus siricidicola* Bedding, 1968, of wood wasps. Another method utilized is the eversion of the reproductive system as seen in the Sphaerularidae. The ovaries in the latter are bathed by the haemocoelic fluid of the insect just as in a culture bath. The mechanism of transfer of nutrients is not known.

Indirect nematode parasites of insects. Development of insect parasitism by nematodes has been limited to fungus-feeding ancestral types. Current evidence (*28*) indicates that bacterial-feeding nematode associates of insects are rarely direct tissue parasites of insects. Some of these are found in the digestive tract, reproductive system, or the haemocoel. Others traditionally thought to be parasites, such as *Neoaplectana* spp, actually introduce pathogenic bacteria into the insect. The bacteria are the true disease agents and the nematode feeds on the bacteria rather than on the tissues of the insect. Such forms are not true insect parasites because development to sexual maturity and oviposition occurs only in dead insects or their environment.

There are true parasites among bacterial feeders or their relatives, which indicates that insect parasitism in the Rhabditida is a recent development.

The concern here has been for the evolutionary development of parasitism and not the exceptions. Perhaps even more important is the evolution of insects as vectors of nematode parasites of vertebrates. All such parasites ancestrally originated from bacterial-feeding nematodes. Unfortunately, we lack any information that could reasonably be applied to an hypothesis for the origin of the vector-parasite relationship.

The role of larval stages in the development of parasitism. Much has been made of the role of larval stages in the evolution of parasitism (*12, 23*). Among tylenchid parasites all stages are important in the transition from one environment to another. However, it should be understood that such a statement applies only to the family Tylenchidae and suborder Aphelenchina, the very same groups thought to be ancestrally important in the development of aboveground plant parasitism and insect parasitism. The larval stages among root parasites do not enter into such a role. Among root parasites the second stage is generally considered infective, but we have seen that all mobile stages have host contact and that the survival stages are the egg, second larva, or the adult. In the Tylenchidae and Aphelenchina this is not the case. For example, the transfer ("dauer") stage for *Anguina* is the second-stage larva; in *Rhadinaphelenchus* it is the third-stage larva; in *Ditylenchus* it is the fourth-stage larva; and in *Cynipanguina* it is the adult male and female. Therefore, the transfer function of the third stage in the insect parasites does not, as suggested by Inglis *(12)*, represent an independently developed adaptation reflecting an inherent Rhabditid tendency. The third stage is important only among insect parasites of Aphelenchina; among Tylenchina the contact stage is often the adult. If there is any reflection of development it is with Diplogasterida.

Transition from one environment to another by a larval form is important among vertebrate parasites. The third stage, as pointed out by Inglis (*12*), is utilized in the transfer among vertebrate hosts from an intra-host environment to another host or to a free-living environment; or from a free-living stage to an intra-host and not just the dauer stage or just, as Osche (*23*) states, the stage of contact with the definitive host. The latter two are too restrictive. The danger in such generalities is to fall into the trap of contemplating a progressive development. This reasoning implies a single taxonomic origin for parasitism among vertebrates rather than a multiple biological origin from different taxa as suggested by Inglis (*12*).

Developmental steps in the phylogeny of vertebrate parasitism by nematodes. The steps in the development of vertebrate parasitism can be surmised by examining the adaptations which have taken place among Rhabditida. Many of these are bacterial feeders in such diverse media as decaying animals, decaying plants, and feces. In such substrates, Rhabditida must withstand variable oxygen tension, great osmotic fluctuations and high temperatures (*23*). In addition, the ability of Rhabditida to withstand desiccation and their behavioral adaptations favoring transport to new environments may be advantageous. Among the free-living taxa an association with insects for transport is important, but this plays only a small role in higher vertebrate parasitism. However, the utilization of third-stage larvae remains important either because of penetrating ability, positive geotropism, or a behavioral stance that enhances the phoretic relationship. For example, there are intestinal nematodes of cattle, *Dictyocaulus viviparus* (Bloch, 1782) Dougherty, 1946, whose behavioral activity causes them to climb to the tops of fungus (*Pilobolus* spp) sporangia. When the sporangia are discharged from the sporangiophore the nematode is catapulted away from the dung pat, where cattle will not feed, to fresh grass, often three meters away, where cattle do feed (*5*). Another nematode, *Rhabditis strongyloides* (Schneider, 1860) Oerley, 1880, frequently lives its early larval stages in the subcutaneous tissue of rodents or under the eyelids. Upon the death of the transport host the nematode leaves the carcass and develops to adulthood on saprobiotic substrates (*23*). These two examples illustrate that behavioral pattern and paratenic host utilization are important steps in the development of parasitism. The effect of the development is usually to diminish the number of exposed life stages of the parasite.

Four steps can be outlined in the development of parasitism. The first step is to facultative parasitism, such as that of *Strongyloides stercoralis* (Bavay, 1876) Stiles and Hassall, 1902, that can complete its entire life cycle in the free-living state, in the parasitic state (autoinfection), or by a combination of both. A second step could be illustrated by hookworms, in which only two larval stages can live free as bacterial-feeders. The third-stage larva (filariform) must find a host to complete its life cycle. It is interesting to note that the third-stage larva normally invades the body by direct penetration and then undergoes a migratory route through the venous system, heart, and lungs before reaching the alimentary canal where it matures in the duodenum. It is further significant that the migration route is not necessary to development. If the third-stage larva is swallowed it can directly develop in the duodenum. The significance is that the normal migration is a behavioral response developed to allow successful attack of the host and not an intrinsic necessity as it is in *Ascaris*. *Ascaris* represents the third step where only the egg stage is outside the host.

Phylogenetic significance of Ascaris migration. If *Ascaris* is to develop to adulthood it must make the migration through the body and spend a developmentally specified period of time in the lungs. *Ascaris* is almost an anomaly in the group because of the direct life cycle. Most representatives of the order use paratenic hosts or intermediate hosts. Perhaps the answer is to be found in the anomaly itself. The necessary migration and need to develop certain larval stages in the lungs may be a substitute for what evolutionarily was the role of the intermediate host. It may be a mechanism of escape from conditions in the intestine that are detrimental to the survival of the early larval stages.

The fourth and final step is known only for Adenophorean parasites, such as *Trichinella spiralis* (Owen, 1835) Railliet, 1895, which never leave the body of the host. The definitive host must be eaten by another definitive host if the parasite is to survive and develop.

Host evolution versus phylogeny of parasite. Another approach to the study of the development of parasitism by nematodes is to examine the evolution of the hosts. Osche (*23*) presents the traditional listing of the Ascarida parasites matched to their hosts in an evolutionary sequence. This list, reproduced below, and modified to include the Anisakidae, presents more problems than it solves, as Osche (*23*) recognized. However, it serves to show that host evolution and parasite evolution need not follow, and indeed may seldom have followed, each other.

PARASITE	HOST
(Acanthocheilidae)	
Acanthocheilinae	Chondrichthyes
(Stomachidae)	
Rhapidascaridinae	Osteichthyes
Stomachinae	Piscivorus Aves, marine Mammalia, marine Reptilia
(Angusticaecidae)	
Angusticaecinae	Amphibia and Reptilia
Ophidascaridinae	Reptilia
(Toxocaridae)	
Porrocaecinae	Aves
(Ascaridoidea)	
Ascarididae	Mammalia
Anisakidae	marine Mammalia with intermediates

This listing ignores a very significant fact: All nematode parasitism had its developmental origin terrestrially among bacterial-feeders. One cannot even place confidence in the taxonomic groupings of the parasites that appear to be based primarily on the hosts. On the basis of morphology, one might consider placing the Angusticaecinae, Porrocaecinae, and Ascarididae together. Since this development was terrestrial, then the further evolution of nematode parasitism among Ascarididae was most likely taken to the marine environment by marine Mammalia and aquatic Aves. This does not exclude the role that may have been played by anadromous fish that may acquire a parasitic burden during their freshwater existence. However, such parasitism is generally among spirurids which are closely related to the Thelaziidae, parasites of birds. This relationship is not surprising since insects make up a large portion of the diets of birds and freshwater fish. This does not, however, explain Ascarid parasites in the marine environment. Sharks may have obtained their parasites from fish or, as Inglis (*12*) suggested, they may have retained parasites acquired by freshwater ancestors.

Sprent (*31*) is of the opinion that Ascarid parasitism began in the aquatic environment and moved terrestrially. As evidence he used the complex life cycles seen among Anisakinae in marine mammals. The implication is that these are newly acquired parasites from the marine environment. Osche (*23*) states that the transition from land to sea caused the normally terrestrial nematode parasites to become extinct. However, indirect life cycles appear more and more to be the norm within Ascarididae. Sprent (*31*) interprets the indirect life cycle as primitive. However, it would be extremely difficult for Anisakids to survive in the marine environment if the direct cycle persisted. Even if the eggs did not require an incubation period it would be very difficult for the eggs to be picked up directly by another definitive host. In view of this it is a natural development for marine parasites to have indirect life cycles and it does not necessarily mean that their development was from already marine parasites and that the original terrestrial parasites were lost. A further aspect that apparently was overlooked by Sprent (*31*), Osche (*23*), and Inglis (*12*) is the geologic time span involved in the transition of nematode parasitism from terrestrial mammals to the marine environment. During the transition there certainly was sufficient time for the parasites to adapt. It is not unlikely that a time existed when the parasite was adapted to a terrestrial and aquatic cycle.

The development of vertebrate parasitism through geologic time. We are now at the point in this discussion where the development of parasitism must be evaluated in the light of geologic time, climate, and vertebrate evolution. Mammals have been in existence since the late Triassic period, 190 million years ago. For the first 140 million years of their existence they were unable to produce any carnivore larger than cat size, or herbivore larger than rat size (7). For the most part they were small shrew-like insectivores. Their life style placed them in a position to be introduced to the preadapted bacterial-feeding nematodes associated with feces, and with the carrion of reptiles, of dinosaurs, and of other mammals.

Inglis (*12*) emphasized that most nematodes are not markedly host specific and that the evolution of most groups of parasitic nematodes has been largely independent of the evolution of their hosts and has tended to occur in groups of hosts with similar ecological requirements and feeding habits. This idea may be linked with that of Desmond (7), who argues that the convergence in shape among the dinosaurs, mammals, and birds reveals that their ecological requirements and feeding habits were similar. The linkage of the thoughts of these two men has great significance to parasitology.

Nematode parasites of dinosaurs possible contemporary parasites of birds and mammals. If Desmond (7) and other proponents of the warm-blooded dinosaur are correct, is it not possible that many avian and mammalian parasites are descendants of the parasites of dinosaurs? This hypothesis may provide an explanation for the gaps seen in the developmental steps of vertebrate parasitism. Further, it explains the lack of relationship between bird parasites and those of lizards and snakes. Birds are closer to dinosaurs than they are to other reptiles and their evolution from dinosaurs is accepted. Since the dinosaurs had to be warm-blooded to survive (7), and probably had a chambered heart to separate oxygenated blood from spent venous blood, they would have provided the necessary requirements for preadaptation of their parasites for birds and mammals. Not all mammalian and avian parasites may have evolved in this manner. At any rate the extinction of the dinosaurs at the end of the Cretaceous must have had a definite impact on the world of parasites, because it is highly improbable that they were free from parasites.

End of Cretaceous marks a revolution in life forms. The end of the Cretaceous saw a change in plant and animal life unequaled in geologic history. There was revolutionary turnover; life forms as diverse as dinosaurs and ammonites, plesiosaurs, and many types of plankton, pterosaurs, and numerous land plants—all were eliminated simultaneously in what was one of the most all-embracing mass extinctions in geologic time (7). Certainly, such mass extinctions had a formidable effect upon the then-existing parasites. Several hypotheses have been proposed for the forces that could have caused such a wholesale catastrophe in a relatively short period of time. The magnitude of the devastation cannot be overestimated. The major theory is that it was caused by outbursts of direct cosmic radiation from the explosion of a nearby supernova star (7), which irradiated the ozone layer and ionosphere with X-rays. But this theory does not explain the survival of mammals, crocodiles, and birds, all of which were exposed just as was the dinosaur.

I propose an hypothesis that involves knowledge gained from the atomic clock. This hypothesis not only accounts in part for the rapid extinction of diverse life forms, but for the equally rapid evolution of plant, insect, and mammalian species thereafter. The atomic clock depends upon electrical oscillations, as in quartz crystal, which are regulated by natural vibration frequencies of an atomic system. It verifies that the earth is decelerating in angular velocity. Deceleration, even constant, would have had, throughout geologic time, a profound effect on the life of the earth. However, linear deceleration would not account for extreme changes in seasons or day length. Abrupt changes in seasonality and day length would result if the earth's inclination were changed, or if there were a rapid deceleration of the earth's angular velocity over a relatively short period of time. Either or both of these phenomena would result from the passing of an asteroid close to the earth.

Without collision, no physical evidence of the asteroid would remain; however, great changes would be evident in the fauna and flora of the earth. We know that such profound changes did occur on the earth at the end of the Cretaeceous. Prior to a change in inclination or before rapid deceleration, temperatures over the land masses would have been relatively more constant. After such changes the earth would be subjected to seasonality and to significant changes in daily temperatures. These are the very factors that would encourage rapid speciation among flowering plants, insects, and mammals. Furthermore, such drastic climatic changes would account for the extinction of planktonic life so delicately controlled by temperature. These same factors would result in the ultimate demise of the dinosaurs. This may explain why we seem to understand the phylogeny of parasitism in plants and insects, the greatest speciation of which postdates the Cretaceous, better than that in vertebrates, which predate the Cretaceous catastrophe.

CONCLUSIONS

Because of a lack of verified information that could substantiate a relationship among pseudocoelomates, the nematodes are treated as an independent phylum: Nematoda. The phylum is divided into two classes: Secernentea and Adenophorea. The latter is further subdivided into two subclasses: Enoplia and Chromadoria.

The order Diplogasterida is proposed as ancestral to the evolution of plant and insect parasitism among Tylenchida.

Subterranean plant parasitism is hypothesized as originating from free-living fungus-feeders. The gradual development of a stronger, stouter feeding apparatus (spear) eventually isolated these nematodes from the fungus-feeding habit. The evolution of subterranean plant parasitism encompasses development from free-living fungus-feeders, to facultative parasites of fungi and higher plants, to obligate ectoparasites, to migratory endoparasites, and culminates in sedentary endoparasitism.

Insect and aboveground plant parasitism has developed, for the most part, from fungus-feeding insect associates.

Vertebrate parasitism is proposed as having evolved terrestrially from bacterial-feeding nematodes associated with saprobiotic substrates. The nematode parasites of marine mammals are viewed as being descendants of the original parasites of marine mammals which carried through the transition of the terrestrial habitat to the marine habitat.

It is proposed that some of the parasites of land mammals and birds are descendants of the nematode parasites of warm-blooded dinosaurs. An hypothesis is put forth suggesting that the rapid speciation of mammals, birds, flowering plants, and their parasites followed a change in the earth's inclination and a rapid deceleration of the earth's angular velocity as a result of the close passing of an asteroid. The results of such a phenomenon provide the mechanisms that, at least in part, could explain the extinction of many Cretaceous plants and animals, including dinosaurs, in a short geologic time span.

LITERATURE CITED

1. Bedding, R. A. 1969. *Deladenus wilsoni n. sp. and D. siricidicola n. sp. (Neotylenchidae), entomophagous-mycetophagus nematodes parasitic in siricid woodwasps.* Nematologica 14: 515-525.

2. Bird, A. F. 1971. *The structure of nematodes.* Academic Press, New York, London. 318 pp.

3. Cobb, N. A. 1919. *The orders and classes of nemas.* Contrib. Sci. Nemat. 8: 213-216.

4. Chitwood, B. G. 1958. *The designation of official names for higher taxa of invertebrates.* Bull. Zool. Nomencl. 15: 860-895.

5. Croll, N. A. 1966. *Ecology of parasites.* Harvard University Press, Cambridge, Massachusetts. 136 pp.

6. De Coninck, L. 1965. *Systématique des nematodes.* Pages 586-731 *in* P.-P. Grasse, ed. *Traité de Zoologie.* Masson, Paris.

7. Desmond, A. J. 1975. *The hot-blooded dinosaurs.* Blond and Briggs, London. 238 pp.

8. Engelmann, G. F., and E. O. Wiley. 1977. *The place of ancestor-descendant relationships in phylogeny reconstruction.* Syst. Zool. 26: 1-18.

9. Fisher, J. M., and W. R. Nickle. 1968. *On the classification and life history of Fergusobia curriei (Sphaerulariidae: Nematoda).* Proc. Helminth. Soc. Wash. 35: 40-46.

10. Hirschmann, H. 1952. *Die Nematoden der Wassergrenze mittelfränkischer Gewässer.* Zool. Jahrb. Syst. 81: 313-436.

11. Huang, C. S. and A. R. Maggenti. 1969. *Mitotic aberrations and nuclear changes of developing giant cells in Vicia faba caused by root knot nematode, Meloidogyne javanica.* Phytopathology 59: 447-455.

12. Inglis, W. G. 1965. *Patterns of evolution in parasitic nematodes.* Pages 79-124 *in* A. E. R. Taylor, ed. *Evolution of parasites.* Brit. Soc. Parasit. (Third Symposium) Blackwell Sci. Pub., Oxford.

13. Linford, M. B., and J. M. Oliveira. 1940. *Rotylenchulus reniformis nov. gen. n. sp., a nematode parasite of roots.* Proc. Helminth. Soc. Wash. 7: 35-42.

14. Maggenti, A. R., and M. W. Allen. 1960. *The origin of the gelatinous matrix in Meliodogyne.* Proc. Helminth. Soc. Wash. 27: 4-10.

15. Maggenti, A. R. 1962. *The production of the gelatinous matrix and its taxonomic significance in Tylenchulus (Nematoda: Tylenchulinae).* Proc. Helminth. Soc. Wash. 29: 139-144.

16. Maggenti, A. R. 1963. *Comparative morphology in nemic phylogeny.* Pages 273-282 *in* E. C. Dougherty, ed. *The lower metazoa.* Univ. of California Press, Berkeley.

17. Maggenti, A. R. 1970. *System analysis and nematode phylogeny.* J. Nematol. 2: 7-15.

18. Maggenti, A. R. 1971. *Nemic relationships and the origins of plant parasitic nematodes.* Pages 65-81 *in* B. M. Zuckerman, ed. *Plant parasitic nematodes.* Academic Press, New York, London.

19. Maggenti, A. R. 1977. *Taxonomic position of nematoda among the pseudocoelomate bilateria.* Pages 1-10 *in* N. A. Croll, ed. *The organization of nematodes.* Academic Press, London, New York.

20. Mamiya, Y. 1972. *Pine wood nematode, Bursaphelenchus lignicola Mamiya and Kiyohara, as a causal agent of pine wilting disease.* Rev. Plant Protect. Res. 5: 46-60.

21. Mamiya, Y., and N. Enda. 1972. *Transmission of Bursaphelenchus lignicola (Nematoda: Aphelenchoididae) by Monochamus alternatus (Coleoptera: Cerambycidae).* Nematologica 18: 159-162.

22. Mamiya, Y., and T. Kiyohara. 1972. *Description of Bursaphelenchus lignicola n. sp. (Nematoda: Aphelenchoididae) from pine wood and histopathology of nematode-infested trees.* Nematologica 18: 120–124.

23. Osche, G. 1963. *Morphological, biological, and ecological considerations in the phylogeny of parasitic nematodes.* Pages 283–302 *in* E. C. Dougherty, ed. *The lower metazoa.* Univ. of California Press, Berkeley.

24. Paramanov, A. A. 1952. *(On the ecological classification of plant parasitic nematodes).* Trudy Gel'mint. Lab. Helminth. Akad. Nauk. SSSR. 6: 338-369.

25. Paramanov, A. A. 1967. *A critical review of the suborder Tylenchina (Filipjev, 1934) (Nematoda: Secernentea).* Trudy Gel'mint. Lab. Helminth. Akad. Nauk. SSSR. 18: 78-101.

26. Pearse, A. S. 1942. *An introduction to parasitology.* Charles C. Thomas, Springfield, Baltimore. 375 pp.

27. Poinar, G. O., Jr. 1969. *Diplogasterid nematodes (Diplogasteridae: Rhabditida) and their relationships to insect disease.* J. Invertebr. Pathol. 13: 447-454.

28. Poinar, G. O., Jr. 1975. *Entomogenous nematodes.* E. J. Brill, Leiden. 317 pp.

29. Riding, I. L. 1970. *Microvilli on the outside of a nematode.* Nature 226: 179-180.

30. Rühm, W. 1956. *Die nematoden der Ipiden.* Parasitol. Schriftenr. 6: 1-437.

31. Sprent, J. F. A. 1962. *The evolution of the Ascaridoidea.* J. Parasit. 48: 818-824.

32. Thorne, G. 1961. *Principles of nematology.* McGraw-Hill, New York, Toronto, London. 553 pp.

33. Tokushige, Y., and T. Kiyohara. 1969. *Bursaphelenchus sp. in the wood of dead pine trees.* J. Jap. For. Soc. 51: 193-195.

11] Biosystematics of Helminthosporium: Impact on Agriculture

BY E. S. LUTTRELL*

ABSTRACT

The generic name *Helminthosporium* was originally applied to relatively obscure deuteromycetous fungi and later generally accepted in misapplication to a large and heterogeneous group of common and economically important fungi. Morphological characters of conidiophores and conidia have been used to separate the resulting miscellany into several more consistent genera. *Helminthosporium* in the restricted sense comprises a compact group of saprotrophs on woody substrates. *Spiropes* comprises hyperparasites on Meliolaceae and other superficial Ascomycetes. Parasites causing numerous diseases of Gramineae, including the Victoria blight epidemic of oats in the 1940's and the corn leaf blight epidemic of 1970, are segregated in *Drechslera, Bipolaris, Exserohilum,* and *Nakatea.* The conidial state characters used to delimit these genera predict the ascigerous states. This system consequently represents a classification of perfect fungi since it is based on the ascigerous state-conidial state combinations *Pyrenophora-Drechslera, Cochliobolus-Bipolaris-Curvularia, Setosphaeria-Exserohilum,* and *Magnaporthe-Nakatea.* It also indicates the attainability of a theoretically sound hierarchial classification for Deuteromycetes that will serve the needs of applied disciplines for organization as well as retrieval of information.

INTRODUCTION

Fungi characteristically reproduce by spores. Spores generally are of two kinds: spores produced in asexual reproduction (the asexual or imperfect state) and spores produced in sexual reproduction (the sexual or perfect state). Asexual spores typically have the function of dissemination. They are associated with the active, assimilatory phases in the life cycle of the fungus. Sexual spores typically have the function of survival during unfavorable periods. In leaf spot diseases of crop plants, for example, the

*Department of Plant Pathology, University of Georgia, Athens, Georgia 30602.

asexual state is associated with lesions on the living plant, and the spores serve as the inoculum for secondary cycles of infection during the growing season. The sexual state develops in the litter of dead leaves during the dormant season. Even when this sexual state is a consistent or obligatory phase in the life cycle, the fungus in the sexual state must be separated from the variety of other fungi sporulating on dead leaves, and the genetic connection between the sexual and asexual states must be established experimentally. When a fungus has developed alternative means of survival, the sexual state may occur only sporadically in nature, may be induced in culture only by special methods, or may remain unknown.

Applied mycologists, pathologists, agronomists, horticulturists, food scientists, and animal scientists, who are concerned with the activities of fungi in parasitic or symbiotic relationships with plants or animals, in deteriorations of agricultural products and formation of toxins, in industrial fermentations, or in maintaining ecological balances, deal primarily with fungi in their asexual states. In part because of this practical necessity, dual systems of classification are recognized. Fungi whose sexual states are known are distributed among the classes Phycomycetes, Ascomycetes, and Basidiomycetes. When the sexual states are unknown or obscure, fungi in the asexual state are placed in the class Deuteromycetes. Asexual states of Phycomycetes generally can be assigned to genera in this class since the classification is based in large measure on asexual state characters. Rusts are recognizable as rusts in all spore states; in other Basidiomycetes the function of dissemination is performed by sexual spores, and asexual states often are obscure or lacking. The class Deuteromycetes, for practical purposes, is composed primarily of the imperfect (conidial) states of Ascomycetes. The convention has been to base the classification of Ascomycetes on the perfect (ascigerous) states and to consider the Deuteromycetes a temporary form class for form species which can simply be inserted into the existing classification of Ascomycetes when they have been connected with their proper ascigerous states. Nomenclature of Deuteromycetes therefore is considered to be provisional.

The following points in the classification of Deuteromycetes and Ascomycetes should be made:

(1) The only perfect fungus is one known in all of its states (*27*). The classification of Ascomycetes must be based on such perfect fungi; it must reflect characters of conidial as well as ascigerous states.

(2) A paradox: development of a theoretically sound hierarchical system of classification of Deuteromycetes is an attainable goal.

(3) Both nomenclature and a theoretical system of classification of Deuteromycetes are of practical importance. The classification must incorporate data from applied disciplines and must serve the needs of these disciplines for organization as well as retrieval of information.

These points are illustrated by a consideration of the deuteromycete genus *Helminthosporium* Link ex Fr. and its segregates.

ECONOMIC IMPORTANCE

Species of *Helminthosporium* cause numerous and diverse diseases of gramineous crops throughout the world. Among the most serious diseases are brown spot (*H. oryzae* B. de Haan) of *Oryza sativa* L. (rice), eyespot (*H. sacchari* Butl. in Butl. & Hafiz) and brown stripe (*H. stenospilum* Drechs.) of *Saccharum officinarum* L. (sugarcane), northern (*H. turcicum* Pass.) and southern (*H. maydis* Nisikado) leaf blights of *Zea mays* (corn), leaf blight (*H. avenaceum* Curt. ex Cke.) of *Avena sativa* (oats), and stripe (*H. gramineum* Rab. ex Schlect.) of *Hordeum vulgare* L. (barley) (*5*). *Helminthosporium* spp have been responsible for two of the major national epidemics of crop plants in the United States in this century, the Victoria blight of oats in the late 1940's and the corn leaf blight epidemic of 1970. Both were notable because these epidemics ironically were consequences of breeding for crop improvement and of extensive planting of similar cultivars. The common factor among oat cultivars susceptible to the new species *H. victoriae* Meehan & Murphy was Victoria parentage, the Victoria oats having been introduced from Argentina as a breeding source for resistance to another disease, crown rust (*28*). The common factor in corn cultivars highly susceptible to the new race of the southern corn leaf blight pathogen, *H. maydis* race T., was Tcms, the Texas male sterile cytoplasmic factor (*14*). This cytoplasmic factor, in combination with fertility restorer genes, had been used to replace the less efficient mechanical detasselling of corn in the production of hybrid cultivars in one of the cleverest technological applications of genetic principles to agricultural production.

The problem in both the Victoria blight and corn leaf blight epidemics was solved by application of the first principle of plant disease control: eliminate the crop; in the absence of the host there will be no disease. The cost of this often disastrously expensive control measure was reduced because it required only elimination of preferred cultivars rather than the crop itself. Costs of the continuing plant breeding research required can be prorated over large acreages and many years. Costs may also be offset by the value of lessons learned in plant breeding and epidemiology. Further offsetting benefits derive from the fact that Victoria blight and corn leaf blight are the major examples of diseases whose specificity can be explained almost entirely on the basis of host selective toxins (*21*). The hosts are major crop plants whose genetics and physiology are at least partially understood. Both are annuals that can be handled in greenhouse or laboratory experiments. The pathogens can be cultured readily. Since they are heterothallic and produce their sexual states quickly in culture, their genetics is open to investigation. Their disease induction mechanism is the relatively straightforward process of toxin production. Long after their field ravages have become of historical interest, *H. victoriae* and *H. maydis* race T continue to furnish primary materials for research aimed at an understanding of the nature of the disease process in plants (*45*). For many of the same reasons and also because of the availability of large populations subjected to changing selective pressures, agricultural systems offer some of the best materials for taxonomic studies of fungi. It would be as appropriate to speak of the impact of agriculture on taxonomy as of the impact of taxonomy on agriculture. Taxonomy and the applied biological sciences are complementary.

DRECHSLERA, BIPOLARIS, AND EXSEROHILUM

The commonly encountered species of *Helminthosporium* have been referred to as "the graminicolous species" (*6*) because their hosts are primarily, although not exclusively, Gramineae. Characteristic of the graminicolous species is the production of porogenous, phragmosporous conidia on the successive apices of simple, sympodially proliferating conidiophores (*23*). The conidia are porospores; an apparent pore in the conidiophore wall characterizes the conidial scar that marks the point of attachment of the conidium, and a corresponding pore is apparent in the hilum at the base of the conidium (Fig. 11.1). Conidia arise singly at the tips of the conidiophores. The conidiophore, however, proliferates to one side of the conidium and forms a new tip on which another terminal conidium is produced (Fig. 11.8). The laterally displaced conidia consequently appear to be pleurogenous, that is, produced on the sides of the conidiophore.

Within the graminicolous group two species clusters were early recognized: the subgenera *Cylindro-Helminthosporium* Nisikado, in which conidia typically are cylindrical and germinate from intercalary as well as polar cells (Fig. 11.4), and *Euhelminthosporium* Nisikado, in which conidia typically are fusoid and germinate only from the polar cells (Fig. 11.7) (*34*). Within *Euhelminthosporium* a convenient primary key character for separating species has been the presence of a protuberant hilum (Fig. 11.2) as opposed to an inserted hilum (Fig. 11.1). In most species of *Cylindro-Helminthosporium* and *Euhelminthosporium* the hilum at the base of the conidium, like the corresponding conidial scar on the conidiophore, is a ringed pore (Fig. 11.1). The wall of the conidium in a circular area surrounding the pore is thickened and darkened. The thickening may become conspicuous, but the hilum essentially is inserted within the contour of the wall of the basal cell. In some species of *Euhelminthosporium* the dark, thickened area around the pore protrudes abruptly from the base of the conidium as a truncated cone. The lateral walls of the cone as well as the wall across the truncated tip are thickened and darkened (Fig. 11.2), and the entire structure may be considered to constitute the hilum. This is the protuberant hilum (*23, 27*).

On the basis of these characters of conidium shape and germination and of hilum structure it is possible to distinguish three groups of species. These three groups have been recognized as the genera *Drechslera* Ito (*4, 11, 12, 17, 36, 37*) (Fig. 11.4) for species in *Cylindro-Helminthosporium*, *Bipolaris* Shoemaker (*4, 11, 12, 36*) (Fig. 11.7) for species in *Euhelminthosporium* with an inserted hilum, and *Exserohilum* Leonard & Suggs (*12, 19, 20*) (Fig. 11.2) for species in *Euhelminthosporium* with a protuberant hilum. The emphasis placed on these apparently minor characters is undoubtedly influenced by their value in predicting the ascigerous states of these fungi. For species of *Drechslera* in which the ascigerous state has been discovered it is in *Pyrenophora* Fr. (*1*); for species of *Bipolaris* it is in *Cochliobolus* Drechsler (*7*); for species of *Exserohilum* it is in *Setosphaeria* Leonard & Suggs (*20, 21, 27*).

Although all of these ascigerous states belong in the Pleosporaceae (Loculoascomycetes), they are clearly distinct. Ascospores of *Pyrenophora* (*1, 29, 44*) are large and dictyosporous. Ascocarps are large, irregularly globose, and beakless. Ascocarps develop slowly over a period of months and require low temperatures for development. In many species they constitute a regular phase in the life cycle. The ascocarps develop in stubble and dead culms during the dormant season, whereas the *Drechslera* conidial states are involved in cycles of infection of green leaves, culms, and inflorescences during the growing season. Ascospores of *Cochliobolus* are scolecosporous and coiled in a helical fascicle within the ascus. Ascocarps are globose with long, filiform beaks. They are rarely found in the field but may be produced in a few weeks when compatible strains are mated in culture under appropriate conditions (*30, 42, 43*). Ascospores of *Setosphaeria* are phragmosporous and are surrounded by a gelatinous matrix that extends from either end of the ascospore as a long tubular appendage after the ascospores are discharged. Ascocarps morphologically more nearly resemble those of *Pyrenophora* but, like those of *Cochliobolus*, develop quickly in culture when appropriate matings are made and thus far have not been found in the field (*19, 20, 22*).

CURVULARIA

Associated with the graminicolous species now placed in *Drechslera*, *Bipolaris*, and *Exserohilum*, and often confused with them, are the species of *Curvularia* (*9, 11, 12*). *Curvularia* species differ from species of *Bipolaris* with three-septate conidia in that the three- or four-septate conidia typically contain conspicuously swollen cells (Fig. 11.5). Usually the third cell from the base is swollen and darker colored, and the conidia may be strongly curved with most of the curvature occurring in the swollen third cell. In a few three-septate species both of the middle cells (second and third) are swollen, and the conidia are straight. In many species the hilum appears to be protuberant. The protrusion, however, is an extension of the wall of the basal cell of the conidium, as it is also in *Bipolaris micropus* (*27*). Often this protrusion seems to be cut off by a septum as a tiny appendiculate cell.

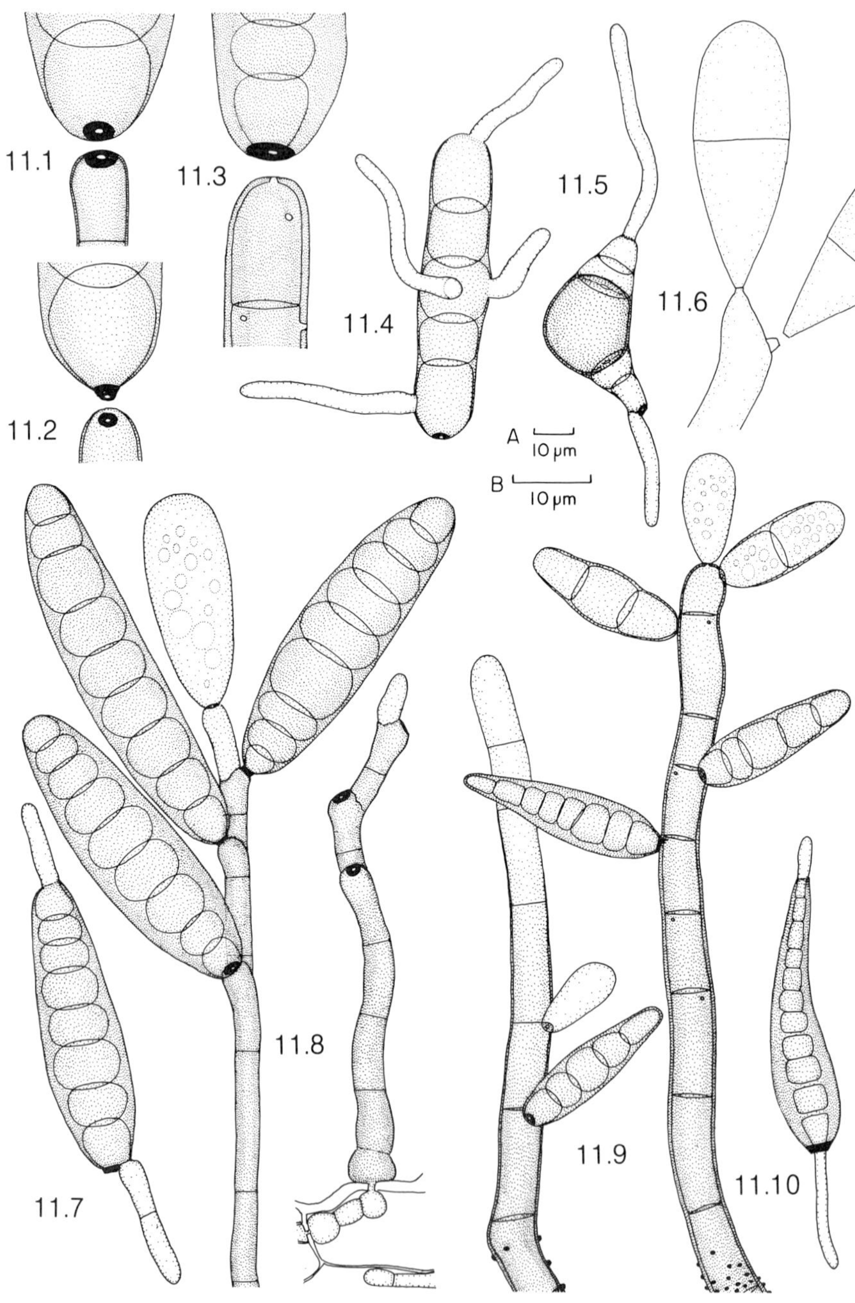
11.1
11.3
11.4
11.5
11.6
11.2
A 10 μm
B 10 μm
11.7
11.8
11.9
11.10

The thickened and the darkened hilum is inserted in the flat tip of the protrusion. The ascigerous states of the four species of *Curvularia* in which they have been produced in culture belong in *Cochliobolus* (*13, 31, 32, 33*).

HELMINTHOSPORIUM AND OTHER SEGREGATES

While attention was focused on the graminicolous species, the typical species of *Helminthosporium* (*8, 11, 15*), occurring on dead woody substrates, were largely ignored. *Helminthosporium* is an example of a name applied originally to relatively obscure fungi and later generally accepted in misapplication to common and economically important species. In fungi that agree with the generic type, *H. velutinum* Link ex Ficinus & Shubert, conidia are pleurogenous. They originate from the sides of the conidiophore behind the continuously growing tip (Fig. 11.9). A conidium may be formed at the tip only after growth of the conidiophore ceases. The conidia tend to be obclavate and the walls of the conidial cells are conspicuously thickened, a condition found only in older conidia of some

---

Figures 11.1–11.10. Diagnostic characters in *Helminthosporium* and segregated genera (*23*). Figure 11.1. *Bipolaris sorokiniana*; ringed pores of hilum at base of porogenous conidium and of conidial scar at apex of conidiophore. Figure 11.2. *Exserohilum turcicum*; protuberant hilum at base of porogenous conidium and ringed pore of conidial scar at apex of conidiophore. Figure 11.3. *Helminthosporium velutinum*; ringed pore of hilum at base of porogenous conidium and simple pores of conidial scars in sides and apex of conidiophore. Figure 11.4. *Drechslera avenacea*; pale, cylindrical, porogenous conidium with germ tubes from intercalarary as well as polar cells; germ tube from basal cell lateral. Figure 11.5, *Curvularia robusta* Kilpatrick & Luttrell; porogenous conidium with third cell from base swollen and darkened and with polar germ tubes; basal germ tube emerging to one side of the ring-pored hilum and extending along the axis of the conidium (semiaxial). Figure 11.6. *Nakatea sigmoidea*; blastogenous conidia formed as buds from the conidiophore wall and leaving the neck of the bud as a denticle on the conidiophore after dehiscence. Figure 11.7. *Bipolaris sorokiniana*; dark, obclavate-fusoid conidium germinating by polar germ tubes; germ tube from basal cell semi-axial. Figure 11.8. *Bipolaris sorokiniana*; sympodial conidiophores resulting from production of single conidia at tips of conidiophores (acrogenous) and lateral proliferations of conidiophores to form new growing tips on which successive conidia are produced. Figure 11.9. *Helminthosporium velutinum*; conidiophores with indeterminate growing tips and conidia originating on sides (pleurogenous), typically below conidiophore septa. Figure 11.10. *Helminthosporium velutinum*; dark, thick-walled, obclavate conidium germinating by polar germ tubes; germ tube from basal cell emerging through hilum (percurrent). Figures 11.1, 11.3, and 11.6 are to scale B; Figures 11.4, 11.5, and 11.7–11.10 are to scale A.

species of *Bipolaris* such as *B. sorokiniana* (Sacc. in Sorok.) Shoemaker. The hilum, although inserted, is conspicuously thickened and darkened. Although germination is polar, the germ tube from the basal cell emerges through the pore in the center of the hilum (percurrent) (Fig. 11.10) rather than from the wall of the basal cell adjacent to the hilum (semiaxial) (Fig. 11.7) as in *Bipolaris*. Contrasting with the conspicuous hilum, the conidial scars are simple pores (Fig. 11.3) rather than the ringed pores characteristic of *Bipolaris*. The ascigerous states of species of *Helminthosporium* in this restricted sense are unknown.

Although the conidia in *Drechslera, Bipolaris, Exserohilum,* and *Curvularia* are borne singly at the tips of the conidiophores, repeated sympodial proliferations of the conidiophore push the successive terminal conidia aside, and the arrangement of the conidia on the conidiophore consequently shows a superficial resemblance to that in *Helminthosporium* proper. Other fungi with acrogenous conidia on sympodially proliferating conidiophores that have been placed in *Helminthosporium* differ even more fundamentally in conidiogenesis. The conidia are blastogenous rather than porogenous (*23*). They arise as buds from the conidiophore wall and are cut off by a septum in the neck of the bud (Fig. 11.6). A group of such species growing mostly as hyperparasites on Meliolaceae and other superficial Ascomycetes have been segregated as the genus *Spiropes* (*10, 11, 12*). In *Spiropes* the neck of the bud that develops into the conidium is short. Consequently, when the conidium dehisces by splitting of the septum in the neck, flat circular conidial scars are left on the conidiophore, and the conidial hilum is correspondingly flat. The septum that forms these scars has a central perforation, and these septal pores must be distinguished from the pores in the conidiophore wall that result from the formation of porospores. The ascigerous states of *Spiropes* species are unknown.

A final example of the many fungi lumped in *Helminthosporium* that have been transferred to new or more appropriate genera is *Nakatea sigmoidea* (Cav.) Hara (*Helminthosporium sigmoidea* Cav.), the cause of culm rot of rice (*5*). As in *Spiropes,* the conidia are blastogenous. The necks of the buds, however, are denticular protuberances; and when the conidia secede from the apices, the denticles remain on the conidiophores as unmistakable evidence of the blastogenous nature of the conidia (Fig. 11.6) (*23*). This species also produces a sclerotial state. The ascigerous state is *Magnaporthe salvinii* (Cattaneo) Krause & Webster (*18*), a fungus originally classified in the Diaporthaceae but later (*3*) placed in the Physosporellaceae (Euascomycetes).

CLASSIFICATION

The relationships between *Helminthosporium* and the genera segregated from it are shown in Table 11.1. Characters distinguishing these genera are summarized in the following key:

KEY TO GENERA

1 Conidia pleurogenous, borne on the sides of the conidiophore behind the growing tip, porogenous (tretic), typically obclavate, with several transverse septa, thick-walled; hilum a dark ringed pore, inserted but often conspicuously thickened; conidial scars simple pores; germ tubes polar, germ tube from basal cell emerging through hilum (percurrent); ascigerous state unknown; primarily saprotrophs on woody substrates ...... *Helminthosporium*

1′ Conidia acrogenous, borne singly on the successive tips of sympodially proliferating condidiophores, the terminally produced conidia pushed aside by the proliferating conidiophore tip and superficially appearing pleurogenous 2

2(1′) Conidia porogenous, delicately attached at the base to an apparent pore in the conidiophore wall; conidial hilum and conidial scars ringed pores; mostly parasitic on Gramineae 3

2′(1′) Conidia blastogenous (blastic), connected to the conidiophore by an obvious extrusion of the conidiophore wall 6

3(2) Conidia germinating from any or all cells, germ tubes lateral; conidia usually cylindrical or subcylindrical, straight, often pale; septum delimiting basal cell formed first during conidium maturation; ascigerous state *Pyrenophora*, with ascospores large, pale, oblong, with transverse and longitudinal septa, surrounded by a thin gelatinous sheath; ascocarps large, irregularly globoid, sclerotioid, covered with hairs or conidiophores, beakless or with a short beak .... *Drechslera*

3′(2) Conidial germination polar, basal germ tube semiaxial; conidia usually fusoid and dark, often curved; middle septum usually formed first in conidium maturation 4

4(3′) Conidial hilum protuberant, in the form of a truncated cone with lateral walls as well as the wall across the truncated tip thicked and darkened; ascigerous state *Setosphaeria*, with fusoid, transversely septate ascospores surrounded by a gelatinous sheath that expands to form a tubular appendage at either end of the discharged ascospore; ascocarps globose, beakless, with dark spines in the region surrounding the ostiole .... *Exserohilum*

4′(3′) Conidial hilum inserted within the basal wall of the conidium, flat, sometimes at the tip of a conical protrusion of the basal wall or a hyaline appendiculate basal cell; ascigerous state *Cochliobolus*, with filiform ascospores coiled in a helical fascicle in narrowly cylindrical asci; ascocarps globoid, with a long cylindrical beak 5

5(4′) Conidia without conspicuously swollen cells, often long and many-septate .... *Bipolaris*

5′(4′) Conidia usually three-or four-septate with the third cell from the base conspicuously swollen and darkened, often strongly curved with most of the curvature in the swollen cell, some with straight three-septate conidia and the two middle cells swollen and darkened .... *Curvularia*

6(2′) Conidia produced on denticles which remain on the conidiophores after abscission, fusoid; mycelium producing sclerotia; ascigerous state *Magnaporthe* with three-septate, fusoid ascospores, asci with a refractive apical ring, and globose, long-beaked ascocarps; parasitic on Gramineae .... *Nakatea*

6′(2′) Conidia not on denticles; conidial scars flat rings; conidia obclavate; ascigerous state unknown; usually hyperparasitic on other fungi .... *Spiropes*

Segregation of *Drechslera, Bipolaris, Curvularia, Exserohilum,* and *Nakatea* leaves *Helminthosporium* in the restricted sense a compact, more precisely defined group in which morphology is correlated with habit. All species are saprotrophs on woody substrates. *H. solani* (*8*), cause of silver scurf of *Solanum tuberosum* L. (potato), is the only plant pathogen in the genus, and this species grows on the corky skin of the potato tubers. The genus *Spiropes* likewise comprises a group of species in which morphology and habit show correlation; for the most part they are hyperparasitic on other superficial fungi. The separation of *Drechslera, Bipolaris, Curvularia, Exserohilum,* and *Nakatea* results in no such correlation between morphology and ecology. All of these genera typically are parasites on Gramineae, and all include economically important plant pathogens. Plant parasitism, however, is so gross a character that it could hardly be expected to distinguish genera. Even when subcategories of parasitism are recognized, very similar types of parasitism recur in widely separated orders of fungi (*26*). Recognition of natural groupings based on morphology, nevertheless, may lead to an awareness of the possibilities of differences in behavior and direct investigations on physiology to a more refined understanding of parasitism.

Confidence that the classification that has been arrived at is natural derives from the fact that it is based on perfect fungi, as indicated by the conidial state-ascigerous state combinations *Pyrenophora-Drechslera, Cochliobolus-Bipolaris-Curvularia, Setosphaeria-Exserohilum,* and *Magnaporthe-Nakatea.* Although many specific ascigerous state-conidial state connections remain to be discovered, sufficient progress has been made to furnish a basis for extrapolating to these unknowns. Presence of a *Bipolaris* conidial state is used as a key character in distinguishing *Cochliobolus* from similar genera such as *Leptospora* and *Acanthophiobolus* (*2, 25*) and was a major character in the original separation of *Cochliobolus* from *Ophiobolus* in its former broad sense (*7*). Presence of an *Exserohilum* conidial state is a key character in distinguishing *Setosphaeria* from genera such as *Keisleriella* and *Leptosphaeria.* This reflects a general trend toward use of conidial state characters in the classification of genera of Ascomycetes, and, to some extent, families and orders.

Use of conidial state characters is of even greater importance at the species level. In his classification of *Pyrenophora,* Wehmeyer (*44*) included the ascigerous states of *Drechslera avenacea, D. bromi, D. graminea, D. teres,* and *D. tritici-repentis* (*1, 24*) in *P. trichostoma,* a species in which the conidial state is unknown. Wehmeyer's classification may represent a logi-

Table 11.1. *Helminthosporium* and its segregates—representative species.

Helminthosporium	Current status	Ascigerous state
	Helminthosporium (*Helmisporium*) Link ex. Fr., 1832 Hughes, 1958	——
H. velutinum Link ex Ficinus & Shubert, 1823	*H. velutinum*	——
H. solani Dur. & Mont., 1849 [=H. *atrovirens* (Harz) Mason & Hughes apud Hughes, 1953]	*H. solani*	——
H. mauritianum Cooke, 1883	*H. mauritianum*	——
H. microsorum D. Saccardo	*H. microsorum*	——
	Drechslera Ito, 1930 (= *Helminthosporium* subgen. *Cylindro-Helminthosporium* Nisi., 1928)	*Pyrenophora* Fr., 1849
H. gramineum Rab. ex Schlecht., 1857	*D. graminea* (Rab. ex Schlecht.) Shoemaker, 1959	*P. graminea* Ito & Kurib. in Ito, 1930
H. avenaceum Curt ex Cke., 1889 (= *H. avenae* Eid., 1891)	*D. avenacea* (Curt, ex Cke.) Shoemaker, 1959	*P. chaetomoides* Speg., 1899 (= *P. avenae* Ito & Kurib. in Ito, 1930)
H. poae Baudys, 1916	*D. poae* (Baudys) Shoemaker, 1962	——
H. cyclops Drechs., 1923 [*Podosporiella verticillata* O'Gara, 1915; *Bipolaris verticillata* (O'Gara) Sprague, 1962; *Angiopoma campanulata* Lév., 1841]	*D. campanulata* (Lév.) Sutton, 1976 [= *D. verticillata* (O'Gara) Shoemaker, 1966	*P. seminiperda* (Brittlebank & Adam) Shoemaker, 1966
H. tuberosum Atk., 1897	*D. tuberosa* (Atk.) Shoemaker, 1959	*P. japonica* Ito & Kurib. in Ito, 1930
	Bipolaris Shoemaker (= *Helminthosporium* Subgen. *Euhelminthosporium* Nisi., 1928	*Cochliobolus* Drechs., 1934
H. maydis Nisi., 1926	*B. maydis* (Nisi.) Shoemaker, 1959	*C. heterostrophus* (Drechs.) Drechs., 1934
H. victoriae Meehan & Murphy, 1946	*B. victoriae* (Meehan & Murphy) Shoemaker, 1959	*C. victoriae* Nelson, 1960
H. sorokinianum Sacc. in Sorok., 1890 (= *H. sativum* Pamm., King & Bakke, 1910)	*B. sorokiniana* (Sacc. in Sorok.) Shoemaker, 1959	*C. sativus* (Ito & Kurib.) Drechs. ex Dastur, 1942

Table 11.1 Continued

Helminthosporium	Current status	Ascigerous state
H. sacchari Butl. apud Butl. & Khan, 1913	*B. sacchari* (Butl. apud Butl. & Khan) Shoemaker, 1959	——
(*Clasterosporium iridis* Oud., 1898)	*B. iridis* (Oud.) Dickinson, 1966	——
H. oryzae B. de Haan, 1900	*B. oryzae* (B. de Haan) Shoemaker, 1959	*C. miyabeanus* (Ito & Kurib.) Drechs. ex Dastur, 1942
	Curvularia Boed., 1933	*Cocliobolus* Drechs., 1934
H. geniculata Tracy & Earle, 1896	*C. geniculata* (Tracy & Earle) Boed., 1933	*C. geniculatus* Nelson, 1964
(*Acrothecium lunatum* Wakker, 1898)	*C. lunata* (Wakker) Boed. 1933	*C. lunatus* Nelson & Haasis, 1964
(*Brachysporium trifolii* Kauff. apud Bonner, 1920)	*C. trifolii* (Kauff.) Boed., 1933	——
H. cymbopogonis C. W. Dodge, 1942	*C. cymbopogonis* (Dodge) Groves & Skolko, 1945	*C. cymbopogonis* Hall & Sivanisan, 1972
	Exserohilum Leonard & Suggs, 1974	*Setosphaeria* Leonard & Suggs, 1974
H. turcicum Pass., 1876	*E. turcicum* (Pass.) Leonard & Suggs, 1974	*S. turcica* (Luttrell) Leonard & Suggs, 1974
H. monoceras Drechs., 1923	*E. monoceras* (Drechs.) Leonard & Suggs, 1974	——
H. rostratum Drechs., 1923	*E. rostratum* (Drechs.) Leonard & Suggs, 1974	*S. rostrata* Leonard, 1976
H. pedicellatum Henry, 1924	*E. pedicellatum* (Henry) Leonard & Suggs, 1974	*S. pedicellata* (Nelson) Leonard & Suggs, 1974
	Spiropes Ciferri, 1955	——
H. guareicola F. L. Stevens, 1918	*S. guareicola* (Stev.) Cifferri, 1955	——
H. capense Thüm.,	*S. capense* (Thüm) M. B. Ellis, 1968	——
H. dorycarpum Mont., 1842	*S. dorycarpus* (Mont.) M. B. Ellis, 1968	——
	Nakatea Hara, 1939 (= Vakrabeeja Subramanian, 1956)	*Magnaporthe* Krause & Webster, 1972
H. sigmoideum Cav., 1889	*N. sigmoidea* (Cav.) Hara, 1939	*M. salvinii* (Catt.) Krause & Webster, 1972

cal conclusion based, as it was, on a consideration of ascigerous state morphological characters alone. On a practical basis it is unacceptable because it does not distinguish between fungi that cause entirely distinct diseases. Although *D. graminea* and *D. teres*, for example, parasitize the same host, *Hordeum vulgare*, *D. teres* causes a local lesion disease known as netblotch and survives on plant residues, whereas *D. graminea* causes a systemic infection and is carried within the grain (*5*). It must be admitted that even when all available morphological characters of both conidial and ascigerous states are employed, these two fungi are difficult to distinguish, and whether they should be recognized at the species level or at some subspecific level is a debatable question (*1*). They must, however, be recognized. It is further apparent that it is futile to attempt to classify *Pyrenophora* species unless their conidial states are known. The problem has not arisen in the classification of *Cochliobolus* and *Setosphaeria* because most of these ascigerous states have been discovered by mating conidial state isolates in culture. Nevertheless, the ascigerous states are relatively uniform; most of the variation occurs in the conidial states, and the conidial state characters must be relied upon in recognizing species.

If conidial state characters are good characters in classifying fungi in which the ascigerous state is known, it follows that they are good characters when the ascigerous state is unknown. Taxonomists may find it convenient to consider the Deuteromycetes a form class to serve as a repository in which fungi known only in their conidial states may be left until their ascigerous states are discovered. These conidial fungi, however, are of practical importance, and a hierarchical system of classifying them is a practical necessity if classification is to serve its purpose as a means of organizing data in a system that will permit extrapolation. Although neither approach is ideal, the need for classifying "imperfect fungi" known only in the conidial state is as great as that for classifying "imperfect fungi" known only in the ascigerous state. Certainly, an a priori decision that a classification based on morphological, physiological, biochemical, and ecological characters of conidial states is necessarily inferior to one based solely on morphological characters of ascigerous states is arbitrary.

In supporting the opinion (*4, 11, 12, 40*) that species of *Bipolaris* and *Exserohilum* should be included in the single genus *Drechslera*, Ellis (*12*) has made the point that *D. campanulata* (*41*) was considered a species of *Bipolaris (B. verticillata)*, until its ascigerous state was found to be a *Pyrenophora (38)*. Additional points are the classification in *Bipolaris* and *Curvularia* of conidial states that later proved to have *Cochliobolus* as a common ascigerous state (*13, 31, 32, 33*) and the tortuous reinterpretation of structure in species of *Curvularia* with an apparently protruding hilum (*27*) when the ascigerous state of one of these species proved to be *Cochliobolus* rather than *Setosphaeria* (*13*). Such impingements, however, are convenient checkpoints in the classification of both conidial states and ascigerous states. With such frequent checks taxonomists attempting to develop a useful, theoretically sound classification of Deuteromycetes (*39*) should be able to proceed with a reasonable degree of confidence.

NOMENCLATURE

To preserve the familiar combinations in *Helminthosporium* for the "graminicolous species" Luttrell (*24*) made two suggestions: (1) Maintain

Helminthosporium in a broad sense and recognize distinctive groups as subgenera. (2) Redefine *Helminthosporium* with *H. maydis* as the type species for the "graminicolous species," and conserve this name against *Helminthosporium* Link ex Fr. Although the first suggestion was extended by Rapilly (*35*), neither suggestion has been accepted; and the second one is too absurdly complicated to be considered. A third suggestion (*24*), that the "graminicolous species" be included in the single genus *Drechslera,* has been adopted by Subramanian and Jain (*40*); Ellis (*11, 12*); and Chidambaram *et al.* (*4*). Since combinations in *Bipolaris* (*36*) for species in *Euhelminthosporium* had already entered the literature, new combinations in *Drechslera* (*40*) simply increased the confusion. Furthermore, *Drechslera* Ito, 1930, is antedated by *Angiopoma* Lév., 1841, and acceptance of Sutton's (*41*) proposal for conservation of *Drechslera* against *Angiopoma* is uncertain. Reversion to Link's original spelling *"Helmisporium,"* rather than the spelling *"Helminthosporium"* accepted by Fries (*15*), is based on Hughes' (*16*) proposal to change the starting point for nomenclature of Hyphomycetes from 1821 (Fries) to 1801 (Persoon). Since there is no possibility of conserving the combinations in *Helminthosporium* for species of economic importance, emphasis might better be placed on recognition of natural relationships in classification (*20*) and on acceptance of the changes in nomenclature that are required.

TAXONOMIC TREATMENTS

Ellis has produced monographic studies of *Helminthosporium* (*8, 11*), *Curvularia* (*9, 10, 11*), and *Spiropes* (*10, 11*). *Drechslera* has been treated comprehensively by Shoemaker (*37*), and *Exserohilum* has been outlined by Leonard (*19, 20*). Although Shoemaker (*36*) has given an outline of nomenclature, there is no comprehensive treatment of *Bipolaris,* the major genus in the group. Descriptions and illustrations of many species of *Drechslera, Bipolaris,* and *Exserohilum* have been provided by Ellis (*11, 12*) and by Chidambaram *et al.* (*4*), but the usefulness of these treatments is reduced by failure to recognize any subgeneric groupings of species that would correspond with the genera *Drechslera, Bipolaris,* and *Exserohilum.*

CONCLUSIONS

The general conception of the relationship between taxonomy and the agricultural sciences is that the function of taxonomy is to provide a service to the applied sciences. It supplies names to apply to organisms and essential identifications of organisms. This creates a division between producers (taxonomists) and users (applied scientists). Taxonomists are more likely to consider themselves creators and to adopt the role of servant with reluctance. This attitude is demonstrated in their reaction to criticism—for example, in irritation with applied scientists for their resistance to changes in nomenclature that advances in classification produce. This attitude overlooks the fact that the problem of the applied scientist is not simply one of learning a new name. These names are key words in information retrieval. Multiplication of synonyms multiplies the number of key words required.

More importantly, this conception overlooks one of the primary functions of classification: to provide a system for the organization of information. Applied scientists have an interest in this information system equal to that of taxonomists and a need to see that their data are incorporated into the system. A large part of the data on genetics, physiology, and ecology of fungi essential to the development of an adequate system of classification is generated by the applied sciences. Applied scientists are producers as well as users. The gaps in Table 11.1 under "ascigerous states" for *Helminthosporium* and *Spiropes* may be worth contemplation. Are species in *Drechslera, Bipolaris, Curvularia, Exserohilum,* and *Nakatea* more amenable organisms, or does the knowledge of life cycles and genetics in these genera reflect the stimulus and funding provided by practical need? Taxonomists would function more effectively if they recognized applied scientists as partners in the development of classifications rather than as uninvolved users of their products.

NOTE

Bertault (Bertault, R. 1970. Deux espèces du genre *Strossmayeria* Schulzer. Rev. Mycol. 35: 130-140.) has described as species of *Helminthosporium* two conidial states of fungi associated with apothecia of *Strossmayeria* spp (Helotiales). Bertault's descriptions indicate, however, that the conidia in his species are acrogenous. This would exclude these species from *Helminthosporium,* in which conidia are pleurogenous. Furthermore, the conidia illustrated by Bertault more nearly resemble those of *Sporidesmium* spp. The evidence presented is against an inference that ascigerous states of *Helminthosporium* spp may be found in *Strosmayeria.*

LITERATURE CITED

1. Ammon, H. U. 1963. *Über einige Arten aus den Gattungen Pyrenophora und Cochliobolus mit Helminthosporium als Nebenfruchtform.* Phytopathol. Z. 47: 244-300.

2. Arx, J. A. von. 1974. *The genera of fungi sporulating in pure culture.* 2nd Ed. J. Cramer, Vaduz. 315 pp.

3. Barr, M. E. 1977. *Magnaporthe, Telimenella, and Hyponectria (Physosporellaceae).* Mycologia 69: 952-966.

4. Chidambaram, P., S. B. Mathur, and P. Neergaard. 1973. *Identification of seed-borne Drechslera species.* Friesia 10: 165-207.

5. Dickson, J. G. 1956. *Diseases of field crops.* 2nd Ed. McGraw-Hill, New York. 517 pp.

6. Drechsler, C. 1923. *Some graminicolous species of Helminthosporium. I.* J. Agr. Res. 24: 641-739.

7. Drechsler, C. 1934. *Phytopathological and taxonomic aspects of Ophiobolus, Pyrenophora, Helminthosporium, and a new genus Cochliobolus.* Phytopathology 24: 953-983.

8. Ellis, M. B. 1962. *Dematiaceous Hyphomycetes: III.* Mycol. Pap. 82: 1-55.

9. Ellis, M. B. 1966. *Dematiaceous Hyphomycetes. VII: Curvularia, Brachysporium, etc.* Mycol. Pap. 106: 1-57.

10. Ellis, M. B. 1968. *Dematiaceous Hyphomycetes. IX. Spiropes and Pleurophragmium.* Mycol. Pap. 114: 1-44.

11. Ellis, M. B. 1971. *Dematiaceous Hyphomycetes.* Commonwealth Mycol. Inst., Kew. 608 pp.

12. Ellis, M. B. 1976. *More Dematiaceous Hyphomycetes.* Commonwealth Mycol. Inst., Kew. 507 pp.

13. Hall, J. A., and A. Sivanesan. 1972. *Cochliobolus state of Curvularia cymbopogonis.* Trans. Brit. Mycol. Soc. 59: 314-316.

14. Hooker, A. L. 1972. *Southern leaf blight of corn—present status and future prospects.* J. Environ. Quality 1: 244-249.

15. Hughes, S. J. 1958. *Revisiones hyphomycetarum aliquot cum appendice de nominibus rejiciendis.* Can. J. Bot. 36: 727-836.

16. Hughes, S. J. 1959. *Starting point of nomenclature of Hyphomycetes.* Taxon. 8:96-103.

17. Ito, S. 1930. *On some new ascigerous stages of the species of Helminthosporium parasitic on grasses.* Proc. Imp. Acad. Tokyo 6: 352-355.

18. Krause, R. A., and R. K. Webster. 1972. *The morphology, taxonomy, and sexuality of the rice stem rot fungus, Magnaporthe salvinii (Leptosphaeria salvinii).* Mycologia 64: 103-114.

19. Leonard, K. J. 1976. *Synonomy of Exserohilum halodes with E. rostratum, and induction of the ascigerous state, Setosphaeria rostrata.* Mycologia 68: 402-411.

20. Leonard, K. J., and E. G. Suggs. 1974. *Setosphaeria prolata, the ascigerous state of Exserohilum prolatum.* Mycologia 66: 281-297.

21. Luke, H. H., and V. E. Gracen, Jr. 1972. *Helminthosporium toxins.* Pages 139-168 *in* S. Kadis, A. Ciegler, and S. J. Ajl, eds. *Microbial toxins,* Vol. 8. *Fungal toxins.* Academic Press, New York.

22. Luttrell, E. S. 1958. *The perfect stage of Helminthosporium turcicum.* Phytopathology 48: 281-287.

23. Luttrell, E. S. 1963. *Taxonomic criteria in Helminthosporium.* Mycologia 55: 643-674.

24. Luttrell, E. S. 1964. *Systematics of Helminthosporium and related genera.* Mycologia 56: 119-132.

25. Luttrell, E. S. 1973. *Loculoascomycetes.* Pages 135-219 *in* G. C. Ainsworth, F. K. Sparrow, and A. S. Sussman, eds. *The Fungi, Vol. IV A.* Academic Press, New York.

26. Luttrell, E. S. 1974. *Parasitism of fungi on vascular plants.* Mycologia 66: 1-15.

27. Luttrell, E. S. 1977. *Correlations between conidial and ascigerous state characters in Pyrenophora, Cochliobolus, and Setosphaeria.* Rev. Mycol. 41: 271–279.

28. Meehan, F. L., and H. C. Murphy. 1946. *A new Helminthosporium blight of oats.* Science 104: 413-414.

29. Müller, E. 1951. *Die Schweizerischen Arten der Gattungen Clathrospora, Pleospora, Pseudoplea, und Pyrenophora.* Sydowia 5: 248-310.

30. Nelson, R. R. 1959. *Cochliobolus carbonum, the perfect stage of Helminthosporium carbonum.* Phytopathology 49: 807-810.

31. Nelson, R. R. 1960. *Cochliobolus intermedius, the perfect stage of Curvularia intermedia.* Mycologia 52: 775-778.

32. Nelson, R. R. 1964. *The perfect stage of Curvularia geniculata.* Mycologia 56: 777-779.

33. Nelson, R. R., and F. A. Haasis. 1964. *The perfect stage of Curvularia lunata.* Mycologia 56: 316-317.

34. Nisikado, Y. 1928. *Studies on the Helminthosporium diseases of Gramineae in Japan.* Ohara Inst. Agr. Res., Special Rept. 4: 1-394. (In Japanese: English summary in Ber. Ohara Inst. landwirtsch. Forsch. Kurashiki 4: 111-126. 1929).

35. Rapilly, F. 1966. *Limites proposées au genre Helminthosporium, relations avec les genres voisons.* Bull. Soc. Mycol. France 82: 221-240.

36. Shoemaker, R. A. 1959. *Nomenclature of Drechslera and Bipolaris, grass parasites segregated from 'Helminthosporium.'* Can. J. Bot. 37: 879-887.

37. Shoemaker, R. A. 1962. *Drechslera Ito.* Can. J. Bot. 40: 809-836.

38. Shoemaker, R. A. 1966. *A pleomorphic parasite of cereal seeds, Pyrenophora semeniperda.* Can. J. Bot. 44: 1451-1456.

39. Subramanian, C. V. 1962. *The classification of the Hyphomycetes.* Bull. Bot. Surv. India 4: 249-259.

40. Subramanian, C. V., and B. L. Jain. 1966. *A revision of some graminicolous Helminthosporia.* Curr. Sci. 35: 352-355.

41. Sutton, B. C. 1976. *Angiopoma Lév., 1841, an earlier name for Drechslera Ito, 1930.* Mycotaxon 3: 377-380.

42. Tsuda, M., and A. Ueyama. 1975. *Identity of Helminthosporium-leaf spot fungi attacking Oryzoideae plants growing in Japan Island (preliminary note).* Trans. Mycol. Soc. Japan 16: 93-94.

43. Ueyama, A., and M. Tsuda. 1975. *Formation in culture of Cochliobolus miyabeanus, the perfect state of Helminthosporium oryzae.* Ann. Phytopathol. Soc. Japan 41: 434-440.

44. Wehmeyer, L. E. 1961. *A world monograph of the genus Pleospora and its segregates.* Univ. Michigan Press, Ann Arbor. 451 pp.

45. Wheeler, H. 1975. *Plant pathogenesis.* Springer, Berlin. 106 pp.

four

RECENT DEVELOPMENTS IN TAXONOMIC TECHNOLOGY

12] Automation of Biosystematics Data: Current Needs

BY T. GARY GAUTIER*

ABSTRACT

The first fifteen years of computer data banking in biosystematics were marked by formulation of basic approaches to using electronic data processing (EDP), and by development of suitable software systems. New challenges face data banking in biosystematics now. With the reluctance to use EDP rapidly declining and the availability of EDP equipment and software systems increasing, we are seeing an accelerating growth in the number and variety of data banks. These data banks are becoming a massive information resource which will require organization and management for it to be as useful as possible. Standards for the representation of data in files, for the exchange of data, and for the description of data banks should be adopted by the biosystematics community. Procedures need to be developed to ensure that the right data will be available to basic and applied research, collection management, libraries, education, and to the public. Basic research is needed in virtually every facet of the field, from the new procedures that will be required to cope with large data sets, to the potential of combining data banks from several disciplines to answer complex questions. The field is wide open for newcomers to make significant contributions.

INTRODUCTION

It has been over fifteen years since biosystematics and curators began looking to electronic data processing (EDP) to store and retrieve the mounting deluge of systematics data. Pioneers in the field saw in automation the opportunity to greatly increase the scientific value of systematics data by making it much more accessible in many more useful forms. Early efforts to build data banks met with skepticism and resistance, however. Some biosystematists felt that data banking would not bring enough benefit to

*Automatic Data Processing Program, National Museum of Natural History, Smithsonian Institution, Washington, D. C. 20560.

warrant the great expense. Others doubted that automated techniques could be developed that would significantly replace or enhance the scientific tasks that had always been done by hand. For these and other reasons, data banking was used by relatively few biosystematists until recently.

The outlook for data banking in biosystematics is now brighter because:

(1) Data processing costs are steadily being driven down by advances in technology and by improvements in methods of handling systematics data.

(2) Data processing equipment and services are becoming more available.

(3) The repertoire of proven procedures is growing rapidly, and the systems are being made easier to understand and simpler to operate.

(4) The number of biosystematists proficient at using data processing is growing.

(5) The practical benefits of data processing are being demonstrated in an increasing number of projects.

(6) National and international organizations are encouraging the construction and use of data banks in biosystematics.

(7) The data problems of biosystematics are becoming more difficult to solve by manual methods alone.

Today, more biosystematists and curators than ever before are looking at data processing as a practical and necessary tool, and are planning to use it in their research and in collection management.

The proliferation of automated data banks being used for an increasing variety of purposes by an increasing portion of the biosystematics community poses new challenges which must be met if the data-bank resource is to be optimally useful. The purpose of this paper is briefly to discuss some of the current problems and needs of biosystematics data banking. Crovello (*3*) has recently presented information on this subject from the botanical viewpoint.

THE PROBLEM OF RAPIDLY GROWING DATA BANKS

The chief value of EDP over manual systems is that it allows us to use electronic machinery to perform virtually all of the functions required to enter, store, manipulate, and retrieve and analyze data from files. Automatic recorders or portable data terminals can be used in the field or in the laboratory to collect data in machine readable form, eliminating the need to transcribe the data later. New data can be interfiled quickly, and errors can be detected by computer programs. Virtually any kind of rearrangement of the records is possible as are massive file-wide changes in data such as taxonomic names that have been superseded. Records can be retrieved on the basis of virtually any combination of data items to answer questions. After retrieval, the records can be printed out in a wide variety of formats, or they can be passed on to other computer programs that perform statistical analyses, or draw maps and charts. The net result is that where we once dealt with hundreds of items of data to solve problems we can now deal with tens of thousands.

A wide variety of data can be put into data banks for a wide variety of purposes. Data banks are being built to index biosystematics resources and legislation of concern to biosystematics. Bibliographic data are being entered to provide better access to scientific literature. We have data banks on natural areas, short-lived biologic phenomena, and on endangered species of animals and plants. We use data banks to keep track of specimen loans from collections and to aid other collection management functions, and we have data banks on marine and terrestrial collecting localities and on sites of marine mammal strandings. These data banks are being built by individual scientists, museums, associations, and government agencies. Their purpose may be to perform the specific research or management function of their sponsors, or to serve the general needs of a larger community of users.

These data banks already contain millions of items on specimens, observations, literature, resources, concepts, and other subjects of interest to biosystematists. The amount of data being automated appears to be growing at an accelerating rate. At the National Museum of Natural History alone, we are entering two million items of data on upwards of 200,000 specimens each year. In another five years it may not be surprising if systematics collections will have entered data for over ten million specimens.

This information explosion is also an information crisis in that technology and the rapid growth of data banks are quickly outstripping our ability to manage and exploit the information to its full potential. Biosystematists cannot easily discover what data banks are available or what data they contain. Very little has been done in biosystematics to set up systems and protocols for exchanging data between files in different projects. If such systems were available, the dissimilarity of data standards between files would probably prevent, or make difficult, their simultaneous use. Once access to files is gained, we have on hand only enough insight and procedures to ask but a small portion of the questions that the data banks might answer.

It is clear that in order to keep up with this rapidly growing information resource, we need to become more aware of where data banking is going in biosystematics. We need to develop management and communication systems to make the data fully available, and we need to carry out more basic research and development on how data banks can be exploited. Aspects of these topics will be discussed more fully in the following sections.

NEED FOR RESEARCH AND DEVELOPMENT

Data banking in biosystematics has made great strides in the few years of its existence. Many systems and files have passed the development stages and are now being used on a routine basis to solve practical problems. But it seems that we have only scratched the surface when we consider how much more is being done in industry, medicine, and other sciences, and when we imagine all the things we could be doing with computer technology. We could certainly think of far more problems to solve and applications to try than have been solved or tried in biosystematics so far.

The field of data banking is wide open for biosystematists and technicians to make highly significant contributions. There are areas of need to meet all tastes—hardware problems for the electronics-oriented person, programming and analysis tasks for those who like to develop systems and procedures, and political and administrative problems for the organizers and leaders. Above all, there are plenty of opportunities for biosystematists to pose challenging scientific questions to data banks to demonstrate what can or cannot be done with them. This input is needed to help administrators and managers refine their objectives and priorities, and to maximize the usefulness of this vast information resource. The remainder of this section will be devoted to brief discussions of a few examples of the areas needing research and development.

Data capture/entry techniques. One of the major factors that discourages biosystematists from building data banks is the high cost in time and money of capturing data and of entering them into computer files. At the National Museum of Natural History, for example, as much as half of the total cost of building specimen data banks is in the capture and entry of data. Numerous methods have already been developed to reduce the expense: more efficient data entry devices have been brought on line; computer systems have been organized so as to reduce or eliminate the need for redundant data handling; automatic data recorders have been employed in laboratories; and the path of data flow has been analyzed and revised to eliminate unnecessary transcriptions of data and to develop efficient recording forms and data formats.

There are still many opportunities for improvement, however. Experiments should be conducted to test the adaptability of voice-data-entry equipment which is now entering the market. Field workers should look into the use of the portable, battery-operated data terminals with large memory capacities. Devices that can read handwriting or pencil marks should be explored further. The plans for research projects that generate data should be reviewed to make sure the methods for handling the data are as efficient as possible, and greater use should be made of existing data banks to obtain data items that would otherwise have to be retyped to build new data banks. There are many other possibilities; any improvements would be significant considering the relative expense of data capture/entry.

Poor data. Data that are poor or entirely wrong are a problem in utilizing automated data banks, especially if the data sets are old or if they were collected by inexpert persons or under inexact conditions. In museum records, for example, one should expect to find: misidentifications, obsolete geographic terms, vague terms, transcription errors, missing or incomplete data, and measurements that are imprecise. Some bad data can be discovered and dealt with when the files are being built, but others may be inserted into the data bases inadvertently or because they cannot be improved.

Many biosystematists may avoid making substantial use of data banks because of the potential harmful effects of bad data, but at the same time they will be cutting themselves off from the good data also contained in the files. It would be more useful to develop a better understanding of the nature and potential impact of bad data and of the ways that might be used to minimize their harmful effects while still being able to extract the useful information. Also, those who build data banks should continue to look for means to detect and eliminate bad data.

Scientific value of automated specimen data. With the possible exception of bibliographic files, computer files of specimen data built during the course of cataloging specimens or of taking inventories of collections probably represent the largest portion of biosystematics data automated so far. They have already proven to be useful in collection management, but the extent of their scientific value is still a matter of conjecture. We know that highly useful indexes to collections can be produced from them which will give scientists easier access to the specimens, but we don't know how far scientists can go in using the data alone—without the specimens—to answer scientific questions and to develop lines of thought.

New emphasis needs to be placed on testing the scientific value of these specimen files. Can the files be used to study associations between species? How much can they tell us about the distributions of animals and plants? Can they show us how distributions and associations have changed through time? Can they help us detect where collecting bias has resulted in erroneous scientific conclusions? How much can they tell us about reproductive cycles, migration patterns, or variation within populations? Answers to these questions might help us determine how future collections should be made and automated in order to minimize sampling bias and to ensure that valuable information is not overlooked and lost.

Retrieving and analyzing information from data banks. Biosystematics data banks would probably be used more extensively if more well-tested and well-documented techniques were available to aid retrieval and analysis of information. Computer graphics, text analysis procedures, and statistical methods are being used far more in other fields that in biosystematics data banking. Biosystematists cannot fully utilize or understand their data banks by using only computer lists of data. More sophisticated techniques for displaying the data in graphic form, for using the computer to find trends, and for discovering subtle relationships in data need to be employed.

Many techniques already exist and simply need to be introduced into systematics, but there is plenty of room to devise new procedures as well.

Interdisciplinary uses of data banks. Using automation, biosystematists have enough capacity for processing data realistically to consider using the data of other fields to carry out research. More ground-breaking projects are needed to explore such possibilities as combing botanical and zoological data banks to discover patterns in the association of animals and plants, or obtaining geographic coordinates from geographic data banks and automatically to insert them into specimen data banks so that computer mapping of distributions can be done. We should experiment with combining weather data banks, earth-satellite data banks, and biosystematics data banks to see if we can pick out relationships between species distributions and environmental factors. Such interdisciplinary uses of data banks will also help in another area of need, that of data standards, discussed below.

NEED FOR COORDINATION OF DATA STANDARDS

One of the first things that a biosystematist must do when starting a data bank is determine what data standards to follow. Data standards specify such things as what data elements will be recorded, the definitions of those data elements, what rules of spelling, format, and syntax will be followed, which data elements are mandatory and which are optional, and what values will be entered in place of missing data. If the data bank is for the exclusive use of one biosystematist, the decisions on data standards can be made on the basis of personal preference, but if the file is to be combined with other files or if it is to be used by other biosystematists, great care must be taken to make sure that the data standards are well documented and that they are compatible with the data standards of the other files.

Coordination of data standards among biosystematics data banks is essential if we are able to exchange and combine files. Computer programs can be written to transfer data from one kind of computer to another or from one data-bank system to another, but it will be far more difficult, or even impossible, to overcome differences in data standards.

The long history of biosystematics research has already done a great deal to bring out coordination of data standards. One of the best examples of this is the Linnean system of taxonomy itself, which has been in use for over two hundred years on a world-wide basis. Nevertheless, a new level of knowledge about the data we collect and use is needed to establish the greater precision and consistency required by data banking. Much useful research and development could be done on questions such as:

What data standards were followed in the past that would influence our interpretation of older data?

What data standards are currently in use?

What benefits would be gained from the coordination of data standards, and how could those benefits be measured?

How can difficult standards problems such as the automation of locality data or the ever-changing taxonomic hierarchy be solved?

Answers to such questions will help promote coordination of data standards.

Several scientists have expressed fear that data standards are a form of regimentation which will constrain free thinking and scientific flexibility. This is not the case. In the first place, in a voluntary data environment such as biosystematics, data standards will be used only if they are beneficial; if the net result of a standard is detrimental, it will not be used. In the second place, the data standards that computers require are not much more constraining than the standards we already follow in biosystematics in expressing bibliographic references and taxonomic names.

Task forces, committees, and certification of standards through organizations such as the American National Standards Institute (ANSI) and the National Bureau of Standards have been used by industry and by the Federal Government to impose data standards or guidelines on their members. It seems unlikely that coordination of data standards in biosystematics will happen in the same way, though certification of biosystematics standards by ANSI might be beneficial after the standards have been proven useful and effective. Data standards in biosystematics are being set today by the users of data banking systems, and to some extent by workers in disciplines such as mammalogy, who are organizing the National Information Retrieval Network for Mammalogy. These people and organizations are setting standards because they have a practical need for them. Coordination of standards between data banks may be fully possible only when the practical need arises to exchange and combine files. We should experiment with such interfile activity as soon as possible in order to discover where data-standards problems are being introduced into the files.

The National Bureau of Standards has published an excellent report (*8*) for readers who wish to learn more about how to set data standards.

NEED FOR MORE ORGANIZATION AND COMMUNICATION

Data banks are becoming as large and as difficult to fully know as are the scientific literature and specimen collections. In order to make them as accessible as possible we need to design and implement management and communication systems that can perform the following functions:

(1) Provide standards for describing the attributes of data banks. These attributes include such things as the data standards, the physical structure of the data on the storage medium, the availability of the file, the conditions under which the data were captured and entered, and the date when the file will become obsolete, if known.

(2) Provide perpetual safe storage of data banks that are no longer active but are still useful. The storage facility should be able to give users access to the data banks—in essence it will function as a library.

(3) Provide a reference system that biosystematists could easily use to discover what data banks are available and where they are located.
(4) Provide computer hardware and software capable of accessing files from different systems in different places.

Crovello (2) has made a great stride forward in this area in compiling his EDP-IR index to data-banking projects in biosystematics, but the level of effort that will be required in the future will be more than one person can provide in his spare time. Further, biosystematics probably cannot provide enough support internally to maintain such management systems. We should do as much as possible through organizations such as the Association of Systematics Collections, but we should also investigate how we can join similar systems that are already being set up in industry and in government.

For more background on these matters the reader should consult several useful reports that have recently been made available by the National Bureau of Standards (*1, 4, 5, 6, 7, 8*).

ACKNOWLEDGMENTS

Richard H. Foote and Lloyd V. Knutson were of great help during the preparation of this paper.

LITERATURE CITED

1. Leong-Hong, B., and B. Marron. 1977. *Technical Profile of Seven Data Element Dictionary/Directory Systems.* Natl. Bur. Standards Spec. Publ. 500-3.

2. Crovello, T. J., and R. D. MacDonald. 1970. *Index of EDP-IR Projects in Systematics.* Taxon 19: 63-79.

3. Crovello, T. J. 1976. *Botanical Data Banking.* Proceedings of the Fifth Biennial International CODATA Conference, Pergamon Press, New York (in press).

4. Hochman, A. (chairperson of the task force). 1977. *Executive guide to the management of data resources* (preliminary draft). Available from the Natl. Bur. Standards Office of ADP Standards Management, Washington, D.C. 20234.

5. Hochman, A. (chairperson of task force). 1977. *Guidelines for the Management of Data Resources* (preliminary draft). Available from the Natl. Bur. Standards Office of ADP Standards Management, Washington, D.C. 20234.

6. McEwen, H. (coordinator of task force). 1977. *A Survey of Eleven Government-Developed Data Element Dictionary/Directory Systems.* Natl. Bur. Standards Spec. Publ. 500-16.

7. Plagman, B. K. 1977. *Criteria for the Selection of Data Dictionary/Directory Systems* (preliminary draft). Available from the Natl. Bur. Standards Office of ADP Standards Management, Washington, D.C. 20234.

8. White, H. S. 1976. *Guide for the Development, Implementation and Maintenance of Standards for the Representation of Computer Processed Data Elements.* Federal Information Processing Standards Publ. 45. (National Bureau of Standards, U.S. Dept. of Commerce).

13] Molecular Phylogenetics

BY LYNN H. THROCKMORTON*

ABSTRACT

Different molecular approaches to phylogenetic analysis are illustrated and evaluated. The principle of genetic indeterminism is formulated, and the restrictions it places on phylogenetic methodology are made explicit. Molecular clocks, and the uses and misuses of genetic distance measures, are commented upon, as are the so-called maximum parsimony methods. Covariation analysis is emphasized as a sound analytical tool; it is defined, and examples are provided from protein studies of *Drosophila*.

INTRODUCTION

Some years ago, when the current bandwagon got rolling, one of the common arguments for molecular approaches to phylogenetic analysis was that they permit almost direct investigation of the genetic material. DNA hybridization can measure genetic divergence, the amino acid sequences of proteins are but one step away from the genetic code itself, and so on. The implication was, of course, that the closer we are to the genetic material the closer we are to pure phylogenetic information, and that if we could read the DNA we could read off phylogeny. That is not true, and I shall begin by noting some properties of the genetic system that severely limit phylogenetic analysis. This will establish what I shall call *genetic indeterminism.* Then I shall discuss some procedures of phylogenetic method and how these apply to molecular data. This will define and illustrate *covariation analysis,* which is the core of standard phylogenetic method and widely accepted as such, although it is not generally called by that name. Finally I shall consider some of the current fads: DNA hybridization, immunological and genetic distance, common ancestor and "maximum parsimony" methods, and so on. These can be compared with standard methods, and

*Department of Biology, University of Chicago, Chicago, Illinois 60637.

with the realities of scientific method, and their advantages and disadvantages for systematics can be appraised.

My overall strategy will be first to emphasize and illustrate theoretical considerations, and second to evaluate current phylogenetic fashions against these principles. Most of my illustrations will be drawn from the virilis group of *Drosophila* (*27, 28,* and unpublished data). This is a holarctic species group with about half a dozen nearctic and half a dozen palearctic species. It probably originated in middle or late Miocene times, perhaps about fifteen million years ago, and its genealogy is quite well established (*25, 29*). The phylogeny is based on information from salivary gland chromosomes, proteins and reproductive behavior, but I shall not try to recount that evidence at this time. There are uncertainties regarding some of the most recently described species, though these do not affect the examples or the conclusions illustrated by them.

GENETIC INDETERMINISM

We have long been aware of genetic variation within natural populations, but it was only during the last few decades that this became directly measurable. Protein electrophoresis made that possible and early studies produced remarkably high estimates of the mean proportion of heterozygous loci within species gene pools. For *Drosophila* this was about 53%, and values for other insects were similar. It was somewhat higher in marine invertebrates (59%) and rather lower in vertebrates (down to 15% in birds). The average heterozygosity per individual ranged from 17% in plants through 15% in *Drosophila* and other insects to about 4% in large mammals (*21*). These estimates are only for structural gene loci. Control loci and elements of other systems may behave differently, but this is still an extraordinarily large amount of genetic variation.

Large as it is, however, recent discoveries find still more (*1, 22, 30*). The first investigations depended on electromorphs, on protein molecules that showed mobility differences during electrophoresis, and it was thought that more than two-thirds of the total genetic variation could be measured in this way. When that was actually checked within populations by using two criteria, mobility plus heat sensitivity, the average electromorph was found to be a collection of proteins with about three different heat stabilities. At the XDH locus in the virilis group, for example, 11 electromorphs became 32 thermoelectromorphs. And when three or more criteria were used to assess differences, the number of morphs increased four or five fold (*4, 5, 23*). This total now approaches theoretical expectations for the number of variants that might be hidden within a single electromorph, so it may be nearly correct. In the virilis group the number of electromorphs per locus per species ranges from 1 to 15, with a mean near 8. If these were increased four to five fold the mean number of alleles per locus per species could be as high as 30 or 40.

The point of all this is that at the level of the gene pool the system is encountered at one of its least predictable levels. Variability is high and natural selection can winnow it in many ways. And since heterozygous gene pools persist through time and many different species bud off from them (*26*), individual genic variants may be apportioned to descendent species almost haphazardly relative to the genealogy of the group. On the average, three electromorphs per locus (12 to 15 alleles?) have been segregating in the gene pools of the virilis group since its origin. Some of the consequences of this are illustrated in Fig. 13.1. First there is a two-allele polymorphism that has persisted in all but one line for about 15 million years (HP-2, lower right); then a three-allele pattern shows the origin of alleles within phylads and their subsequent haphazard distribution from then on (LAP, lower left). The upper examples (MDH, HP-1) show combinations of these patterns. If some single locus were being followed and distribution of its alleles used as evidence for phylogeny, no one of these would give the correct result.

This segregation is quite real, but operationally its effects are confounded with another aspect of genetic indeterminancy. Mutation occurs at predictable rates and individual genic variants may recur quite regularly so long as appropriate parent variants persist in the gene pool. At the upper right in Fig. 13.1 (HP-1) electromorph 6 is found almost throughout one phylad, but it is also found once in the other phylad. Is this convergence or parallelism? We cannot sharply separate the two. This instance might be mutational. The distribution of alleles 3 and 4 of HP-2 (lower right) almost certainly is not. Insofar as phylogenetic analysis is concerned it makes no difference. Whichever it is, it is noise, and any competent phylogenetic analysis must cope with it.

Granting that single alleles have great limitations, one might suggest that perhaps analyses combining information from many loci will overcome the drawback. That can be true or not, depending on how the results are analyzed. If a similarity index is calculated (phenetic distance, genetic distance, etc.) it is not true. High levels of similarity, large numbers of alleles in common, can also arise in parallel. Perhaps the most obvious and most common example of this is evolutionary conservatism, where two or more species on different evolutionary pathways remain very similar through retaining a large proportion of ancestral alleles in their genotypes. Alternately, from a common heterozygous heritage they may independently generate highly similar genotypes through responding in similar ways to similar environmental challenges. Again these last two alternatives cannot be distinguished operationally, but that does not change the conclusion. So long as mechanisms exist by which high levels of similarity arise independently, degree of similarity, a blind number, cannot reliably stand as evidence of propinquity of descent.

Table 13.1 illustrates the above. The matrix to the left shows the distribution of similarities attributable to shared primitive states. In this instance

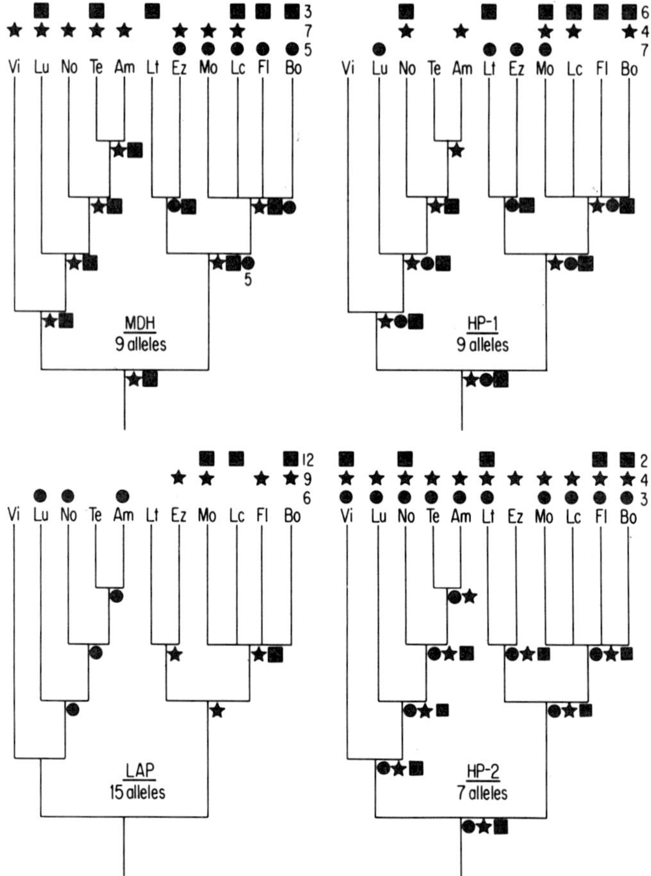

Figure 13.1. Genetic indeterminism. The segregation of different electromorphs is illustrated by symbols on the genealogy of the virilis group of *Drosophila* species. The electromorphs are numbered to the upper right. The locus and the total number of electromorphs at each locus are indicated between the phylads. MDH, malic dehydrogenase; LAP, leucine amino peptidase; HP-1, hemolymph protein 1; HP-2, hemolymph protein 2; Vi, *D. virilis*; Lu, *D. lummei*; No, *D. novamexicana*; Te, *D. a. texana*; Am, *D. a. americana*; Lt, *D. littoralis*; Ez, *D. ezoana*; Mo, *D. montana*; Lc, D. lacicola; Fl, D. flavomontana; Bo, *D. borealis*.

the primitives were detected operationally as those electromorphs that were present in at least one species of each of the two cytological phylads. Derivatives were recognized as those electromorphs restricted to the species of only one phylad. On the average, primitive states account for about 78% of the genetic similarity between species in the virilis phylad (upper left corner of the matrix) and 59% of the similarity between species in the

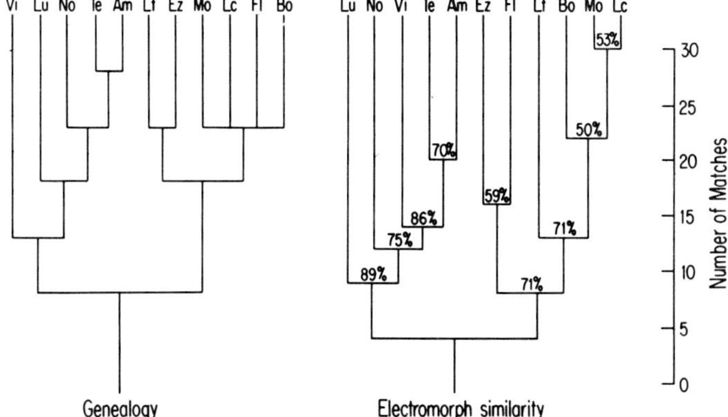

Figure 13.2. Genetic indeterminism. A contrast is shown between the phylogeny of the virilis group of *Drosophila* species (left) and the dendrogram of genetic similarity derived from protein data (right). The number of matches is shown on the right side. Numbers on crossbars are the percent contributed to that similarity by primitive electromorphs. Abbreviations are as in Figure 13.1.

montana phylad (to the lower right in the matrix). Some consequences of this for phylogeny are illustrated in Fig. 13.2. On the left I show the genealogy of the virilis group again. On the right I show a dendrogram of relationship derived by clustering species on the basis of the number of matches between them for electromorphs at 10 loci. The numbers on the crossbars indicate the percent of matches that in each case result from sharing of ancestral genes. This ranges from 50% to nearly 90%, and the overriding effect it can have on similarity, and on phylogenies blindly based on similarity indexes, is obvious.

Heterozygosity, its persistence, and its predictably unpredictable consequences over long periods of evolutionary time, cannot be avoided. Genetic indeterminism is a fact of life, and the next question is whether, or how, it can be dealt with. To my knowledge there is only one method that minimizes its effects, and that is covariation analysis.

COVARIATION ANALYSIS

Covariation and correlation are not the same thing. Attributes that covary are correlated, but things that are correlated do not necessarily covary. *Character states can be said to covary when any one of the states is not observed unless another state of the set, from at least one other independent character of the set, is present in the same unit also.* Table 13.2 shows the covariation of sets of inversions, one set for *Drosophila lummei* through *D. a. americana,* the other for *D. ezoana* through *D. borealis.* Note that *D. virilis* is not included in either set. Within the first set there is a sequential

Table 13.1 Genetic indeterminism. The distribution of similarities generated through the retention of ancestral electromorphs (left) and those resulting from segregation of derivative electromorphs (right). Abbreviations for species names are as in Fig. 13.1.

	Shared primitive states											Shared derivative states										
	Vi	Lu	No	Te	Am	Ez	Lt	Fl	Mo	Lc	Bo	Vi	Lu	No	Te	Am	Ez	Lt	Fl	Mo	Lc	Bo
Vi		9	11	12	14	7	9	10	11	8	14		2	1	2	5	0	0	0	0	0	0
Lu			8	9	9	7	9	7	12	9	8			1	1	3	0	0	0	0	0	0
No				9	12	4	8	8	11	9	10				4	5	0	0	0	0	0	0
Te					14	8	8	9	13	10	11					6	0	0	0	0	0	0
Am						7	7	9	13	10	13						0	0	0	0	0	0
Ez							6	10	11	8	8							2	7	10	6	9
Lt								9	11	10	11								2	4	4	4
Fl									13	10	12									10	6	8
Mo										16	14										14	16
Lc											12											11

Table 13.2. Covariation analysis of gene sequences. Species are indicated to the left and abbreviations are as in Fig. 13.1. The gene sequences of salivary gland chromosomes are across the top. Individual sequences are not named since only the pattern of covariation is emphasized here. Unique sequences, those found only in one species, are indicated in the column to the right.

Species	Character states—chromosome sequences			U
Vi				1
Lu	+ + + +			?
No	+ + + + + + + + +			0
Te	+ + + + + + +			0
Am	+ + + + + + + + + + + +			3
Ez		+ + + + + + + +		4
Lt		+ + + + + + +		6
Fl		+ + + + + +	+ + + + + + + + +	4
Mo		+ + + + + +	+ + + + + + + + +	33
Lc		+ + + + + +	+ + + + + + + + +	13
Bo		+ + + + + +	+ + + + + + + + +	7

Table 13.3. Covariation analysis of electromorphs. Species are indicated to the left and abbreviations are as in Fig. 13.1. Different electromorphs are shown horizontally. The upper section shows the ancestral electromorphs and uniques. The lower section shows derivative electromorphs and should be compared with the pattern seen in Table 13.2.

Character states—ancestral electromorphs

Species																															U
Vi	+	+	+	+	+	+	+				+	+			+	+	+	+	+								+	+	+	+	3
Lu		+	+	+		+		+	+		+	+			+											+		+	+	+	0
No	+	+	+	+							+	+			+		+					+	+	+	+		+	+	+	+	2
Te	+	+		+		+			+		+	+	+	+	+		+	+	+	+	+							+	+	+	0
Am	+	+	+	+		+				+	+	+	+	+	+	+	+	+	+			+	+					+	+	+	4
Ez					+	+		+		+	+	+			+				+	+	+					+		+	+	+	1
Lt	+	+		+	+	+		+	+						+										+	+	+	+	+	+	4
Fl				+	+	+	+	+		+		+			+				+	+	+	+			+		+	+	+	+	5
Mo		+	+	+	+	+	+	+	+	+	+	+	+	+	+					+	+		+	+	+	+		+	+	+	6
Lc		+		+	+	+		+	+	+	+			+	+						+		+	+	+			+	+	+	3
Bo	+	+	+		+	+	+	+	+	+					+	+	+	+	+	+			+		+		+	+	+	+	5

Character states—derivative electromorphs

Species																																
Vi	+	+	+	+	+																											
Lu	+	+				+																										
No			+			+	+	+	+		+																					
Te	+			+			+	+	+	+	+																					
Am	+	+	+	+	+	+	+	+	+	+																						
Ez												+	+	+	+	+	+	+	+			+	+									
Lt												+	+							+	+											
Fl												+	+	+	+			+	+				+	+	+	+						
Mo												+	+	+	+	+	+	+	+	+	+	+	+	+	+	+	+	+	+	+	+	+
Lc												+	+	+	+	+	+			+	+			+	+		+			+	+	+
Bo												+	+	+	+	+	+	+	+	+	+	+		+	+		+	+	+			

subdivision, with three taxa, *D. novamexicana, D. a. americana,* and *D. a. texana,* sharing inversions not yet seen in *D. lummei.* The second set is subdivided dichotomously, with *D. ezoana* and *D. littoralis* partitioned off from the other four species. This happens to be the cytological evidence for the phylogeny of the virilis group, and it should be pointed out that inversion phylogenies are nondirectional. In themselves they contain no information as to the position of the "root" of the tree. Non-directionality is a peculiar property of inversion data and not a general property of covariation. In this instance information from chromosome fusions, and also other evidence, indicates that the tree is "rooted" in such a way that *D. virilis* is included phylogenetically with *D. lummei, D. novamexicana,* and *D. americana.*

A more typical set of data is given in Table 13.3. This shows some of the proteins from the virilis group. Species are again listed to the left and alleles are shown horizontally. In the top section ancestral alleles and the uniques are given. At the bottom the derivative sets are shown. The pattern seen for the ancestral alleles gives some indication of the "noise" in the system and of the haphazardness of distributions generated through evolutionary time as a consequence of heterozygosity. The sets of derivative and primitive electromorphs were, of course, intermingled in the original data set. Ninety-five electromorphs are present; 30 (31%) are primitive, 32 (34%) are derivative, and 33 (35%) are unique. Since species boundaries are known, the uniques are easy to detect. Of the remainder about half are "information" and half are "noise," and identifying which is which can be difficult. It follows that I am emphasizing covariation analysis as an effective phylogenetic method, not as an easy one.

In a sense it is misleading to refer to the ancestral electromorphs as noise. This is a matter of perspective. Insofar as intragroup relations are concerned, they are noise. But within the genus these are the information that sets the group apart from all other groups. In an early study of sibling species from all parts of the genus, it was found that less than six percent of electromorphs are

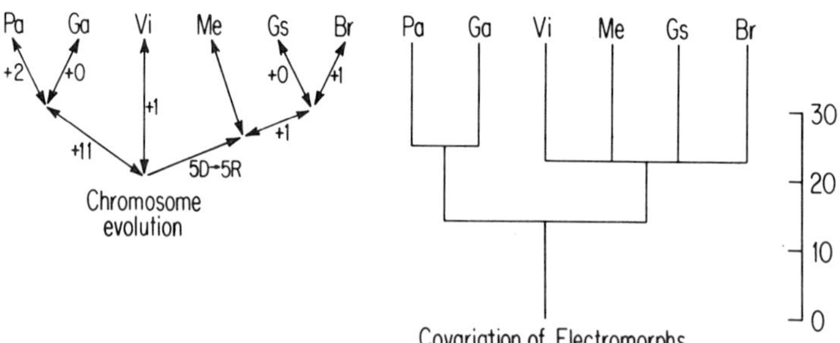

Figure 13.3. Covariation analysis of the mesophragmatica group of *Drosophila* species. To the left is given the diagram of chromosome relationships. Most arrows are bidirectional. In one case, 5D to 5R, the dot chromosome was increased by heterochromatin to become a small rod. This may have been a directional change since the dot is primitive for the genus. Numbers along the right side indicate the size of the covariation set that establishes each group. Chromosome data are recalculated from Brncic *et al.* (*3*) and protein data are recalculated from Nair *et al.* (*18*). Pa, *D. pavani*; Ga, *D. gaucha*; Vi, *D. viracochi*; Me, *D. mesophragmatica*; Gs, *D. gasici*; Br, *D. brncici*; U, uniques.

Table 13.4. Covariation analysis of the mesophragmatica group. Species are indicated to the left and character states across the top. Chromosome data are recalculated from reference *13* and protein data are recalculated from reference *14*.

Chromosomes Species															U
Pa	+	+	+	+	+	+	+	+	+	+	+	+			10
Ga	+	+	+	+	+	+	+	+	+	+	+	+			0
Vi	+														1
Me													+		4
Gs													+	+	6
Br													+	+	0

Ancestral electromorphs Species															U
Pa	+		+	+	+	+	+	+	+	+	+		+	+	5
Ga	+	+		+			+	+	+		+	+			2
Vi					+	+		+	+	+	+				7
Me			+	+		+	+			+		+	+	+	3
Gs	+	+				+	+					+		+	2
Br	+		+			+	+	+	+	+	+	+			0

Derivative electromorphs Species																				
Pa	+	+	+	+	+	+	+	+	+	+	+									
Ga	+	+	+	+	+	+	+	+	+	+	+									
Vi												+	+	+			+			+
Me												+		+		+		+	+	+
Gs													+		+	+	+	+	+	+
Br													+		+	+	+			+

shared between species of different species groups (*14*). If we use 94% (100%-6%) as a guide, approximately 28 of the 30 ancestral electromorphs are distinctive relative to those of other species groups in the genus; which gives some idea of the power of this approach for partitioning taxa.

Over the years I have gotten the impression that some systematists envy *Drosophila* their salivary gland chromosomes. Or at any rate they envy the systematists who make use of them. Protein electrophoresis can, to some degree, give all taxonomists a tool as powerful as salivary gland chromosomes. Each has its advantages and disadvantages, as is true for almost all methods. The covariation patterns of the chromosomes (Table 13.2) and of the derivative states of the electromorphs (Table 13.3, bottom) can be contrasted. That of the chromosomes is a much tighter pattern, with fewer gaps in it, indicating that these chromosomes were less sensitive, on the average, to the indeterminism of heterozygosity than were the electromorphs. Thus these inversions are somewhat better evidence of genealogy than are the proteins. On the other hand, covariation patterns of proteins are directional while those of the inversions are not, so what is gained in one way can be lost in another.

Not all chromosomes give such definitive patterns as those of the virilis group. In Table 13.4 and Fig. 13.3 an example is given from the mesophragmatica group (*3, 18*). These six species are split into two clusters, a sibling species pair, *D. gaucha* and *D. pavani,* and four other species separated from it by at least 11 inversions. In one case (5D to 5R, Fig. 13.3) the dot chromosome has been converted to a small rod by the addition of heterochromatin. Since the dot-like chromosome 5 is primitive for the genus, this is probably a directional change, but it and the few remaining inversions fixed between species cannot help much to delineate further the genealogy of these species. Their numbers are so low that they could easily have segregated in this pattern from heterozygous ancestral gene pools with no relationship to phylogeny.

The protein data from the mesophragmatica group parallels the chromosome data and resolves some of the phylogenetic problems (Fig. 13.3, right). The "root" of the tree appears to be between the sibling species pair and the other four species. *D. viracochi* belongs to the phylad on the right and the origin of the rod-like fifth chromosome probably occurred during the early evolution of that phylad. We still cannot tell how to apportion the eleven paracentric inversions that separate the two clusters, and we do not know the details of the relationships of the four species in the viracochi phylad. This is the parsimonious answer that can be derived from these data, and, while we might like to have more, we certainly cannot complain of what we do have.

One last example of covariation can be noted (Table 13.5). This does not come from *Drosophila,* but it is of a kind familiar to almost everyone. I picked, almost at random, a 30-residue segment from amino acid sequence data published by Dayhoff (*6*). It happens to be from some vertebrate polypeptide hormones. The different covarying sets are indicated by underlining and spacing. The point of the example is to show the second major way covariation patterns are seen in molecular data. Processing is the same as for electromorph data. Here the amino acid residue at a particular site in a protein is the character state. There the particular electromorph at a specific locus is the character state. In both cases direction of evolution is resolved by reference to outside groups, with states encountered outside the group regarded as operational primitives of no value for determining within-group relationships. Since, on the average, fewer than one in ten of the electromorphs within a group are found outside a group, I have not emphasized that aspect of the analysis for electromorphs. For other types of data, anatomical as well as molecular, this is a very important consideration.

Before continuing to the fads and fashions that we encounter today in molecular phylogenetics, it is necessary to comment on the scientific, or epistemological, justification for covariation analysis and for regarding the product of such an analysis as a nonarbitrary and parsimonious indication of the true genealogy of the species involved.

The argument is brief, but all the more potent for being concise. First

Table 13.5. Covariation of amino acid residues. A 30-residue section of some vertebrate hormones is shown. Horizontal lines separate the different covarying sets. 1, cow prolactin; 2, sheep prolactin; 3, human lactogenic hormone; 4, human growth hormone; 5, cow growth hormone; 6, sheep growth hormone. Data are from Dayhoff *(6)*.

Species	Amino acid residues							
1	L L F A H L E F	R M S N S E M D	R V I	D H	V	S V F	D Y K	S N
2	L L F A H L E F	R M S N S E M D	R V I	D H	V	S V F	D Y K	S N
3	L L F A H L E F	S L A Q A D T E	H Q A	D H	V	P M Y	R R E	I Q
4	L L F A H L E F	S L A Q A D T E	N R L	D H	I	P M Y	R R E	F E
5	L L F A H L E F	S L A Q A D T E	N R L	A Q	M	S V F	G H R	A K
6	L L F A H L E F	S L A Q A D T E	N R L	A Q	M	S V F	G H R	A K

recall the definition of covarying character states. *Character states are said to covary when any one of the states is not observed unless another state of the set, from at least one other independent character of the set, is present in the same unit also.* An analysis based on covariation is phylogenetic. A highly improbable distribution (covariation) of character states is observed. Such an association of character states cannot be excused on the grounds of chance and some explanation is required for it. Only a limited number of alternatives is available. Covariation can be a consequence of pleiotropy, or it can be due to independent adaptation to common environmental factors (convergence), or it can be due to common descent uncomplicated by other elements. If no positive evidence for pleiotropy or convergence can be adduced, common descent is accepted as the parsimonious inference. Alternatives involve *additional assumptions* (and so are less parsimonious). Special creation is one example of an additional assumption, if you do not mind calling it that, and independent origin through mutation is another. But whatever the case, so long as there is no *concrete* evidence for other factors, the inference of common descent is justified and so is the conclusion of phylogenetic accuracy that derives from it.

It would be wrong to decide that this implies a method of absolute accuracy. Those seem to be rare in this universe and covariation analysis is no paragon. All attributes that are accessible to covariation analysis, whether anatomical or molecular, are influenced by genetic indeterminancy. The only way the consequences of this can be minimized is through the use of large sets of characters. Choice between conflicting sets of covarying states, and these do appear frequently, is based on the nature and relative sizes of the sets concerned. Naturally, the larger and "tighter" sets are the weightier sets. If alternative sets are nearly the same size, no choice is possible and the details of relationship must remain unresolved.

In brief, then, covariation analysis is not necessarily easy, but it can be effective in producing a nonarbitrary and parsimonious phylogeny. As we all know, parsimony does not guarantee truth. What it does guarantee is the best answer from the existing data. It is neither right nor wrong. It is just very correct!

FADS AND FASHIONS

In criticizing methods it is necessary to recognize at the outset that almost every method is good for something, although it may not have the usefulness its proponents claim for it. Some methods produce useful preliminary results and are fine for exploration. Others produce definitive results, but they may be too cumbersome or expensive for general use. There are four chief sources of data for molecular phylogenetics. I can describe none of them in detail, but can only direct attention to each method, indicating generally its approach, and occasionally noting technical or practical difficulties. Then I consider the manner in which these data have been, or can be, processed and the ways in which this restricts the systematic uses to which they can be put. Here the emphasis must be not so much on the steps in the procedure as on the premises that underlie it. It is these premises that validate an analysis, that give it whatever scientific merit it can have, hence they are of the utmost importance.

Obtaining the data. Electrophoresis, amino acid sequence analysis, immunology and DNA hybridization have been most frequently exploited to provide molecular data for phylogenetic analysis, and each is optimistically reviewed elsewhere (*6, 11, 19, 24, 31*). Electrophoresis makes use of samples of body fluids, homogenized tissues, homogenized whole organisms, and so on. For my own work a single fly is homogenized per sample, which gives an indication of the amount of material required. Some of the results I gave earlier were taken from a study of the virilis group involving more than 60,000 samples (flies) run in acrylamide gel over a period of about five years. Once in the gel the samples are subjected to a charge differential for varying lengths of time, then they are removed to a suitable reaction mixture to visualize the proteins. The gels can then be photographed to record the data. This method assays only products of structural gene loci, and of these it deals only with those that produce soluble products, products that are detectable by available assay systems, and so on. It is generally thought that they represent a random sample of structural gene loci. At least there is no evidence that they comprise a sample biased in some evolutionarily significant way, and they are evaluated accordingly. Without special additional procedures this method detects electromorphs, not necessarily alleles, so similarities calculated from electrophoretic data are as much phenetic distances as they are genetic distances. I have already given examples of results from electrophoresis, and since one dendrogram is much like another, I shall not give additional specific examples for the other methods.

The second important method is that of amino acid sequence analysis (*6, 11*). Probably the best known examples of this are the cytochromes *c* and the hemoglobins (*9*). This method requires a purified protein as its starting point, and there must be suitable amounts of protein from each of the organisms to be investigated—sometimes not an easy requirement to meet. The cytochrome *c* of a gastrotrich will probably be quite hard to acquire, for example. Much of the procedure for sequencing proteins has been automated, but it is still not simple and relatively few laboratories are equipped to do it. Once the sequence has been determined, it is possible to infer the DNA sequence that produced it, to compare DNA sequences, calculate mutation distance, etc. An example of amino acid sequence data is given in Table 13.5, and it can be regarded as true genetic data, not just approximately genetic data, as was the case for the electromorphs.

Immunology is the third important method, the most common procedure being that of microcomplement fixation (*20*). This method also requires samples of purified protein, although usually not in the amounts needed for sequence analysis. An antiserum is produced against this protein and reacted against serum from the same (homologous) and different (heterologous) species. The factor by which the serum concentration from heterologous samples must be raised to give a reaction equal to that of homologous serum provides an index of dissimilarity. The higher the required concentration of heterologous serum the greater the dissimilarity. The log of the index of dissimilarity is approximately linear with respect to the proportion of different amino acids between the sequences of the proteins used in the reaction if the proteins are not too different, at least for a few proteins that have been tested so far. To obtain approximate linearity, however, it is necessary to make full reciprocal comparisons between species. This is a warning that the procedure is less trustworthy than might be wished. From a study of bird transferrins (*12*), for example, I calculated the mean percent difference between reciprocal comparisons, and for more than half (52.5%) this difference was 20 percent or greater. Thus, microcomplement fixation measures an approximate genetic distance. The distance is roughly comparable to that obtained for single proteins by amino acid sequence analysis. It is not an accurate measurement, however.

The fourth method is DNA hybridization (*2, 10, 13, 15, 16, 17, 19*). For this the DNAs are hybridized and the thermal stability of the hybrid DNA is measured. At the temperature at which one-half the hybrid DNA dissociates, a 1.5% mismatch between strands lowers the thermal stability by 1°C. This is, accordingly, a convenient measure of differences between DNAs, the lower the thermal stability of the hybrid DNA the greater is the distance between the parent DNAs (*16*). Several years ago DNA phylogenies, mostly based on a less accurate technique, were much in evidence. Then serious complications became apparent (*19*). It is now known that the DNA of an organism is complex; exactly how complex is not yet certain. There are at least two fractions of DNA, one unique and presumably coding for structural gene loci, the other repetitive and acting as spacer DNA, etc. Different species may have different relative quantities of these types of DNA, and the different ways these different kinds of DNA might evolve are scarcely understood at all (*10*). In consequence, DNA comparisons between species do not necessarily involve homologous DNAs, or involve homologous DNAs only in very uncertain proportions, even when care is exercised to use just what are hoped to be the unique sequences. If it were *known* that only unique sequences were involved, the genetic distance calculated from the DNA might be roughly comparable to the genetic distances calculated from amino acid sequence data. It would estimate genetic distance between total genomes, however, rather than over just one structural gene as in the case of genetic distances between molecules of, say, hemoglobin. Since the DNAs involved in such comparisons cannot be sharply defined yet, the significance of the difference values is uncertain.

Analyzing the data. Phylogeny means different things to different people. I use phylogeny in the sense of genealogy, to refer to the sequence by which different groups arise. A construction that cannot accurately represent genealogy cannot be regarded as a phylogeny, and a method that does not accurately reproduce genealogy is not a phylogenetic method, as I use the term here. Of course, a phylogeny does not need to be completely detailed to be accurate. Accuracy is required only for the details that are shown. There are three fundamentally different ways of approaching phylogenetic analysis. One of these is covariation analysis, which I have already considered briefly. A second is the "clock" approach, and the third is one based on some concept or another of "minimal" evolution.

The concept of a molecular clock has been with us now for almost a generation (*6, 8, 19*). It grew out of the observation that the degree of amino acid sequence difference between certain proteins is nearly linearly related to the times of divergence between them. Different proteins have evolved at different rates, and hence it should be possible to use slowly evolving proteins to clock ancient events and rapidly evolving ones to clock recent ones. Unfortunately, there are serious reservations regarding just how such a clock can be used.

First, it is now realized that constancy of rate is a consequence of the long periods of time involved. A protein can change at quite different rates over short periods of time and its average rate can still be quite constant. Evidence for inconstant rates comes from a variety of sources, a convenient one for me being the data on protein evolution in the virilis group. The two phylads of this group have been evolving for the same length of time, but one phylad added 20 new electromorphs while the other added 48. Of the 10 loci investigated, three evolved faster in the virilis phylad, six evolved faster in the montana phylad, and one did not change in either phylad. The net

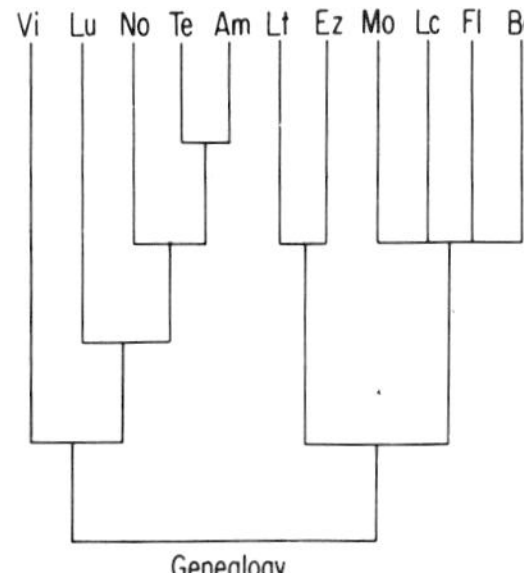

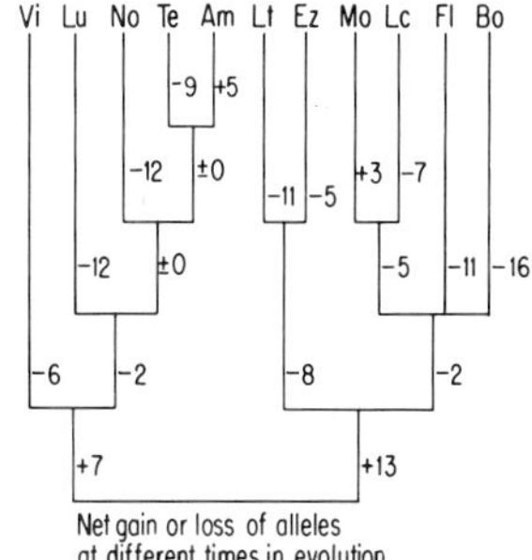

Figure 13.4. The irregular molecular clock. The genealogy of the virilis group of *Drosophila* species is given to the left. On the right a slightly modified version is shown with the net gain or loss of alleles during each interval shown beside the internodes. In total the virilis phylad gained 20 electromorphs while the montana phylad gained 48.

result is that the montana phylad evolved 2.4 times faster than the virilis phylad. The general pattern for gain and loss of alleles is shown in Fig. 13.4. The net gain or loss of electromorphs is given for each internode of the dendrogram and ranges from +13 to -16, with two internodes showing no net gain or loss. The rate of molecular evolution is not constant here. The ecologically conservative phylad changed slowly; the one encountering evolutionary opportunity in new environments changed rapidly. Hence, there is no molecular clock that can be used for timing short intervals like the branches of most phylogenetic trees. It is, in fact, a very poor clock since, as we all know, an inaccurate clock can be worse than none at all. It may be correct sometimes, but one never knows when those times are.

The uses of this kind of clock are very limited, since they must be compatible with uses of an *average*. The restriction is very simple. One average can be used to estimate another, with more or less success depending on circumstances. But one cannot take an average and from it discover the particular individual observations that contributed to it. One cannot, for example, use the height of the average adult U.S. male and assert that *that* will be the height of the next man one meets. But that is effectively what is done when genetic distance is used to specify the individual internodes on a dendrogram. For the length of an internode (= time) is established in proportion to genetic distance. Genetic distance and time are presumed to be equivalent and the rate of evolution is treated as constant over the entire dendrogram. The data from the virilis group illustrate an instance where that is not the case; which thus introduces a constraint that eliminates all genetic distance measures as effective means of constructing phylogenies. Genetic distances can be used in phylogenetic analysis only if molecular change occurs with clocklike regularity, and present evidence shows that that is not true.

One might suppose that there is a situation where genetic distance can be used, as an average, to time divergence between groups. If one had, say, two groups of several species each, one could make pairwise measurements of genetic distance between them, and from these observations calculate a pretty fair average genetic distance between them. Then, if one could somehow calibrate this genetic distance and determine how much time, on the average, was related to each distance unit, one might have an estimate of the time his groups diverged. One still would not know how much evolution occurred in the line leading to one group, and how much occurred in the other, but that might not be important. But there is another problem, and that is calibration. This genetic distance must be measured between groups for which times of divergence are known from fossil, or other evidence. And those are hard to come by. Calibration must usually be carried out on one group of organisms and the results applied to others. This creates an epistemological problem. What guarantees that the rate observed in one group actually applies in others? After all, if I want to find out what is in my pocket I do not look in yours, and if I measure the rate of evolution in frogs or bacteria I cannot have too much assurance that it applies for *Drosophila*. In short, if the method cannot be properly calibrated it cannot be used, and if it can be calibrated, it is not needed, since the required times of divergence are already available from other more reliable sources.

It follows from this that phylogenetic methods depending on molecular clocks, as do all varieties of genetic distance, whether they be calculated from DNA, electrophoretic data, amino acid sequences, or immunological data of one kind or another, are inadequate. At least they cannot produce non-arbitrary and parsimonious phylogenies. They can, however, be very useful for exploratory work, and they should not be discounted for systematic research of that kind.

The minimal distance methods (*7, 9*), including "common ancestor" methods (*6*), present a somewhat different scientific problem. Those who have worked much with arranging dendrograms know that many different trees may often be created by slight changes in clustering methods, sequence of analyzing the data, and similar trivial variations. How does one choose among the trees that can be produced? In fact, there is no nonarbitrary choice, and that in itself is sufficient to eliminate these methods as definitive phylogenetic methods. If the choice of dendrogram is arbitrary, there can be no necessary relation between it and the genealogy of the species involved. And there is an additional problem. A favorite method is to try, by any one of many ingenious means, to discover the "shortest" possible tree that is compatible with the data. The argument for this is that it reflects "minimal evolution" of some kind and is therefore the "best" tree. The difficulty with this approach should be obvious to everyone. Formal evolution theory does not assert, and never has, that evolution proceeds from one particular species to another by the shortest possible route. What it does say

Table 13.6. Relationship between kind of data and method of analysis. In the body of the table a "yes" indicates that data from a given source is suitable for analysis by a given method. Only two kinds of data are suitable for covariation analysis, and covariation analysis is the only epistemologically sound method of analysis by which genealogies can be inferred.

	Method of analysis		
Source of data	Genetic distance	Minimum distance[a]	Covariation
Electrophoresis	Yes	Yes	Yes
Amino Acid Sequence	Yes	Yes	Yes
Immunology	Yes	Yes	No
DNA Hybridization	Yes	Yes	No

[a]Minimum mutational distance, maximum parsimony, etc.

is that change from one generation to another will be small (but not necessarily the smallest possible change), and that one character state will evolve into another gradually. That is a far cry from saying, for some haphazard assortment of extant (or fossil) species, that the lot of them must have evolved from each other, or from annectant ancestors, by the shortest possible route. To regard such a procedure as parsimonious completely misconsceives the intent and use of parsimony in science.

Parsimony is a means of choosing between answers on the basis of the inferences employed in arriving at those answers from the available data. One asks which of the analyses involves the fewest unnecessary inferences, the fewest presumptions not directly dictated by the data themselves. It is most definitely not a method that involves looking at two "answers" and deciding which is the least "extreme." Minimum distance methods force data to conform to a specific preset bias. Data are not permitted to "disclose the structure of the universe," as is the intent of scientific method. Instead they must produce an answer as nearly as possible in conformity with a predetermined model. Far from being parsimonious, minimum distance methods are thoroughly procrustean, and they succeed only in giving inherently arbitrary choices quite undeserved dignity and standing.

CONCLUSIONS

The gene pool carries a great deal of variation, and while its distribution from generation to generation can be predicted with some facility, its course through longer periods of evolutionary time is quite unpredictable. This indeterminancy, reflecting persistent heterozygosity, haphazard retention of ancestral alleles among descendent species, convergent mutation, and parallel generation of similar genotypes, places severe restrictions on methods of phylogenetic analysis. And the closer the material is to the gene itself, the more sensitive it is to these perturbing influences. Thus, the distribution of alleles at one locus must be quite uncertain, while the joint distribution of many alleles at many loci (covariation) can, if properly analyzed, accurately evidence genealogy.

Genetic distance methods depend on the reliability of unreliable molecular clocks, and so are not adequate for phylogenetic analysis. The minimum-distance methods force data to conform to preset bias, hence are procrustean, and so are wholly nonparsimonious and also unsuited to produce phylogeny. Only covariation analysis (Table 13.6), which searches for covarying character sets and hence can in principle discover nonarbitrary partitions within the data, can produce sound evidence from which parsimonious phylogenetic inferences can be drawn. Present immunological and genetic distance methods cannot produce data suitable for covariation analysis. Only electrophoretic data and amino acid sequence data are suitable, and definitive molecular phylogenies must come from them. At best, other so-called phylogenetic methods are suitable only for exploration, to provide tentative evidence of probable lines of descent. Such preliminary work can precede more discriminating study by some form of covariation analysis, but it cannot replace it.

ACKNOWLEDGMENTS

This research was supported in part by grants GM 11216 and GM 23007 from the National Institutes of Health.

LITERATURE CITED

1. Bernstein, S. C., L. H. Throckmorton, and J. L. Hubby. 1973. *Still more genetic variability in natural populations.* U.S. Natl. Acad. Sci., Proc. 70: 3928–3931.

2. Britten, R. J., and D. E. Kohne. 1968. *Repeated sequences in DNA.* Science 161: 529–540.

3. Brncic, D., P. S. Nair, and M. R. Wheeler. 1971. *Cytotaxonomič relationships within the mesophragmatica species group of Drosophila.* Univ. of Texas Publ. 7103: 1–16.

4. Cochrane, B. J. 1976. *Heat stability variants of esterase-6 in Drosophila melanogaster.* Nature 263: 131–132.

5. Coyne, J. A. 1976. *Lack of genic similarity between sibling species of Drosophila as revealed by varied techniques.* Genetics 84: 593–607.

6. Dayhoff, M. O. 1972. *Atlas of protein sequence and structure.* Natl. Biomed. Res. Found., Washington, D.C.

7. Farris, J. S. 1972. *Estimating phylogenetic trees from distance matrices.* Amer. Nat. 106: 645–668.

8. Fitch, W. M. 1976. *Molecular evolutionary clocks.* Pages 160–178 *in* F. J. Ayala, ed. *Molecular Evolution.* Sinauer, Sunderland, Mass.

9. Fitch, W. M., and E. Margoliash. 1967. *Construction of phylogenetic trees.* Science 155: 279–284.

10. Galau, G. A., M. E. Chamberlin, B. R. Hough, R. J. Britten, and E. H. Davidson. 1976. *Evolution of repetitive and nonrepetitive DNA.* Pages 200–224 *in* F. J. Ayala, ed. *Molecular Evolution.* Sinauer, Sunderland, Mass.

11. Goodman, M. 1976. *Protein sequences in phylogeny*. Pages 141-159 *in* F. J. Ayala, ed. *Molecular Evolution*. Sinauer, Sunderland, Mass.

12. Ho, D. Y.-K., E. M. Prager, A. C. Wilson, D. T. Osuga, and R. E. Feeney. 1976. *Penguin evolution: protein comparisons demonstrate phylogenetic relationships to flying aquatic birds*. J. Mol. Evol. 8: 271-282.

13. Hoyer, B. H., B. J. McCarthy, and E. T. Bolton. 1964. *A molecular approach to the systematics of higher organisms*. Science 144: 959-967.

14. Hubby, J. L., and L. H. Throckmorton. 1968. *Protein differences in Drosophila. IV. A study of sibling species*. Amer. Nat. 102: 193-205.

15. Kohne, D. E. 1970. *Evolution of higher-organism DNA*. Quart. Rev. Biophys. 3: 327-375.

16. Kohne, D. E., J. A. Chiscon, and B. H. Hoyer. 1972. *Evolution of primate DNA Sequences*. J. Hum. Evol. 1: 627-644.

17. Laird, C. D., and B. J. McCarthy. 1968. *Magnitude of interspecific nucleotide sequence variability in Drosophila*. Genetics 60: 303-322.

18. Nair, P. S., D. Brncic and K. Kojima. 1971. *Isozyme variations and evolutionary relationships in the mesophragmatica species group of Drosophila*. Univ. of Texas Publ. 7103: 17-28.

19. Nei, M. 1975. *Molecular population genetics and evolution*. North-Holland, Amsterdam, and Elsevier, New York. 288 pp.

20. Sarich, V. M., and A. C. Wilson. 1966. *Quantitative immunochemistry and the evolution of primate albumins: micro-complement fixation*. Science 154: 1563-1566.

21. Selander, R. K. 1976. *Genic variation in natural populations*. Pages 21-45 *in* F. J. Ayala, ed. *Molecular Evolution*. Sinauer, Sunderland, Mass.

22. Singh, R. S., J. L. Hubby, and L. H. Throckmorton. 1975. *The study of genic variation by electrophoretic and heat denaturation techniques at the octanol dehydrogenase locus in members of the Drosophila virilis group*. Genetics 80: 637-650.

23. Singh, R. S., R. C. Lewontin, and A. A. Felton. 1976. *Genetic heterogeneity within electrophoretic "alleles" of xanthine dehydrogenase in Drosophila pseudoobscura*. Genetics 84: 609-629.

24. Soule, M. 1976. *Allozyme variation: its determinants in space and time*. Pages 60-77 *in* F. J. Ayala, ed. *Molecular Evolution*. Sinauer, Sunderland, Mass.

25. Stone, W. S., W. C. Guest, and F. D. Wilson. 1960. *The evolutionary implications of the cytological polymorphism and phylogeny of the virilis group of Drosophila*. U. S. Natl. Acad. Sci., Proc. 46: 350-361.

26. Throckmorton, L. H. 1962. *The problem of phylogeny in the genus Drosophila*. Univ. of Texas Publ. 6205: 207-343.

27. Throckmorton, L. H. 1972. *Genetic change and speciation in the virilis group of Drosophila*. Presented at 14th Internatl. Congr. of Entomol., Canberra.

28. Throckmorton, L. H. 1972. *Abstracts,* 14th Internatl. Congr. of Entomol. (Canberra) Page 16 (Abstr.)

29. Throckmorton, L. H. 1975. *The phylogeny, ecology, and geography of Drosophila*. Pages 421-469 *in* R. C. King, ed. *Handbook of Genetics 3*. Plenum, New York.

30. Throckmorton, L. H. 1977. *Drosophila systematics and biochemical evolution*. Ann. Rev. Ecol. Syst. 8: 235-254.

31. Wilson, A. C., S. S. Carlson, and T. J. White. 1977. *Biochemical evolution*. Ann. Rev. Biochem. 46: 573-639.

14] Automated Identification of True and Surrogate Seeds

BY CHARLES R. GUNN* AND LEO LASOTA**

ABSTRACT

A user-computer dialogue has been established at the U. S. National Seed Herbarium to identify isolated true and surrogate seeds and fruits. Four major data processing programs are used to make identifications or to solve taxonomic problems. The compare program (CMPARE) compares taxa within a data matrix. A description for each species is generated by the describe program (DSCRBE). The identification program (IDENT4) enables unknowns to be identified, if sufficient data are in the data bank. A key program (KEY3) issues indented, dichotomous, diagnostic keys. Properly prepared documentation sheets facilitate the recording and numeric coding of diagnostic seed-fruit characters. These coded characters are data banked as dichotomous, multistate, or quantitative observations. Because the manual completion of these documentation sheets may be quite time consuming, systems are being considered to generate computer-compatible data automatically. Elements of one system are the scanning electron microscope (SEM) and an optical scanner with a direct computer feed. SEM photographs can be scanned optically to determine seed shape, size, and surface topography as well as the relationship of the hilum, lens, and aril (if present) to seed shape and size. Seeds of five species of *Ononis* are used to illustrate the advantages of using SEM photographs in interpreting seed coat topography. The computer is programmed to suggest characters for an identification, to recognize character and taxon weights, and to permit correction of user errors while an identification is in progress. Computer identification feasibility has been demonstrated through research and service identification with a completed data bank for seeds and fruits of North American species of the Papaveraceae. Data gathered from tribes and genera of the Fabaceae (Leguminosae) are now being processed.

*Plant Taxonomy Laboratory, Plant Genetics and Germplasm Institute, ARS, U. S. Department of Agriculture, Beltsville, Maryland 20705.
**Department of Horticulture, University of Maryland, College Park, Maryland 20742.
Scientific article number A2311. Contribution number 5312 of the Maryland Agricultural Experiment Station, Department of Horticulture.

INTRODUCTION

Comparative seed and surrogate-seed (seed-like fruit) morphology continues to be generally neglected by biosystematists. The lack of reliable morphological data has hampered the identification of isolated seeds and surrogate seeds and diminished their usefulness in considerations of vascular plant phylogeny. Although there are a few books and papers specifically designed to identify seeds, most vascular plant floras and monographs either ignore seeds or give them summary treatment. This is surprising for two reasons: First, seeds generally exhibit a conservative and stable morphology that often evinces familial characters; second, vascular plants are often transported and used (consumed or planted) in their seed stage, usually with no collateral plant parts to aid in their identification.

The lack of published data on seed morphology was often overcome by workers who accumulated unrecorded insights into seed identification. Obviously, such data are perishable, because the workers are mortal. A conversational (on line) computer provides a means of capturing and (with proper preparation) codifying these data without publication. By data banking accurate, clearly defined, and unprejudiced data, we can rapidly correct past inadequacies in our knowledge of seed morphology. Our short-range goal to bank relevant data and to use seed-fruit characters for mechanized identifications has been achieved. Our long-range goal to automate and standardize the seed-fruit identification process has yet to be attained.

A true seed is usually a fertilized mature ovule that may include an embryonic plant, stored food material (rarely missing), and a protective coat or coats (Fig. 14.1). Application of the term "seed" is seldom restricted to this morphologically accurate definition. Rather, "seed" is usually used in a functional sense, that is with reference to a unit of dissemination, a disseminule. In this sense, the term embraces dry, one-seeded (rarely two- to several-seeded) fruits or surrogate seeds (Fig. 14.1) as well as true seeds. A fruit is a mature ovary which may contain one or more seeds and may include accessory floral parts. In addition to seed-like fruits, there are multiseeded fleshy or dry fruits. The latter fruit type may be dehiscent or indehiscent. Hereinafter, both true and surrogate seeds will be referred to as seeds. Fruits will refer to either papaveraceous capsules or leguminous fruits (mainly legumes and loments).

CONVERSATIONAL COMPUTER SYSTEM

Gunn and Seldin (*10*) demonstrated how a conversational (on-line or time-sharing option) computer system including documentation sheets, concatenated data matrices, a character list, a remote terminal, and several taxonomically useful programs, may be used to identify isolated seeds and fruits. Seed-fruit morphological data gathered from the North American Papaveraceae formed the data base (*9*).

Figure 14.1. Diagrams of a surrogate seed (seed-like fruit) and a true seed.
(A) Longitudinal section of a maize (*Zea mays* L.) caryopsis (seed-like fruit);
(B) bean (*Phaseolus vulgaris* L.) seed, one of the two cotyledons removed;
(C) bean seed, showing micropyle, hilum, and lens.

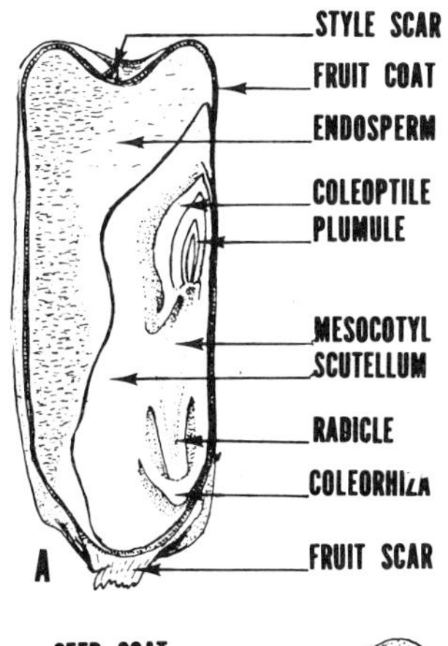

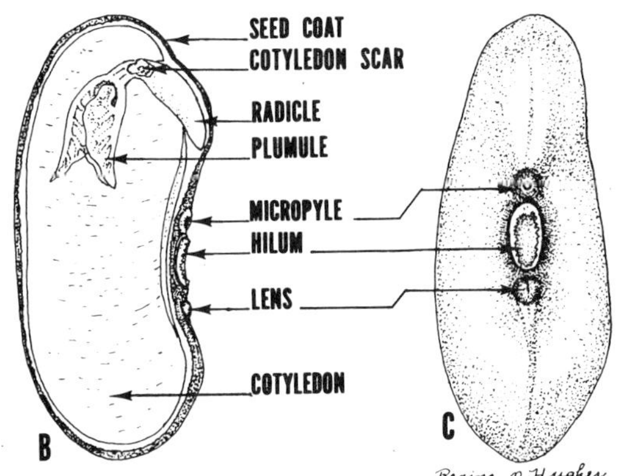

The computer used in the papaveraceous survey and to be used in the legume survey is the U. S. Department of Agriculture's International Business Machine 370/168* conversational computer. The interaction between user and computer may be accomplished by use of a remote terminal, such as the one located in the U. S. National Seed Herbarium, Beltsville, Maryland, which is connected to the computer by a telephone line. This permits a laboratory equipped with a remote terminal to have ready access to a distant computer, except during repair or heavy use. The procedures that we espouse offer few advantages to the laboratory without a terminal and without access to a conversational computer system.

A basic tenet is to assay correctly identified and properly vouchered seed

*Mention of a trademark or proprietary product does not constitute a guarantee or warranty of the product by the U.S. Department of Agriculture and does not imply its approval to the exclusion of other products that may also be suitable.

and fruit samples for characters amenable to taxon identification. In the papaveraceous study, identifications were made to species, whereas in the legume survey the identifications will be made to genus. The reasons for handling these data differently are based on the size of the families: Papaveraceae has 23 genera and 268 species, whereas Fabaceae has 717 genera and 17,600 species. A logical sequence of questions and observations for legume seeds and fruits was incorporated into a first generation documentation sheet. Selected questions and observations are shown in Fig. 14.2. We anticipate that a second generation of sheets will be prepared near the completion of the legume survey.

The basic concept is that data must be expressed numerically for this computer system to function. Once that is acknowledged the process of formulating the documentation sheets becomes routine. The taxonomic principles used in preparing and testing numerically answered questions are identical to those using word answers. Therefore, the preparation of documentation sheets is not an extra step or added burden in data processing. These numerical answers have no intrinsic meaning or value. The documentation sheets were prepared so that language answers are always visible. When answers are requested from the computer, programs use the character list to convert these numbers into their word equivalents, usually as they are recorded on the documentation sheets.

A prerequisite to a functional data bank is that numeric answers be filed by their categories: dichotomous characters (D), multistate characters (M), and quantitative characters (Q). The examples in Fig. 14.2 illustrate that D characters are answered by single-digit numbers; M characters are answered by double-digit numbers, 00, 01, 02, 04, 08, 16, 32 (a binary system, whose total cannot exceed 99, that permits intermediate answers such as 01 + 02 = 03 or 02 + 08 + 32 = 42, with these intermediate answers being created by only one sequence of binary numbers); and Q characters are true measurements that are recorded as two triple-digit numbers, a low and high range.

The last dichotomous character number is D 34 and the first multistate character is M 51. This hiatus permits a seventeen-character expansion of D characters. A similar hiatus occurs after the last M character. Q characters may be expanded by increasing the total number of characters.

Concatenated data matrices form a data bank, and the design of the data bank is described by Gunn and Seldin (*10*), Morse (*18*), and Morse *et al.* (*19*). The design must meet the requirements of the programs and data. It is important that each matrix to be concatenated have the same design, possess parallel data, be completely filled out, and share one character list.

One character list must be used for all concatenated matrices, because the same characters (D, M, and Q) are answered for all studied taxa. A character list is a dictionary for the matrices. Without a character list, program outputs would be cryptic numbers rather than words. Although the character list has as many entries as the documentation sheets and both contain identical data, they are distinct entities. Documentation sheets used during

FABACEOUS SEED-FRUIT SURVEY
C. R. GUNN, MARCH 1977

D QUESTIONS--1:
0 = unknown, 1 = yes, 2 = sometimes,
3 = no, 4 = inapplicable

001. Are seed lengths at right angles to fruit lengths?	0 1 2 3 4
008. Is endosperm present?	0 1 2 3 4
017. Are seeds contained in individual compartments of fruit?	0 1 2 3 4
034. Are valves coherent between seeds?	0 1 2 3 4

FABACEOUS SEED-FRUIT SURVEY
C. R. GUNN, MARCH 1977

M OBSERVATIONS--1:
00 = inapplicable, -9 = unknown
< = less, > = greater

052. Longest seed dimensions are
00 =
01 = < 01 mm
02 = 01- < 03 mm
04 = 03- < 05 mm
08 = 05- < 10 mm
16 = 10- < 20 mm
32 = 20- < 30 mm

053. Longest seed dimensions are
00 =
01 = 30- < 40 mm
02 = 40- < 50 mm
04 = 50- < 100 mm
08 = 100+ mm
16 =
32 =

063. Testa is
00 =
01 = monochromatic
02 = dichromatic
04 = trichromatic
08 = polychromatic
16 =
32 =

064. Testa dichromism caused by
00 =
01 = 2 colored areas
02 = mottles
04 = points
08 = streaks
16 =
32 =

143. Fruits are
00 =
01 = indehiscent
02 = tardily dehiscent
04 = dehiscent
08 = elastically dehiscent
16 = elastically revolute
32 = follicularly dehiscent

144. Fruits dehisce by
00 =
01 = transversely circular slit
02 =
04 =
08 =
16 =
32 =

154. Septa between seeds are
00 =
01 = chartaceous
02 = frass-like
04 = spongy
08 = hard
16 = double walled
32 =

155. Seed bearing sutures are
00 =
01 = not intruded
02 = + intruded
04 = intruded
08 =
16 =
32 =

FABACEOUS SEED-FRUIT SURVEY
C. R. GUNN, MARCH 1977

Q MEASUREMENTS--1:
000.000 = inapplicable
-99.999 = unknown

184. Seed length ________

185. Fruit length ________

Figure 14.2. Selected questions and observations from a modified (for this figure) first generation documentation sheet for scoring legume seed-fruit characters. The top of the first page of the documentation sheet is reproduced along with four dichotomous questions, a modified top of the first multistate observation page with four examples, and a modified top of the first quantitative observation page with the two observations.

data collecting are stored in the computer as a numeric data bank, whereas the character list is stored separately in the computer. When desirable, each data matrix may have its own character list. However, such matrices cannot be concatenated.

In preliminary studies we used two terminals not described here. The remote terminal used in this study was a Texas Instruments Silent 700 ASR Electronic Data Terminal (Fig. 14.3). This terminal has a keyboard similar to a typewriter keyboard. Use of this terminal may preclude the use of key-punched cards and batch processing. Another useful feature of this terminal is its cassette-tape capability. Data may be taped and then transmitted to the computer, or vice versa. The tapes may also be used to store data. The terminal is connected to the computer by a telephone line, by use of an Anderson-Jacobson acoustical coupler shown on the right-hand side of the terminal in Fig. 14.3. A telephone line permits the use of a terminal in a laboratory remote from a computer.

We used four of six main data processing programs designed and discussed by Morse (*18*) to solve seed-fruit identification problems. The four

Figure 14.3. A Texas Instruments Co. Silent 700 ASR Electronic Data Terminal with an Anderson-Jacobson acoustical coupler cradling the telephone receiver.

programs to be discussed separately are compare (CMPARE), describe (DSCRBE), identification (IDENT4), and key (KEY3). One of the two unused main data processing programs is CARDKEY. This program converts the data matrices to punch cards that may be used as field keys. The other program is INVERT, which groups species by characters, rather than characters by species as in the DSCRBE program. Several subprograms, discussed by Morse, may be used by the main programs. A service program (MXCHEK) is designed to check for format errors in a data matrix. During the operation of any program the weights assigned to the taxa and/or characters may be altered. If uniform numbers are assigned to the taxa and characters, weighting is inoperable.

A matrix checking program (MXCHEK), designed to locate and correct format errors, may be operated as a detailed (line by line printout) or as a summary (number of lines printed dependent on size of matrix) search. Printing stops when a format error is located. A data matrix rejected by MXCHEK is not accepted by any other program mentioned in this paper.

To ascertain which taxa within a matrix have similar or dissimilar characters, CMPARE offers three options that may be used separately or sequentially. In the diagnostic powers option, the taxa in the matrix are compared, the taxa ranked by number of differences, and the data summarized. The second option permits the user to compare a selected taxon with one, several, or all other taxa in the matrix. The third option contrasts pairs of species selected by the user.

Taxon descriptions may be obtained using the three options of DSCRBE: sequential, decreasing importance values, and selected characters. The sequential option gives the user a complete numeric or linguistic (if the character list is used) description of the taxa. The decreasing importance values option lists the most important characters for each taxa in decreasing importance. Any number of characters may be selected. If the characters are not weighted, the computer searches for unique answers first and then for characters shared by two taxa. This process continues until the number of characters selected is fulfilled. The third option permits the user to select the characters to be printed out.

The IDENT4 program is a specimen-identification program of exceptional value as a taxonomic tool, because it is designed to identify unknowns. IDENT4 has seven options including ones that list possibilities remaining; delete last character-input set; start new specimen, same matrix; request new matrix for new specimen; and recycle the program. Perhaps the most unusual option is the "delete last character-input set" option that permits the user to move from bottom to top of the matrix. This feature is seldom present in programs.

The key-constructing program constructs an indented dichotomous key that may be issued as words or numbers. The key output may be altered with the edit routine, especially changing character weights, while remaining within the KEY3 program. Coarse to fine keys may be requested on a basis of one for fine to 10 for coarse. Intermediate numbers may be used for corresponding intermediate keys, depending on the size of the matrix and the homogeneity of the data. The only improvement that we would make in the second statement of each couplet is to make it a positive statement rather than a negative one.

FABACEAE

The diversity of this large family is reflected in the several classifications that have been proposed. De Candolle (*5*) divided the Fabaceae into two suborders, the Curvembriae and the Rectembriae. His sytem, which separated the suborders on the basis of curved or straight embryos, foreshadowed the importance that seed characters would have on legume systematics. Bentham and Hooker (*1*) created a third suborder by dividing de Candolle's Rectembriae into the Caesalpineae (Suborder II) and the Mimosae (Suborder III). Taubert (*23*), retaining this tripartite division, considered the suborders to be subfamilies. In the century following 1862, the number of described legume genera nearly doubled. Hutchinson (*14*) offered a pragmatic solution to separating this "large and very natural (homogeneous) assemblage" by elevating Taubert's subfamilies to families comprising the order Leguminales: Caesalpiniaceae, Mimosaceae, and Fabaceae (Papilionaceae).

Seed characters support a return to the concept of one family (the Fabaceae) as advocated by de Candolle over 150 years ago. The entire family is characterized by seed coats that contain Malpighian and osteosclereid cells (*3, 4, 20, 21*). Malpighian cells develop from the outer epidermis of the outer integument and constitute the palisade layer. They are more or less hexagonal in cross section, contiguous, and thick-walled. Osteosclereids, also known as hour-glass cells, occur adjacent to the inner surface of the palisade layer. Separated by wide intercellular spaces, the hour-glass cells comprise either the hypodermal layer or the inner epidermis of the outer integument. As the term implies, an hour-glass cell is narrowed in the middle. Hour-glass cells may be rudimentary or absent in seeds of some legume species.

De Candolle's bipartite subdivision of Fabaceae is also supported by seed characters (*6*). Hilum structure, ovule form (anatropous or campylotropous), radical length, vascularization, and general shape of seed and seed parts unite the Mimosoideae with the Caesalpinioideae as a group distinct from the Faboideae (*7, 15*). The lack of the free amino acid, canavanine, in seeds of the mimosoid-caesalpinioid types and its occurrence in most tribes of the Faboideae also reinforces the validity of the broad Curvembriae-Rectembriae subdivisions of de Candolle. The weight of evidence may cause some species to be assigned to new affiliations, while the exact phylogenetic position of such intermediate tribes as the Swartzieae remains uncertain. However, the features that support a basic bipartite division of the Fabaceae appear stable.

The basic subdivisions of de Candolle permit assignment of an unknown legume seed into either a group of thousands or a group of hundreds. Further reduction necessitates more diagnostic characters. The number of discrete characters is not limitless, but the scanning electron miscroscope (SEM) permits a closer approach to the limit by confirming the presence or absence of characters that are not readily discerned by light microscopy. Quantitative measurements are also readily made with the SEM. Moreover, the output is an electrical signal, relating the SEM more closely to the language of a computer than manually recorded descriptors from a stereoscopic microscope are related.

We have developed a data bank to accept such information and would welcome input on SEM seed-fruit research. However, the relative lack of legume seed-fruit studies using SEM (*2*) indicates that the instrument is either an evolving species or an endangered species. We trust that the latter is not the case. The suggestions that follow are based on SEM analysis of seeds from the larger of de Candolle's two groups, the Faboideae (Curvembriae).

Those seed traits that have helped to define the legume subdivisions are of equal importance for seed identification. One character alone can reduce the odds for the correct identification of a legume seed from one in about 17,600 species to one in either 5,600 or 12,000. This gain is paralleled by the not-so-overwhelming reduction of from one in about 717 genera to one in 235 or 482. Combinations of characters decrease further the number of possibilities. If species matrices were created, species could be identified. However, the 17,600 species would strain the value of this operation, especially with the presence of a major seed collection (*8*). Computer assistance yields to herbarium reliance only as fast as the choices of databanked characters are exhausted.

SCANNING ELECTRON MICROSCOPY

Seed preparation. Seeds were obtained from either herbarium sheets or identified seed samples. In the latter case, identification was confirmed by SEM and light microscope comparison of non-vouchered and vouchered seeds. The herbarium symbols cited in Figs. 14.4 to 14.6 are defined in Holmgren and Keuken (*13*).

For the SEM study, seeds of at least two samples for each species were examined. Seeds were cleaned for 10 seconds in absolute ethanol by either a Cole-Palmer Model 8845-3 Ultrasonic Cleaner or by rinsing. Other samples received no pretreatment. Seeds were attached by silver print paint to aluminum specimen stubs and then stored in a desiccator for at least 24 hours before coating.

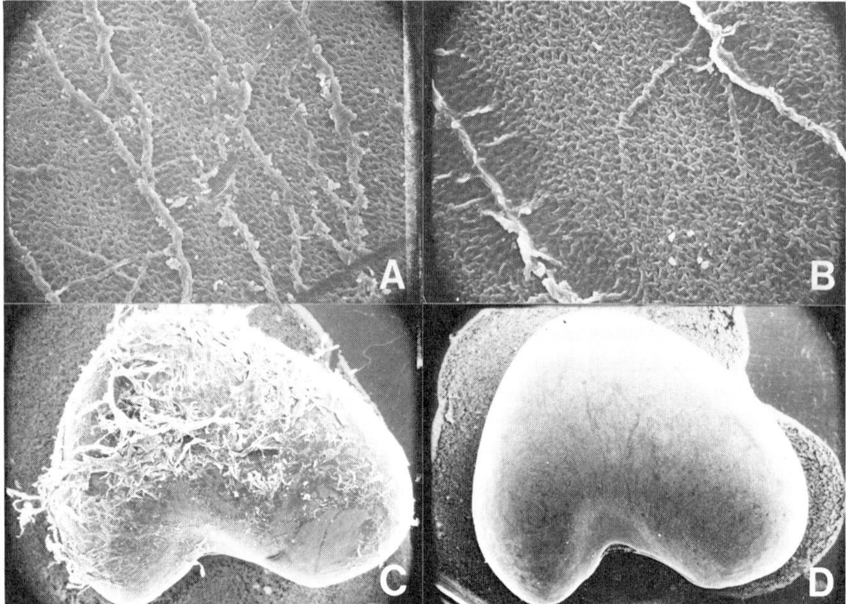

Figure 14.4. Seed cleaning for scanning electron microscopy. Seeds (A) and (C) were not cleaned; seeds (B) and (D) were treated for 10 seconds in absolute ethanol by being ultrasonically cleaned (B) or rinsed (D). (A) and (B) *Robinia pseudoacacia* L. (United States: L. LaSota 123, MARY), 200 ×; (C) and (D) *Clianthus puniceus* Lindley (New Zealand: Millan PQ 61226, BARC), 15 ×.

Artifacts present a major problem for SEM analysis. Ideally, seed should be removed with forceps from fruits because, once handled, seeds may accumulate surface films and attendant particulate matter. Both types of artifacts obscure surface detail. The seed in Fig. 14.4A was transferred between thumb and forefinger from fruit to specimen stub. The seed appears to be lubricated from the residue of this transfer. The seed in Fig. 14.4B was from the same fruit and similarly handled, but the specimen was rinsed for 10 seconds in absolute ethanol before transfer by forceps to a specimen stub. The *Clianthus puniceus* Lindley seed in Fig. 14.4C was manipulated entirely by forceps. The artifacts in this instance are from seed contact with the hairy interior of the fruit. Rinsing of other seeds from the same fruit did not remove the hairs. However, ultrasonic cleaning for 10 seconds in absolute ethanol did effectively clean these seeds (Fig. 14.4D). If topography or shape analysis is to be based on SEM information, the removal of fruit hairs and other obfuscating matter is essential.

For comparison, untreated seeds were mounted on the same stubs as seeds cleaned with ethanol. No differences in the seed coat pattern were detected. Untreated seeds from most fruits with smooth interior walls were essentially identical to cleaned seeds from the same source. Water and detergent were not as effective as ethanol. Cleaning agents such as acetone or toluene cannot be recommended because they tended to deposit more residue than they removed. Cracks did develop in coats of some seeds treated with ethanol. These cracks were probably caused by rapid ethanolic penetration through natural seed openings, because the extent of cracking was comparable in ultrasonically cleaned and rinsed seeds. Cracking neither obscured nor confused topographic pattern, and thus was not considered to outweigh the merits of a rapid preparation technique for service identification. Moreover, at the magnifications used intact areas were always available.

Coating with gold-palladium was done in a Technics Hummer D. C. Sputtering Coater. Chamber pressure was reduced to 20 millitorrs and then flushed five times with argon before chamber stabilization at 100 millitorrs. Coating time was two minutes at an operating potential of 1600–2000 volts and at a current of 10 milliamperes. Although measurements were not made of its thickness, the gold-palladium coat was not deemed excessive for the desired magnifications. Resolution was not improved by operating the Hummer at lower voltages or for shorter coating times, but charging did increase with these changes. At higher magnifications or with different biological material (*17*), artifacts of thick conductive coatings could be both more apparent and more critical than with our relatively low magnification study of legume seeds.

Scanning procedure. Seeds were scanned in a Hitachi SEM, model HHS2R. Accelerating voltage was 10KV, final aperture size 200 μm, and a working distance to specimen of 30 mm. All seeds were scanned at magnifications of 500 X and 1500 X. For some seeds, other magnifications also were used. SEM photographs are reduced by a factor of two in this publication.

All seeds were scanned in side or face view (Fig. 14.4D), because surface patterns were generally uniform in this area. Patterns were much more irregular along the edges of seeds. This increased randomization in topography created what were considered insoluble replication problems for seeds scanned in that plane. Thus, it is recommended that seed coat topography be determined and defined as the repetitive pattern(s) occurring within the area of the seed face. Structural elements, such as the micropyle or hilum, are excluded from this definition, but they could also be scanned for topographic comparisons of specific seed parts.

Magnifications of 500 X and 1500 X seemed the best compromise as standards for all legume seeds. Recurrent, but widely separated, features were occasionally excluded from view at magnifications in excess of 500 X.

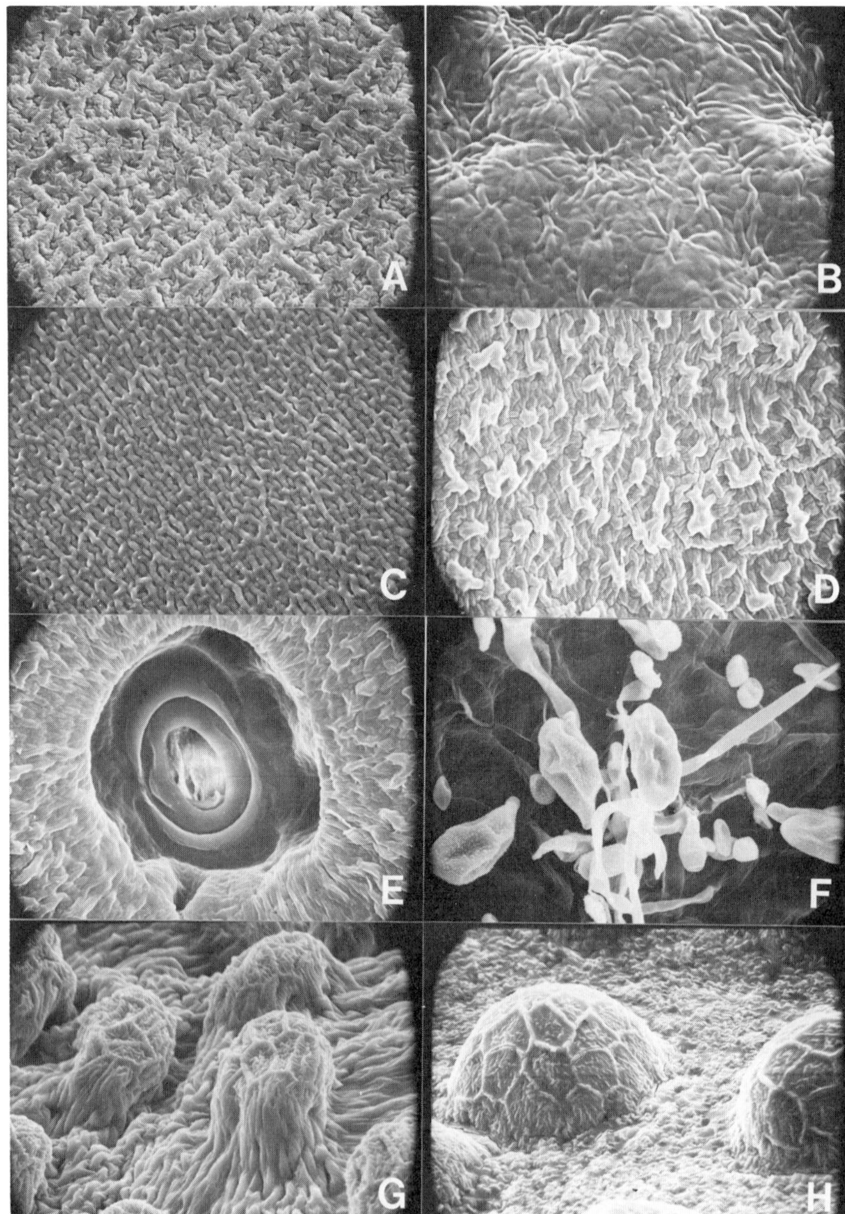
A
B
C
D
E
F
G
H

Similarly, at magnifications above 1500 X the consistent patterns of a species often merged into the variation found among different areas of the same seed.

Seed topography. Forty-one species representing 16 genera were examined. Seed topography permitted consistent separation of all but four of the genera.

Figures 14.5A through 14.5C illustrate the pattern and individuality of three species that would all be described as "smooth" if viewed under a dissecting microscope. Figure 14.5D shows the general seed coat topography and Fig. 14.5E illustrates the remarkable pore-like structures that easily separate *Olneya tesota* A. Gray, a monotypic genus, from other studied species. Figure 14.5F, the surface of a legume fruit, suggests yet another area where the SEM could be used to identify legumes. Morphological detail or even the presence of species specific pathogens on legume fruits could be taxonomically significant.

Regardless of seed size, specimens examined within and among samples of the same species were alike in surface topography. A shrivelled, obviously immature, seed of *Ononis sicula* Gussone (Fig. 14.5G) differed from a mature seed (Fig. 14.5H) in the same sample by degree but not by type of patterning. This sample consistency was generally observed with other species studied. However, pending examination of the extent of variation due to seed maturity in many more species, it is suggested that descriptions of seed topography be broad enough to accommodate pattern variability and be based on fully ripened seeds (*11, 12*).

Seed tuberculation, as occurs in *O. sicula*, is a common character of the genus *Ononis*. Many species, such as *O. spinosa* L., appear granular to the unaided eye and can be definitely so described when viewed through a dissecting microscope. Thus, the pre-SEM monograph by Širjaev (*22*), which describes 67 species, classes approximately half the genera as having tuberculate seeds. The related term granulata also occurs frequently as a seed surface description.

Figure 14.5. Surface topography of seeds and internal surface topography of a legume (fruit). Specimens (A) through (F) were not cleaned; (G) and (H) were cleaned for 10 seconds in absolute ethanol in an ultrasonic cleaner. (A) through (E), (G), and (H) are seeds; (F) is a legume inner surface. (A) *Colutea orientalis* Miller (Netherlands: F. G. Meyer 6249, NA); (B) *Sophora microphylla* Aiton (England: R. Polhill 8.X.76, BARC); (C) *Pueraria lobata* (Willdenow) Ohwi (China: Tak & Chow 3067, NA); (D) and (E) *Olneya tesota* A. Gray (United States: R. H. Peebles 7909, NA); (F) *Anthyllis vulneraria* L. (England: R. Polhill 15.X.76, BARC); and (G) and (H) *Ononis sicula* Gussone (Iraq: Gillett & Rawi 2.IV.47, US). All 750 ×.

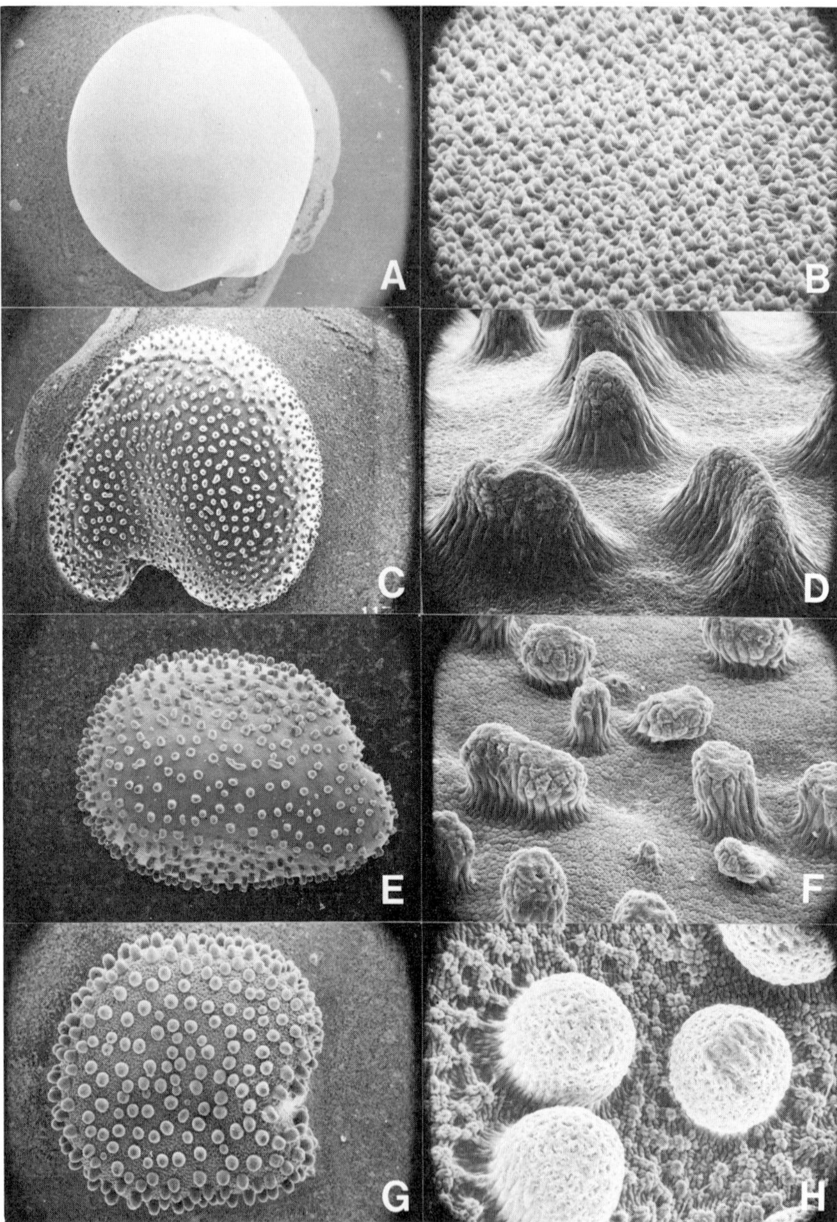

Figure 14.6. Seeds of *Ononis* spp. cleaned for 10 seconds with absolute ethanol in an ultrasonic cleaner. (A) and (B) *O. pubescens* L. (Morocco: A. Faure 18.VI.31, US) 13 × and 750 ×; (C) and (D) *O. spinosa* L. (England: R. Polhill 15.X.76, BARC) 13 × and 250 ×; (E) and (F) *O. oligophylla* Tenore (Italy: N. Guzzino VIII.89, US) 25 × and 250 ×; and (G) and (H) *O. ornithopodioides* L. (Italy: P. Savi, US) 30 × and 250 ×.

Of five *Ononis* species scanned, all could be characterized as tuberculate. Included in this number was *O. pubescens* L., a species that is essentially smooth under the light microscope and has been so described (*16*). With the SEM this species has a distinctly tuberculate surface at 1500 X. A key, based on SEM photographs at 500 X and 1500 X, may be used to identify seeds of five species of *Ononis*. With the instruments then available, Širjaev was able to discern differences and described seeds of these species as: *O. pubescens*, nigra-punctata; *O. oligophylla* Tenore, minute micronulata; *O. spinosa*, tuberculata; *O. sicula*, tuberculata; and *O. ornithopodioides* L., acute tuberculata. The increased depth of field at high magnifications of the SEM permits a more discrete separation.

1. Tubercles dense, occurring with frequency greater than 1/cm^2 at 1500 X .... *O. pubescens*, Fig. 14.6B
1. Tubercles sparse, occurring with frequency less than 1/cm^2 at 1500 X.
   2. Tubercles coalesced into ridges at a frequency of greater than 10 percent at 500 X or less.
      3. Ridges capitate .... *O. oligophylla*, Fig. 14.6F
      3. Ridges prismatic .... *O. spinosa*, Fig. 14.6D
   2. Tubercles coalesced into ridges at frequency of less than 10 percent at 500 X or less.
      4. Tubercles ribbed .... *O. sicula*, Fig. 14.5H
      4. Tubercles ruminate .... *O. ornithopodioides*, Fig. 14.6H

Images from the SEM could be handled in several ways. An on-line terminal adjacent to the SEM monitor would permit a user-computer dialogue while the specimen was being scanned. Through the use of videotape, simultaneous or subsequent remote study would be possible. To reduce bias and to increase the quantitative information extracted from the SEM, micrographs could be traced onto a coordinate digitizer or rotated on a drum and digitized by an optical scanner. The latter method permits a high degree of resolution while increasing the automation of the system. Finally, the SEM signal could be interfaced, without a photographic intermediary, directly to the computer. This would require sophisticated component interaction in such aspects as raster generation and control of signal intensity. A direct feed system has been shown to be practical in other disciplines (personal communication from R. H. Foote). Regardless of input, techniques such as Fourier series expansion could be used to compare an unknown with holograms stored in the computer.

CONCLUSIONS

Seeds and one-seeded fruits often exhibit stable external and internal morphology that has seldom been used to its fullest extent even though seeds are a major unit of dissemination and use. As a means of overcoming the lack of published data, a conversational computer has been programmed and data have been banked to process service identifications and research data. This system is functional. However, the manual completion of documentation sheets, which are the source of the data matrices, may be partially or nearly completely abridged by automation. One approach is to use SEM output and an optical scanner which feeds data to a conversational computer. Seed characters reducible to number, size, shape, and relation of parts are candidates for this type of automation. With appropriate computer interfacing, data can be recorded automatically. We plan to develop this type of system because 1) new seed topography data will be available, 2) excellent illustrations of seeds (SEM photographs) will be available, 3) data banking will achieve a higher degree of accuracy and be more rapid, and 4) descriptors will be clearly defined and uniformly applied. The benefits of a SEM-computer interface appear to outweigh the cost of equipment and seed processing time.

ACKNOWLEDGMENTS

We appreciate the support of Larry Morse, computer programmer at Harvard University, and Margaret J. Seldin, U.S. Department of Agriculture computer programmer, and her colleagues in the Data Systems Application Division of the Agricultural Research Service. We also thank William Wergin and Bruce Engber of the Department of Agriculture's Reproduction Laboratory and Gene Taylor of the University of Maryland Institute for Physical Science and Technology for their assistance with the scanning electron microscopy.

LITERATURE CITED

1. Benthan, G., and J. D. Hooker. 1865. *Genera plantarum*, Vol. 1, pp. 434–600. Reeve and Co., London.

2. Brisson, J. D., and R. L. Peterson. 1976. *A critical review of the use of scanning electron microscopy in the study of the seed coat.* Workshop on Plant Sci. Applications of SEM, Proc. (I. I. T. Research Institute, Chicago, April, 1976) 2: 477–495.

3. Corner, E. J. H. 1951. *The leguminous seed.* Phytomorphology 1: 117–150.

4. Corner, E. J. H. 1976. *The seeds of dicotyledons.* Vol. 1, pp. 161–173. Cambridge University Press, Cambridge, England.

5. de Candolle, A. 1825. *Mémoires sur la famille des Légumineuses.* 525 pp., 70 plates. A. Berlin, Paris.

6. El-Gazzar, A., and M. A. El-Fiki. 1977. *The main subdivisions of Leguminosae.* Bot. Notiser 129: 371–375.

7. Gunn, C. R. 1972. *Seed characters.* Pages 55–143 (Chapter 2) *in* T. T. Kozlowski, ed. *Seed Biology. Vol. 3.* Academic Press, New York.

8. Gunn, C. R. (chairman), 1977. *Systematic collections of the Agricultural Research Service.* U. S. Dept. Agric. Misc. Pub. 1343. 84 pp.

9. Gunn, C. R., and M. J. Seldin. 1976. *Seeds and fruits of North American Papaveraceae.* U. S. Dept. Agric. Tech. Bull. 1517. 96 pp.

10. Gunn, C. R., and M. J. Seldin. 1977. *Preparing seed and fruit characters for a computer with discussion of five useful programs.* Seed Sci. and Tech. 5: 1–40.

11. Gutterman, Y. 1973. *Differences in the progeny due to daylength and hormone treatment of the mother plant.* Pages 59–80 (Chapter 4) *in* W. Heydecker, ed. *Seed Ecology.* Pennsylvania State University Press, University Park.

12. Gutterman, Y., and W. Heydecker. 1973. *Studies of the surfaces of desert plant seeds. 1. Effect of day length upon maturation of the seedcoat of Ononis sicula Guss.* Ann. Bot. 37: 1049-1050.

13. Holmgren, P. K., and W. Keuken. 1974. *Index herbariorium.* Part 1, ed. 6. Regnum Veg. 92: 1-397.

14. Hutchinson, J. 1964. *The genera of flowering plants.* Dicotyledons. Vol. 1, pp. 221-489. Clarendon Press, Oxford.

15. Isely, D. 1955. *Observations on seeds of the Leguminosae: Mimosoideae and Caesalpinioideae.* Iowa Acad. Sci., Proc. 62: 142-145.

16. Ivimey-Cook, R. B. *The phenetic relationships between species of Ononis.* Pages 69-90 *in* A. J. Cole, ed. *Numerical Taxonomy.* Academic Press, London.

17. Leuenberger, B. E. 1976. *Die Pollenmorphologie der Cactaceae.* Dissertationes Botanicae 31. J. Cramer, Vaduz.

18. Morse, L. E. 1974. *Computer programs for specimen identification, key construction and description printing using taxonomic data matrices.* Publs. Mus. Michigan State Univ., Biol. Ser. 5: 1-128.

19. Morse, L. E., J. A. Peters, and P. B. Hamel. 1971. *A general data format for summarizing taxonomic information.* BioScience 21: 174-180.

20. Pammel, L. H. 1899. *Anatomical characters of the seeds of Leguminosae, chiefly genera of Gray's manual.* Acad. Sci. St. Louis, Trans. 9: 91-275.

21. Pitot, A. 1935. *Le développement du tégument des graines des Légumineuses.* Bull. Soc. Bot. France 82: 311-314.

22. Širjaev, G. 1932. *Generis Ononis L. revisio critica.* Beih. Bot. Centr. 49: 381-668.

23. Taubert, P. 1894. *Leguminosae.* Pages 70-388 *in* A. Engler and K. Prantl, eds. *Die Naturlichen Pflanzenfamilien, Bd. 3, Teil 2.* Wilhelm Engelmann, Leipzig.

five

APPLICATIONS OF TAXONOMIC DATA

15] Museums, Museum Data, and the Real World—The Taxonomic Connection

by RICHARD H. FOOTE*

ABSTRACT

Traditionally research in taxonomy has been conducted without regard for the many ways in which it is useful for the control of economically important pests, studies of the environment, and other practical matters. Accurate and precise identifications are indispensable to the conduct of such programs, but the research that makes these determinations possible must take place in museums that offer adequate reference collections, catalogs, and libraries as working tools. These activities, identification and research, must go hand in hand for two reasons—1) the taxonomist delivers information much more effectively to the users if he is thoroughly familiar with the action programs involved, and 2) the user often requires identification-associated information (overall distribution, biology, economic importance, etc.) that the taxonomist is not likely to deliver on a routine basis. The taxonomist, in complete command of his resources, is in a unique position to conduct inventories of the groups on which he specializes. Such inventories will eventually form the backbone of comprehensive information systems that will be used extensively by those engaged in action programs and by museum personnel alike. The resources of museums, devoted perforce mostly to museum objectives, must be augmented in many ways before the full potentials of museums can be brought to bear on problems of the non-museum world, and such augmentation should be provided by those organizations involved most closely with action programs.

INTRODUCTION

The science of systematics, or taxonomy, has been pursued for a very long time by practitioners whose concern for the economic applications of their

*Systematic Entomology Laboratory, IIBIII, Agricultural Research Service, U.S. Department of Agriculture, c/o U.S. National Museum, Washington, D.C. 20560.

work (i.e., the "real world")* has been minimal. They have worked at systematizing knowledge about plants and animals because the classification of living things was interesting for its own sake rather than because there was any practical application of their work outside the realm of systematics. With few exceptions these systematists have restricted themselves to a "museum world," and understandably so, because a museum is the only organization properly equipped to house and care for a collection of any group large enough to support studies in systematics, and because the financial support for systematics has been more or less readily available from museum-centered sources.

At the same time, it has become continually more apparent within the past decade or two that other disciplines of biology have been supported to a much greater extent than has systematics. For example, increasing public awareness that we are irreversibly polluting our environment has resulted in the initiation of ecological studies on a grand scale. New developments in pesticides, coupled with our need to know how to use them more efficiently because of the adverse environmental side effects of some, and the increased costs associated with their use, have induced the expenditures of large sums of money for chemical research and development. Outstanding successes in the biological control of some of our economically important insect and weed pests have been responsible for a large amount of research leading to the search for, and the study and introduction of, many parasites, predators, and pathogens. Also, much attention is being paid to developing integrated control methods, which require input from many fields of biology.

However, until quite recently the size and popularity of these and other programs, produced at least in part by political and economic pressures as well as by the tendency of many systematists to ignore the economic considerations of their work, have caused us to lose sight of this essential fact: *The results of research in systematics—i.e., a strong classification reflecting natural relationships coupled with means for identifying organisms with precision—are essential to the conduct of economically oriented day-to-day activities outside the "museum world."*

A review of relatively recent literature devoted to systematics as vital to the outcome of "real-world" programs reveals that fewer than half of the approximately 40 papers were written before 1968. An awakening need for systematics in support of "real-world" programs is shown by the appearance every year of more written words in its defense. The general need for systematics has been noted by an impressive number of authors (Darlington, *6*; Denmark, *8*; Essig, *10*; Evans, *11*; Frison, *12*; Gahan, *13*; Kim, *17, 18*;

*The terms "museum world" and "real world," as used throughout this paper, are convenient tags to denote two different arenas of scientific activities. However, the use of the term "real world" for extra-museum activities is not intended to imply that the museum milieu is other than "real" or useful.

Kosztarab, *19*; Mahmood, *22*; Murphy, *26*; Oman, *28, 29*; Pechuman, *30*; Robinson, *32*; Sabrosky, *37, 38*; Slater, *42*; and Wilson, *46, 47*), who have argued its value in supporting biological control programs (Biliotti, *4*; Clausen, *5*; Labeyrie, *21*; Rosen and DeBach, *34*; Sabrosky, *36*; Sailer, *40*; and Schlinger and Doutt, *41*), in reinforcing numerous studies in ecology and the environment (Allen, *1*; Davidson, *7*; Grotta, *14*; Hendrickson, *15*; Heywood, *16*; Munroe, *25*; Resh and Unziker,*31*; Sabrosky, *35*; Sailer, *39*; Wiggins, *45*; and Wilson, *46*), in increasing the effectiveness of agricultural quarantines (Muesebeck, *23*), in furthering studies in veterinary entomology (Becklund, *3*), and in promoting research on mosquito-borne disease and disease control (Eldridge, *9*; Oman, *27*; Rogers, *33*).

These papers have a significant commonality: Whether an author discusses this subject in general terms or in relation to a specific discipline, in every case his thesis depends heavily on the need, although often only implied, for reliable field studies or surveys. In turn, the field study or survey cannot possibly succeed in reliability without having as a firm base the accurate identification of the organisms involved. And the achievement of accurate identifications in any well-known group is impossible unless a taxonomic authority with adequate working tools (an extensive, properly identified reference collection and an adequate library) is available. Because some "real-world" activities are so closely dependent upon museum services, then in a very real way those individuals who perform the services must be considered as museum users. In practice, taxonomic specialists either work in a museum or museum-like setting or depend on museum services. Hence the link between the "real world" and the museum is indispensable.

A CASE IN POINT: THREE AGENCIES

In this paper I attempt to illustrate more precisely how museums, museum-related data, and systematic entomologists working in museums, are indispensable in supporting some of the programs I have mentioned—programs the activities and objectives of which appear on the surface to be far removed from those of the museum world. As an example, two organizations within the U.S. Department of Agriculture (USDA) and their relationships to the Smithsonian Institution should clearly demonstrate the "taxonomic connection." In addition, I hope that this example will: (1) lead us to re-examine the role that museums play in the conduct of "real-world" activities; and (2) thereby help establish a firmer economic justification for the existence of modern natural history collections, the data associated with them, and the activities of scientists working in them.

As my example of the museum, I have selected the Smithsonian Institution's United States National Museum of Natural History (USNMNH), not only because of its size and eminence, but also because it houses, as one of its users, the Systematic Entomology Laboratory (SEL) of the USDA's

Agricultural Research Service (ARS). The SEL makes available the results of its taxonomic research and service programs for the support of a host of organizations within and outside the U.S. government. Among these is the USDA's Animal and Plant Health Inspection Service (APHIS). SEL, through APHIS, is deeply involved in American agriculture and to a very real extent the two are mutually dependent. Neither could conduct its most important programs without the other, and neither could manage without access to well identified, well maintained natural history collections, which are the sources of readily available data. Brief descriptions of the composition and activities of these three organizations—USNMNH, SEL, and APHIS—are given below. Lines of communication among them are illustrated in Fig. 15.1.

The United States National Museum of Natural History (USNMNH). Of the nine subject-matter departments in USNMNH, the Entomology Department has ultimate responsibility for one of the largest insect collections in the world: some 25 million specimens. The Department employs 13 full-time scientists who undertake taxonomic research on insect groups lying within their special interests and expertise. This staff is supplemented by more than 20 research associates, collaborators, and affiliated scientists and has a large support staff.

The taxonomic research undertaken by the Smithsonian Institution staff is not mission-oriented: typically the researcher seeks to fill gaps in scientific knowledge wherever he or she finds them; however, Entomology Department scientists do respond to requests for information, some of which may be directly relevant to the solution of economic problems. An important responsibility, for example, is the maintenance of the National Collection of Insects, which is depended upon heavily as a working tool by USDA and Smithsonian Institution scientists. A history of the Entomology Department was published in 1963 (*2*) and a rather complete description by Ward *et al.* appeared in 1976 (*44*).

The Systematic Entomology Laboratory, Agricultural Research Service, United States Department of Agriculture (SEL, ARS, USDA). Intermediary between the Smithsonian Institution's natural history collection in the USNMNH and APHIS is the Systematic Entomology Laboratory of the USDA's Agricultural Reseach Service (Fig. 15.1). The Smithsonian Institution provides space for the research and service activities of 25 of the Laboratory's 29 scientists. The Laboratory is supported entirely by funds from ARS and APHIS of the USDA; thus its mission-orientation is tied quite closely to Department objectives in contrast to the orientation of the Museum's Entomology Department. This mission-orientation requires that the staff engage in taxonomic research related primarily to agriculturally important insects and mites, and that it deliver identifications and

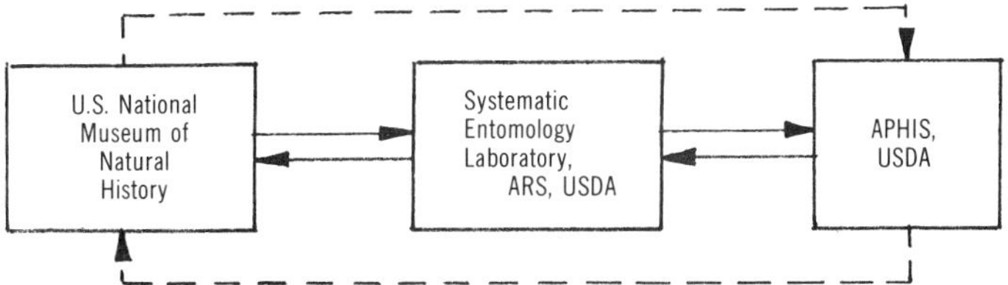

Figure 15.1. Direct (solid) and indirect (dotted) lines of communication among U. S. National Museum of Natural History (USNMNH), Systematic Entomology Laboratory (SEL), and Animal and Plant Health Inspection Service (APHIS), illustrating the central role played by SEL in the exchange of information by these organizations.

provide a wide variety of other services more or less "on demand" to support agricultural programs of U.S. and foreign organizations and individuals. For example, in the calendar year 1976, SEL scientists made 48,753 identifications of 278,240 insect specimens. Because the pursuit of taxonomic research requires the presence of a reference collection, library, etc., as pointed out earlier, each SEL taxonomist can be regarded as a prime user of the Smithsonian Institution's National Collection of Insects. The origins, growth, and principal activities of SEL were well described by Sabrosky (*37*) and more recently by Kim (*17*).

The Animal and Plant Health Inspection Service, United States Department of Agriculture (APHIS, USDA). The mission of APHIS is to detect and control agriculture's most important pests and pestilences, and to exclude from the U.S. those that might be introduced.* Among its many Plant Protection and Quarantine programs are six that are closely associated with entomological museums and these museums' taxonomic expertise. Three of the programs relate to pest survey and control. These are:

(1) Cooperating with various state agriculture departments in furnishing funds, personnel, and equipment to combat five or six of the nation's most important insect pests.

(2) Supporting a widespread domestic insect survey in cooperation with state level departments of agriculture and with certain universities.

(3) Compiling and issuing a weekly status report on insect conditions throughout the country. This newsletter, the *Cooperative Plant Pest*

*Many of the activities of the Plant Protection and Quarantine programs of APHIS were described in a most interesting fashion in an unpublished talk entitled "Quarantine Entomology and its Relationships with Taxonomic Research and Services," presented by Dr. Maynard J. Ramsey at the Eastern Branch Meeting, Entomological Society of America, 1974.

Table 15.1 Significance of programs of the Animal and Plant Health Inspection Service (APHIS), U.S. Department of Agriculture in insect regulation and detection.

APHIS program	(i) What is the extent of problem?	(ii) What information is needed?	(iii) Who needs the information?	(iv) What is best source of the information?	(v) To what extent are taxonomic identification services needed?
Domestic Surveys and Pest Control:					
1) *Program Pests*	5–6 species of primary importance	Extent of spread into new areas (new county records, etc.); extent of damage; effectiveness of control	Control personnel	Survey personnel; field and published records	Verification of specimens from new localities (program pest species only)
2) *Surveys*	All insects	Type and extent of damage; seasonal population fluctuations	Survey and control personnel	Survey personnel; records	Precise (species level) to general (e.g., "grasshoppers")
3) *Cooperative Plant Pest Report (CPPR)*	Up to 10,000 species affecting man & his activities	Type and extent of damage; seasonal population fluctuations	Survey and control personnel	Survey personnel; records	Precise (species level) to general (e.g., "grasshoppers")
Quarantines:					
4) *Interceptions at ports of entry*	Exotic insects worldwide	Point of origin; economic importance at point of origin; biology at point of origin	Customs and USDA quarantine officials	Literature; previous records of importation; specimen data	As precise as possible, especially if insect is known to be important at source
5) *New Pest Detection*	Exotic insects worldwide	Home range; economic importance in home range; biology in home range	Control and survey personnel	Literature; previous records of importation; specimen data	As precise as possible, especially if insect is known to be important at source, and indication of closely related species
6) *Insects Not Known to Occur in the U.S. (INKTO)*	300–400 exotic insects of economic importance worldwide	Home range; economic importance in home range; biology (all aspects) in home range; means for identification	Control and survey personnel	Literature; specimen data	Precise identification (to species level) and indication of closely related species

Report (CPPR) (formerly the *Cooperative Economic Insect Report*) (CEIR), is widely acknowledged as the country's foremost means of forecasting entomological trouble spots in the nation's agricultural economy.

The other three programs are related to quarantines. These are:

(4) Maintaining agricultural quarantines at more than 80 sea, air, and border ports of entry throughout the United States. APHIS works closely with the U.S. Customs Service in prohibiting entry of agriculturally important pests or plant diseases that might become established within the United States.

(5) Conducting surveys, especially in areas surrounding the principal ports of entry into the U.S., in an effort to detect pests that might have escaped quarantines and gained an initial foothold in this country.

(6) Issuing and circulating widely a series of publications entitled "Insects Not Known To Occur in the United States" (INKTO) to facilitate the recognition, accurate identification, and further treatment of insects intercepted at quarantine or detected as newly established.

All of these programs are continually evaluated to improve their effectiveness, and all are subject to expansion or other changes to meet the continually changing needs of American agriculture.

Table 15.1 summarizes the various activities and information needs of individuals working in these six APHIS programs.

APHIS' own in-house programs and a corps of 18 area identifiers permit APHIS to meet its simpler needs for taxonomic services: the identification of five or six important insect pests (Table 15.1, program 1) and a part of the more difficult identification of arthropods intercepted at quarantine (Table 15.1, program 4). The training of APHIS personnel to discharge these responsibilities has been greatly assisted by SEL expertise. Furthermore, as backup for these APHIS programs, the specialized knowledge of SEL taxonomists is called upon to monitor identifications made by APHIS personnel and to identify those specimens beyond the ability of nonresearch personnel to recognize. In addition, taxonomic services in support of surveys and the CPPR (Table 15.1, programs 2 and 3) that are provided by professional nonresearch taxonomists in various states can be supplemented by SEL. Finally, taxonomic research capability is essential for the support of most other APHIS programs (Table 15.1) because of the immense total number of taxa, the obscurity of many, and the large numbers of organisms not yet known to science.

THE TAXONOMIC CONNECTION

From time to time individuals represented as being at point A in Fig. 15.1, and with any or all of the APHIS programs described in the previous section, have need for identification of insect specimens, data associated with insect specimens, or both.

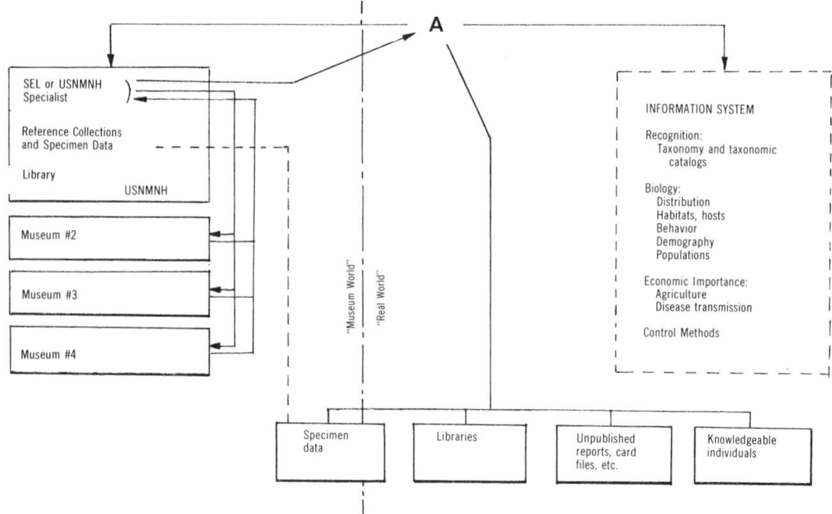

Figure 15.2. Schematic representation of relationships in the "museum" world and the "real" world.

Their first requirement is to know with certainty the organisms with which they are concerned, and to meet this requirement they submit specimens to SEL for identification. Using the reference collection, the library, and any other appropriate working tool, SEL or Smithsonian Institution specialists identify the specimens to the lowest possible taxon consonant with available time, expertise, and the sender's needs. The identifications, and the data associated with them, are used in two important ways: (1) they are returned to the sender, and/or (2) they are cycled back into the collections of the USNMNH, and possibly other museums (museums #2, #3, #4 in Fig. 15.2), where the data contribute to an ever-growing bank of information about the identity, distribution, and biology of those particular organisms. If the sender needs to take action concerning the specimens that have been identified, a requirement for additional information about the insect is generated, since the sender usually received only the name of a taxonomic entity. Senders typically depend upon sources such as libraries, unpublished reports, knowledgeable individuals, and specimen data (Fig. 15.2, bottom boxes) for information on recognition, biology, economic importance, and control methods (Fig. 15.2, dotted box at right) that will assist them in making appropriate decisions. As shown in Table 15.1, the amount and detail of information required varies greatly with each APHIS activity and program, but in *every case* we know that the information the sender receives from the specialist must be fortified with additional facts to support his decision-making process. Sometimes the sender has on hand the information needed, but more often is forced to look elsewhere. He or she can go back to the specialist for supporting specimen data, to the library, to unpublished documentation, or to others who may have the needed information. As a result of this activity and the decisions that are made, the information gained is eventually recycled into the information sources originally consulted, resulting in a gain in knowledge against which any similar kinds of decisions can be more expeditiously and effectively made in the future.

NEEDS OF APHIS FOR TAXONOMIC SERVICES

The "real-world" programs of APHIS are clearly dependent upon natural history collections, upon specimen-related data, and upon the scientists who use them as working tools. First, any APHIS action program can be only as effective as the identification upon which it is based. The misidentification, or delay in identification, of an important program pest in a previously uninfested area (Table 15.1, program 1), or of a potential pest allowed to pass otherwise rigid quarantine inspections (Table 15.1, program 4), or of an insect found for the first time living on or in plants near a quarantine station (Table 15.1, program 5) might delay for weeks the application of effective control measures. During this time the insect might well have the opportunity to become fully established in a new geographic area. The accurate identification of such a previously undetected pest is fully dependent not only upon the ability and knowledge of the taxonomic specialist involved, but also upon the accuracy with which specimens of the species-in-question have been named and labeled in the reference collection.

Second, specimen-related data are vital to APHIS programs. When the sender (at A in Fig. 15.2) consults the specialist for additional information, the specimen-related data for this species must have been accurately recorded, otherwise they have no value. If individuals in APHIS are concerned about reporting an outbreak of a familiar pest in CPPR (Table 15.1, program 3), they may wish to learn the complete distribution of that pest in the United States. Specimen-related data are commonly-used sources of such information which, if kept inaccurately, could prevent concerned agriculturalists from becoming aware of possible outbreaks in their area of the country. And anyone compiling the information for widespread publication about an insect not yet known to occur in the United States (Table 15.1, program 6) is certain to consult data associated with specimens for distributional and biological information such as hosts, habitats, etc. In these and many other cases, misleading specimen data can cause—and actually have caused—undue loss of agricultural revenue, increased costs of pest control, and irreparable damage to the credibility of those who distributed the information.

Third, the specialist working with an insect collection provides an invaluable aid to APHIS in the working library of books, reprints, and unpublished material which he or she has painstakingly accumulated in order to pursue taxonomic research. Although the specialist's main objective is to acquire the published taxonomic record (descriptions, figures, nomenclatural actions, etc.) concerning a specialty, a large amount of non-taxonomic data is also gathered along with it. What better place to begin a search for information on distribution, economic importance, and control methods than this specialist's file, already indexed and categorized taxon by taxon, author by author, year by year?

MAKING THE TAXONOMIC CONNECTION EFFECTIVE

As a result of research and the services provided, the taxonomic specialist becomes the steward of a very large amount of information. Most taxonomists, without sufficient support, are hard put to select and record those items they can predict will be needed and used most widely, to organize them effectively, and to add to them meaningfully. When they are requested to provide information for purposes that lie outside their normal predictions, they can comply only at the expense of much time and effort. This is due partly to a rather widespread lack of capable subprofessional support, and partly to the absence of means for effectively retrieving the information in a variety of different ways. It is certainly possible to handle extremely detailed specimen-related data such as locality, date of collection host, and other information in a variety of ways for a variety of purposes. Where there is a very large amount of data, this organizing process must be facilitated by the ready availability of professional assistance for the input phases and an automated information system for storage and retrieval. Unfortunately, these advantages are currently available to only a comparatively small number of systematists. As a consequence, the specialist has a great deal of information that lies fallow except for the occasional visitor who must try to retrieve data for his special need—data that are often poorly organized or hard to find because the specialist simply has not had the time or help to cope with them.

Despite obstacles to the full use of existing information, the specialist performs minor miracles in marshalling significant data for his or her taxonomic research and service responsibilities. In publishing the results of his research or in delivering critical identifications, the specialist typically demonstrates a competent command of the literature regarding his or her subject and is able to summarize rather effectively not only the purely taxonomic information accumulated during the studies but also additional data concerning biology, distribution, and economic importance.

It is easy to see, then, how a taxonomic specialist becomes the logical person to conduct an inventory of the taxa which make up a special taxonomic group. Likewise, it is easy to see how valuable such an inventory would be to those who are in hand-to-hand combat with day-to-day pest problems in the field. This inventory, of which the taxonomic catalog is a part, is one of the natural objectives of a taxonomist's research; when the effort is well organized on an international scale, as was the case with USDA hymenopterists in the early 1950's (*24*), and later with USDA dipterists (*43*), comprehensive and highly useful catalogs can be produced. Computerization of the North American Hymenoptera catalog has been described by Krombein *et al.* (*20*), and a description of a new catalog of the North American Coleoptera is being prepared for publication. Because any catalog effort must be based from the beginning on data associated with specimens, it is only natural that museum data-banking be designed, at least in part, to support this valuable enterprise.

Taken one step further, the inventory of an insect fauna can form the backbone of a much more comprehensive information system comprising all living things. The organization of this system can and should be based on names of organisms. Such a system could be produced through the cooperation of appropriate specialists in museums and other institutions at many different locations. Adequate funding could provide the building blocks—data gathering, automation, networking, cooperation—needed to construct a comprehensive information system about all living things, a system in which museums would play a central role and which museums themselves could utilize to great advantage.

CONCLUSIONS

Data-banking, as commonly discussed within the "museum world," is seen largely as a means to assist curators and administrators in the management of collections. However, a considerable amount of data, rarely used in such management activities, is generated as a by-product of these collections—data that actually form an important base for "real world" activities. More serious consideration must be given to the interrelationships of these two worlds so that museum data and data systems are used to their full potential.

We must acknowledge that natural history collections are meaningful to the "real world" only when competent, knowledgeable, and interested taxonomic specialists, actively using and adding to the information they contain, can interpret the data for "real-world" purposes. The taxonomist thus becomes the natural liaison between the two worlds.

The value to the "real world" of a museum data-bank is directly proportional to its relevance, volume, and accuracy, and the only insurance for the presence of these qualities is the active participation of specialists with adequate facilities. Support must be increased to make potentially valuable data more readily available and useful to all museum users, and especially to those outside the "museum world" who, like the agencies described in this paper, depend upon these data for the successful conduct of their activities.

Museum objectives often differ so markedly from those of the "real world" (i.e., accreditation of museums primarily for public exhibits, lack of concentration on economically important organisms, etc.) that museum resources are likely to fall short of developing the exciting potential they possess. Because users of museum data, whoever they may be, depend so heavily upon the taxonomic specialist and the museum services he needs to conduct effective research, "real-world" organizations and individuals must seriously consider giving more meaningful support to museum activities. There are many ways in which such support can bring museum programs to bear more directly and effectively on "real-world" objectives, to the betterment and better integration of both "worlds."

LITERATURE CITED

1. Allen, R. T. 1975. *Ecosystematic entomology*. Bull. Entomol. Soc. Amer. 21: 13–17.
2. Anon. 1963. *Smithsonian Institution's Department of Entomology*. Bull. Entomol. Soc. Amer. 9: 273–277.
3. Becklund, W. W. 1969. *The role of the National Parasite Collection in veterinary parasitology*. Biol. Soc. Wash., Proc. 82: 603–610.
4. Biliotti, E. 1961. *Les problemes de systématique dans le recherches écologiques sur les insectes entomophages*. Entomophaga 6: 117–123.
5. Clausen, C. P. 1942. *The relation of taxonomy to biological control*. J. Econ. Entomol. 35(5): 744–748.
6. Darlington, P. J., Jr. 1971. *Modern taxonomy, reality, and usefulness*. Syst. Zool. 20: 341–365.
7. Davidson, J. F. 1952. *The use of taxonomy in ecology*. Ecology 33: 297–299.
8. Denmark, H. A. 1971. *Insect identification: What is it?* Florida Entomol. 54: 101–104.
9. Eldridge, B. F. 1974. *The value of mosquito taxonomy to the study of mosquito-borne diseases and their control*. Mosquito Systematics 6: 125–129.
10. Essig, E. O. 1942. *The significance of taxomy in the general field of economic entomology*. J. Econ. Entomol. 35: 739–743.
11. Evans, H. E. 1973. *Taxonomists' curiosity may help save the world*. Smithsonian Magazine 4(6): 36–43.
12. Frison, T. H. 1942. *The significance of economic entomology in the field of insect taxonomy*. J. Econ. Entomol. 35: 749–752.
13. Gahan, B. 1923. *The role of the taxonomist in present day entomology*. Entomol. Soc. Wash., Proc. 25: 69–78.
14. Grotta, L. D. 1975. *The need for taxonomic information and services by environmental consulting firms*. Assoc. Syst. Coll. Newsletter 3: 3–5.

15. Hendrickson, J. 1965. *The relationship of ecology to systematics.* Pacific Insects 7: 8-13.

16. Heywood, V. H. 1973. *Ecological data in practical taxonomy,* Pages 329-347, *in* V. H. Heywood, ed. *Taxonomy and Ecology.* Syst. Assn. Spec. Vol. 5. Academic Press, London. 370 pp.

17. Kim, K. C. 1975. *Systematics and systematics collections: Introduction.* Bull. Entomol. Soc. Amer. 21: 89-91.

18. Kim, K. C. 1975. *Systematics and systematics collections: A concluding remark.* Bull. mol. Soc. Amer. 21: 95-98.

20. Krombein, K. V., J. F. Mello, and J. J. Crockett. 1974. *The North American Hymenoptera catalog: A pioneering effort in computerized publication.* Bull. Entomol. Soc. Amer. 20: 24-29.

21. Labeyrie, V. 1961. *Taxonomie, écologie et lutte biologique.* Entomophaga 6: 125-131.

22. Mahmood, S. H. 1973. *Why systematics?* Entomol. Soc. Karachi, 2nd Ann. Mtg. (Feb. 22, 1973).

23. Muesebeck, C. F. W. 1942. *Fundamental taxonomic problems in quarantine and nursery inspection.* J. Econ. Entomol. 35(5): 753-758.

24. Muesebeck, C. F. W., K. V. Krombein, and H. Townes. 1951. *Hymenoptera of America North of Mexico. Synoptic Catalog.* U. S. Dept. Agric., Agr. Monogr. 2. 1420 pp. (+ 2 supplements).

25. Munroe, E. 1958. *Systematics as a tool in ecology.* Tenth Int. Congr. Entomol., Proc. 2(1956): 663-667.

26. Murphy, R. C. 1952. *Taxonomy today.* Science 115. 3.

27. Oman, P. W. 1957. *The relation of insect taxonomy to mosquito control.* Mosquito News 17: 147-151.

28. Oman, P. W. 1960. *The relation of insect taxonomy to applied biology.* Bull. Entomol. Soc. Amer. 6: 3-5.

29. Oman, P. W. 1974. *Identification and classification in pest management and control.* Pages 77-86, *in* F. G. Maxwell, and F. A. Harris, eds. Summer Inst. Biol. Control of Plant Insects and Diseases, Proc. Univ. Press of Mississippi, Jackson.

30. Pechuman, L. L. 1975. *University collections.* Bull. Entomol. Soc. Amer. 21: 91-93.

31. Resh, V. H., and J. D. Unziker. 1975. *Water quality monitoring and aquatic organisms: The importance of species identification.* J. Water Poll. Contr. Fed. 47: 9-19.

32. Robinson, W. H. 1975. *Taxonomic responsibilities of nontaxonomists.* Bull. Entomol. Soc. Amer. 21: 157-179.

33. Rogers, A. J. 1974. *The value of mosquito taxonomy to mosquito control.* Mosquito Systematics 6: 121-124.

34. Rosen, D., and P. DeBach. 1973. *Systematics, morphology, and biological control.* Entomophaga 18: 215-222.

35. Sabrosky, C. W. 1950. *Taxonomy and ecology.* Ecology 31: 151-152.

36. Sabrosky, C. W. 1955. *The interrelations of biological control and taxonomy.* J. Econ. Entomol. 48(6): 710-714.

37. Sabrosky, C. W. 1964. *Taxonomic entomology in the U. S. Department of Agriculture.* Bull. Entomol. Soc. Amer. 10: 211-220.

38. Sabrosky, C. W. 1970. *Quo vadis, taxonomy?* Bull. Entomol. Soc. Amer. 16: 3-7.

39. Sailer, R. I. 1969. *A taxonomist's view of environmental research and habitat manipulation.* Pages 37-45, *in* Tall Timbers Conf. Ecol. Animal Control by Habitat Management, Proc.

40. Sailer, R. I. 1972. *The inventory of beneficial insects.* XIV Internat. Congr. Entomol., Canberra, Australia.

41. Schlinger, E. I., and R. L. Doutt. 1965. *Systematics in relation to biological control.* Pages 247-280, *in* P. DeBach, ed. *Biological Control of Insect Pests and Weeds.* Reinhold, New York. 844 pp.

42. Slater, J. A. 1960. *The responsibilities of the insect taxonomist.* Bull. Entomol. Soc. Amer. 6(1): 17–19.

43. Stone, A., C. W. Sabrosky, W. W. Wirth, R. H. Foote, and J. R. Coulson, eds. 1965. *A catalog of the Diptera of America North of Mexico.* U. S. Dept. Agric., Agric. Handb. 276. 1696 pp.

44. Ward, R. A., O. S. Flint, A. S. Menke, and F. C. Thompson. 1976. *The United States National Entomological Collections.* Smithsonian Inst. Press, Washington, D.C. 47 pp.

45. Wiggins, G. B. 1966. *The critical problem of systematics in stream ecology.* Pages 52–58 *in* K.W. Cummins, C. A. Tryon, Jr., and R. T. Hartman, eds. *Organism-substrate relationships in streams.* Pymatuning Symposia in Ecology, Spec. Publ. 4, Pymatuning Lab. of Ecology, Univ. Pittsburgh, Pa. 145 pp.

46. Wilson E. O. 1968. *Recent advances in systematics.* BioScience 18: 1113–1118.

47. Wilson, E. O. 1971. *The plight of taxonomy* (editorial). Ecology 52: 741.

16] Overview of Predictiveness of Agricultural Biosystematics

by L. R. BATRA, D. R. WHITEHEAD, E. E. TERRELL, A. M. GOLDEN, AND J. R. LICHTENFELS*

ABSTRACT

Biosystematics is fundamental to most of agriculture. It provides scientifically sound classifications of organisms and critical data on phyletic relationships, hybridization, host-parasite relationships, geographic distribution, and environmental requirements of taxa. It plays a significant role in the exploration and introduction of germplasm for agronomically improved and disease-resistant biotypes, and in reducing pre- and post-harvest crop losses due to pests. Along with other fields of study, systematics must keep pace with technological advances to answer many questions posed by changing cultural practices designed to enhance crop production, to conserve soil moisture and fertilizer, and to minimize need for chemical pesticides. This overview examines some of the principles governing exploration, collection, identification, introduction, selection, and protection of agricultural biota. It focuses on the predictive values of relatedness or unrelatedness of units of evolution, their recognition and their use to our advantage. Some critical problems and challenges in recognition of variability and coevolution of host-parasite relations are identified. Also, we discuss the role of biosystematics in quarantine, in forecasting epidemics and losses due to pests, in estimating weed potential of undesirable biota, and in biological control.

INTRODUCTION

The basic task of the agricultural biosystematist is to provide evolutionary classifications of plants and animals, with emphasis on ecological, behav-

*The authors, respectively, are staff members of the various laboratories and institutes engaged in biosystematic research at the Beltsville Agricultural Research Center, Beltsville, Maryland 20705: (L. R. Batra) Mycology Laboratory, Plant Protection Institute; (D. R. Whitehead) Systematic Entomology Laboratory, Insect Identification and Beneficial Insect Introduction Institute; (E. E. Terrell) Plant Taxonomy Laboratory, Plant Germplasm and Genetics Institute; (A. M. Golden) Nematology Laboratory, Plant Protection Institute; and (J. R. Lichtenfels) Parasite Classification and Distribution Unit, Animal Parasitology Institute.

ioral, and genetic characteristics of agriculturally beneficial or harmful populations. Evolutionary classifications are tangibly most useful and effective in their contributions toward acquiring, improving, and protecting food resources. Thousands of studies—monographs, revisions, checklists, distribution maps, host indices—have been made, and much more such information will be needed as we select and transport additional useful, more effective crops from one region to another in response to changing and increasing human needs. It may seem that most of the plants related to crop species and that most of the important pathogens or pests of cultivated plants have been discovered, but we cannot afford the conceit or ignorance of thinking that our knowledge of them is adequate (*22*).

We offer here an interdisciplinary consideration of biosystematic problems associated with the modern culture of native and introduced biota and with the management of native and exotic pests. More than 3,000 plant species have been used as food; about 300 of these are widely grown, but just 12 crops (rice, wheat, maize, sorghum, the millets, rye, barley, potato, the sweet potatoes, cassava or manioc, bananas, and coconuts) furnish nearly 90 percent of the food for the world's population (*51*). The agricultural biota and its undomesticated relatives took millions of years to evolve in harmony within diverse ecosystems. However, in the short span of a few hundred years man has selected, transported, hybridized, multiplied, and widely disseminated such biota to diverse climatic regions of the world. The New World, Africa, and Oceania offer the most glaring examples wherein new associations and ecosystems created by man have allowed the introduced biota to interact with the native wild species.

Predictive values are considered here under two major agricultural activities: 1)—*Acquisition of agricultural biota* (plants and animals domesticated for their food, timber, fiber, or chemical value; pollinators or other beneficial insects; desirable symbiotes; and the like). Predictive values are those that aid in increasing yields, such as predicting new sources of energy or products, predicting ecological characteristics, or predicting the outcome of particular hybridization programs. 2)—*Minimizing losses due to pests.* Predictive values are concerned with forecasting results of deliberate and accidental environmental modification, and also with predicting epidemics. A third section is included on opportunities to improve predictiveness.

Agricultural biosystematics might in simple terms be described as the capability to distinguish agriculturally important taxa and the science of structuring this information in such a way that it is useful to agriculture. From these data, reliable predictions and forecasting may be made. Figure 16.1 diagrams data input and analysis for vascular plant taxonomists; this summary differs in detail but not in substance from summaries of other systematic disciplines. Systematics is the synthesizer of information from all fields of biology—organizing the data into a classification that groups related species. A classification is most useful if we can predict with a

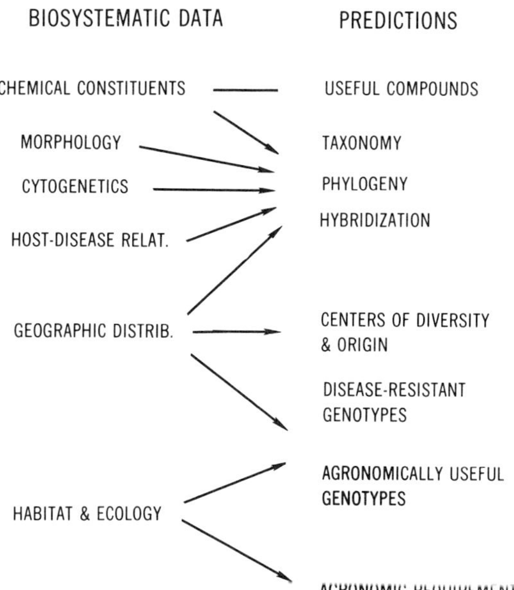

Figure 16.1. Biosystematic data and their predictive values regarding vascular plant attributes.

satisfactory degree of probability relatedness among taxa on the basis of previously uninvestigated character systems, and if a few diagnostic characters of a newly discovered taxon enable us correctly to relate it to previously known taxa (*25*). A natural classification that correctly reflects evolutionary patterns is most predictive simply because it groups organisms that are most closely related. Thus, biologists of all kinds can use a natural classification to learn probable characteristics of their organisms (see Kavanaugh, Chapter 8, this volume).

How can a natural classification be useful in domestication? Wild species that are most closely related to those under cultivation are the most likely to have agronomically useful properties. They might be tested for possible cultivation themselves or might be considered for hybridization.

Though of common knowledge, the following uses of natural classification are commonly overlooked or neglected: 1) Magnitude of similarity generally indicates biological relatedness; this principle has been the basis of all initial hybridization programs. 2) Taxa with intergradient morphological characters may be of hybrid origin, and any unusual characteristics may have come from either of the parent taxa. 3) Habitat and behavioral characteristics of organisms in the wild indicate their adaptive amplitude for certain climatic zones. 4) We predict from the basis of chemical as well as taxonomic data that certain useful chemical constituents are more likely to be found in some taxonomic groups than in others; for example, since

Juglans regia L. has the insect repellent juglone, *J. nigra* L. could be predicted to have this chemical and, in fact, it does (*103*).

However, the agricultural biosystematist is concerned not with evolutionary relationships alone, but in a much broader sense with similarity; phylogenetically unrelated biota may have the same or similar agronomically desirable properties. We might predict agricultural usefulness of certain distantly related taxa from chemical similarity alone; if latex-bearing *Euphorbia* spp can be used to provide petroleum, latex-bearing plants in other families (e.g., Asclepiadaceae) may also provide petroleum. We might make similar predictions from an even more indirect source of data; if wild plants have insect or parasite associates similar to those of cultivated plants, then biochemical properties of these plants are likely to be similar (*81*).

ACQUISITION AND IMPROVEMENT OF AGRICULTURAL BIOTA

Some questions we explore here are: Where are agronomically useful genotypes most likely to be found? What are the desirable or undesirable effects of domestication? What can we predict about undomesticated biota? How can knowledge of ecosystematics be useful in domestication?

Sources of diversity and Vavilov's theory. Creech and Reitz (*24*) reviewed some aspects of acquisition, maintenance, and distribution of germplasm. As stated by them and by numerous others (*67, 72*), present day theories about sources of germplasm and methods to acquire it are mostly refinements of Vavilov's theory of the phytogeographic basis of plant exploration (*116, 117, 124, 125*) in the gene centers (centers of diversity).

Vavilov assumed that host and parasite genotypes coevolved through time, exerting reciprocal selective pressures on each other; that surviving individuals tolerate each new or increasingly virulent pathogen from the area of the primary gene center, maintain low levels of diverse inocula, and are thus well-adjusted or adapted to the genotype of the pathogen. Domestication may alter gene frequencies through selection and hybridization, providing opportunities for development of susceptible hosts which otherwise have desirable attributes and increased build-up of inocula which previously were minimized by resistant wild types or by hyperparasites.

Vavilov's theory predicts that the greatest diversity of agronomically useful genotypes and disease-resistant genotypes will be found in primary and secondary centers of biotic diversity. For example, wheat genotypes that will grow under various conditions and in various soils should be sought at the center of diversity in southwest Asia and the Middle East. Various species of wild wheats (*Triticum* spp) from this region have been crossed with *T. aestivum* L. (bread wheat) to give resistance to rusts and other fungi (*72*). The center of diversity for *Lolium* is southern Europe, the Mediterranean, and southwest Asia; thus turf and forage grass breeders often sample these regions for new genotypes (*108*). At least three wild species of *Lycopersicon* in Andean South America have genes for resistance to diseases and insects and might therefore be crossed with *L. esculentum* Mill. (tomato) (*118*). For additional examples see Table 16.1 and other authors (*2, 24, 36, 39, 42, 56, 57, 59, 87, 89, 93, 104, 105*).

A corollary of Vavilov's theory is that when pest-free hosts from one primary center of diversity are introduced into an ecosystem with related hosts, the pests of the latter may devastate the immigrant. *Prunus* spp (cherries, plums, peaches, and apricots) are severely damaged by the American brown rot fungus *Monilinia fructicola* (Winter) Honey, even though these hosts left behind their European brown rot fungus *M. fructigena*

Table 16.1 Source(s) of desirable characteristics for some crops.

Crop and related taxa [and center(s) of diversity]	Desirable characteristic(s)[a]
Lycopersicon esculentum Mill. (tomato) [South America]	
L. chilense Dunal	pr
L. hirsutum Humb. & Bonpl.	pr
L. peruvianum (L.) Mill.	pr
L. pimpinellifolium (L.) Mill.	pr
Saccharum officinarum L. (sugarcane) [New Guinea and Southeast Asia]	
S. robustum Brandes & Jeswiet ex Grass	ai
S. spontaneum L.	pr
Solanum tuberosum L. (potato) [Central and South America]	
S. acaule Bitter	pr
S. chacoense Bitter	cp
S. demissum Lindl.	pr
S. stenotum Juz. & Buk.	pr
Triticum aestivum L. (wheat) [Eastern Mediterranean to Southwest Asia]	
T. boeticum Boiss.	pr
T. dicoccoides (A. & G.) Aaronsohn	pr
T. monococcum L.	pr
T. timopheevii (Zhuk.) Zhuk.	pr

[a] Abbreviations: ai, agronomic improvement
cp, consumer preference
pr, pest or disease resistance.

(Aderh. and Ruhl.) Honey; the American fungus does little damage to the largely inedible American *Prunus* spp (L. R. Batra, unpublished). Similarly, *Leptinotarsa decemlineata* (Say) (Colorado potato beetle), originally restricted to *Solanum rostratum* Dunal (buffalo bur) and for 30 years of no economic significance, became an important pest of *S. tuberosum* L. (potato) introduced in Colorado in 1874. It moved eastward about 35 miles per year and is now present in most of the Atlantic states (*82*).

Search for germplasm. The use of pesticides for non-specific or saprophytic pests probably will continue, for it is uncommon to find genes for field resistance (also called generalized or horizontal resistance in contrast to specific or vertical resistance). However, for many field crops it is economically not feasible to use pesticides, and we must breed for pest resistance or minimize losses by improved cultural practices. Exploration, utilization, and conservation of gene pools of primitive cultivated plants and their wild relatives are therefore extremely important to the plant breeder (*24*).

Since it is impossible to predict with certainty what diseases or pests will arise or become established in the future, we must preserve diverse genetic materials. This is especially true because modern crops and herds bred for genetic uniformity (Table 16.2) are especially vulnerable to new, more virulent strains (*22, 41*). For example, of the 197 varieties of corn available, just 6 accounted for 71 percent of total acreage in the United States in 1969. The now well analyzed 1970 epiphytotic of *Helminthosporium maydis* Nisikado and Miyake (southern corn blight) was a result of a mutant of the pathogen that built up undetected over a period of time. Unfortunately, sources of diverse germplasm are rapidly dwindling, particularly from large areas of Asia, Africa, and Latin America as wild lands are eliminated and "improved" crops introduced. Vavilov's famous living plant collection of 250,000 accessions is now dispersed or otherwise unsuitable because of inadequate maintenance (*72*).

MINIMIZING LOSSES DUE TO PESTS

The agricultural biosystematist faces many questions, only some of which we attempt to answer here. The following give some indication of the scope of these questions: How can we predict pest potential? How can we predict pest epidemics? What are the undesirable consequences of accidental or deliberate introductions of new cultivars, new pests, or new biocontrol agents; of domestication of new useful biota; or of population shifts in previously innocuous pests? Will biocontrol agents exert selective pressure (as do pesticides) such that the target coevolves to become resistant, and, if so, how long does this take? What factors govern or relate to host-parasite dynamics of well-known pests, and how do they operate in everyday agricultural life?

Table 16.2 Acreage and farm value of major United States crops and extent to which small numbers of varieties dominate crop acreage (1969 figures) (adapted from ref. *22*).

Crop	Total acreage (millions)	Value (millions of dollars)	Total varieties available	Major varieties used	Acreage of major varieties (percent)
Cotton	11.2	1,200	50	3	53
Corn	66.3[a]	5,200	197[b]	6	71
Peanut	1.4	312	15	9	95
Potato	1.4	616	82	4	72
Rice	1.8	449	14	4	65
Sorghum	16.8	795	?	?	?
Soybean	42.4	2,500	62	6	56
Sugar beet	1.4	367	16	2	42
Wheat	44.3	1,800	269	9	50

[a] Corn includes seeds, forage, and silage.
[b] Released public inbreds only.

Host ranges and ecosystematic amplitudes. Host range is extremely important in control and classification of a pathogen or pest. That this was recognized early is evident from the many pathogen-host indices for plants and animals (*11, 46, 99, 100, 111, 112*). The Forest Service and the Agricultural Research Service of the U.S. Department of Agriculture (USDA) for many years have monitored foreign diseases of North American trees and crops planted abroad, and they likewise monitor diseases of foreign plants growing in the United States (*99, 100, 119*). This is an efficient and effective way of obtaining data on foreign diseases without the actual transfer of experimental inocula. Similar detailed monitoring is needed for other plant pests and for their possible biological control agents.

Ecosystematic amplitudes of a species form a basis for predicting which crops and pests are most likely to be associated with each other in any given area or where alien taxa are most likely to find a suitable ecological niche. Thus, data on host range and ecological amplitudes, along with data on taxonomic relationships, form the basis for predicting pest potential—one of the important questions directed to the agricultural biosystematist.

The USDA is preparing a computer-generated "List of intercepted plant pests," including several thousand insect species considered to be potential problems for our agriculture. Several hundred of these are particularly likely to become established, as judged from their ecological requirements and the magnitude of commerce. The USDA also has a major detection program to survey for 124 foreign insect crop pests and many animal diseases (*110*). Similar programs on foreign fungi and bacteria (*119*), animal parasites (*10, 112*), and plant nematodes (*46*) are needed, both here and abroad.

Forecasting of losses and epidemics. Estimates of losses by pests are important in planning for adjustments in production, storage, and transportation. Several methods are available for forecasting percentage loss of a crop due to individual pests; however, a grower needs an integrated system to estimate loss due to the entire pest complex (we cannot just add up individual losses), and we must follow the entire succession of pests—each subsequent pest having less crop available to attack. Critical basic and applied research on the sociobiology and population dynamics of significant pests, with accompanying qualitative and quantitative models, are urgently needed (*120, 123*).

Ideally, management of epidemic outbreaks of pest species should be a matter of prevention rather than suppression. Yet, despite efforts in recent years to reduce complexities of epidemic population booms to simple models, forecasting generally remains a matter of hindsight rather than science. Even in situations involving recurrent, cyclic epidemics, we still know too little to predict with accuracy when and where epidemics will occur for many pest species.

What mechanisms allow pest populations suddenly to rise from innocuous levels to epidemic proportions? Given prior knowledge of (a) biosystematics, bionomics and dynamics of host and parasite populations, (b) physical characteristics of the environment, and (c) incipient populations or inoculum load on wild relatives or bridging hosts, we could predict the probability of an epidemic (even though the effect of weather to predispose hosts or to influence pests may not be within the province of biosystematics). The minimum requirement is an observed *sequence of conditions* of an outbreak—inoculum load, effect of temperature, water conditions, and so on (*85, 86, 114*). Negative forecasts (predictions that significant outbreaks will not occur) are as necessary as positive forecasts, since they may save money and needless pesticide application.

The USDA in cooperation with state agricultural experiment stations has operated a plant disease forecasting and warning service for many years. Most of these forecasts are concerned with fungus diseases that fluctuate greatly and are influenced primarily by temperature, rainfall, relative humidity, and dew at plant level. Currently the following pathogens are covered: *Phytophthora infestans* Debary (late blight of potatoes and tomatoes), *P. phaseoli* Thaxter (downy mildew of lima beans), *Peronospora tabacina* Adam and *Alternaria tenuis* Nees ex Corda (blue mold and brown spot diseases of tobacco respectively), *Cercospora beticola* Sacc. (leaf spot of beets), and *Helminthosporium maydis* (Southern corn leaf blight). Simple climatic factors may serve to trigger outbreaks; for example, Kalkstein (*62*) was able to isolate critical climatic factors sufficiently to predict an outbreak of *Dendroctonus frontalis* Zimmerman (Southern pine beetle) in Texas.

Morphological features of gastrointestinal nematodes of ruminants

(*Nematodirus, Haemonchus,* and *Ostertagia*) are useful in understanding host-parasite population dynamics. When diapause, quiescence, and associated phenotypic variation are more fully described (*37, 83*) it may become possible to predict seasonal outbreaks of nematodiasis in the United States as is done for nematodiasis of sheep in Great Britain (*88*) caused by *Nematodirus battus* Crofton and Thomas.

Other systems are more complex. In North America, much is known about past epidemic outbreaks of *Choristoneura fumiferana* (Clemens) (spruce budworm), yet we still lack sufficient insight into biosystematic aspects to make useful predictions. For example, what role do parasites play? Many parasitoid species are abundant during peak or collapse phases but disappear during endemic phases, whereas others help maintain low population levels of the host during endemicity but are unable to respond to outbreak phases (*84*).

In most areas of the northeastern United States that are subject to infestation by *Lymantria dispar* (L.) (gypsy moth), population levels of the pest are normally at innocuous levels where they might remain indefinitely unless bolstered by external factors. Such bolstering is most likely to occur in suburban areas, where disturbances along the forest edge provide abundant pupation sites; control efforts, including establishment of parasites, might be most effective in such sites (*17*).

Conversely a stable, natural relationship between a pest and its parasites may require very little manipulation to induce an "outbreak" of the desired parasite. Whereas under normal field conditions the tachinid *Lixophaga diatraeae* (Townsend) does not effectively control the pyralid *Diatraea saccharalis* (F.) (sugarcane borer), the pest can be virtually eliminated within one growing season by supplemental releases of the parasite at the appropriate host density (*65*).

In addition to the aforementioned weather-based systems to predict outbreaks, some progress is also being made toward analysis of genetic vulnerability of cereals to rusts, powdery mildews, and selected blights. Here the predictive models are based on surveys and projections of virulent races and susceptible acreage of hosts. Thus during 1934–35 it became apparent that the hitherto rust-resistant wheat *ceres* occupying large acreage must be eliminated at the appearance of race 56 of *Puccinia graminis* Pers. *f. sp. tritici* Eriks. and E. Henn. Similarly, corn lines with Texas (T) male sterile cytoplasm were eliminated after the 1970 outbreak of race "T" of the southern corn blight organism.

Endemic outbreaks and selection of germplasm. Endemic disease or pest outbreaks in primary or secondary centers of diversity offer good opportunities to seek resistant genotypes from surviving populations. Biosystematists must promptly characterize both the susceptible and resistant populations, corelate them with others, and determine potential for improvement by hybridization. The effect of genetic versus environmental factors must be critically delineated as soon as possible, for wild populations are seldom uniform and common, and unchanged strains of pathogens may soon be unavailable.

Often minor diseases suddenly become important because of severity of attack of endemic or epidemic nature. This may be, among other factors, due to the appearance of a previously unknown or undetected pathogen strain. *Helminthosporium victoriae* Meehan and Murphy, which in the 1940's caused the demise of oat varieties derived from otherwise desirable Richland X Victoria crosses, was known as an unimportant and obscure fungus for a long time (*102*), as was the southern corn blight fungus (*22*).

Quarantine activities and biosystematics. Importation of foodstuffs from newly developed growing regions, development of new crops in established growing regions, or increased agricultural exploitation of underdeveloped areas may suddenly alter the picture of agricultural pests: they may be transported by commerce, shift to a new food source, or increase to epidemic numbers as a once scarce food source becomes abundant.

Local populations of pest species often develop a balance with native plants over a long period, causing little or no obvious damage. But, with introduction of a new host (for example, an economic crop) into the area, one or more pest species might cause serious damage. *Dolichodorus heterocephalus* Cobb (awl nematode) was described in 1914 from Silver Spring, Florida, evidently associated with native flora; in the early 1950's it was found as a serious parasite on vegetables and some other crops in the area, and it is now known in other southern states. *Heterodera schachtii* Schmidt (sugarbeet nematode) and *Nacobbus aberrans* (Thorne) (false root nematode) were found by Thorne (*109*) on native *Atriplex canescens* (Pursh) Nutt. (shad-scale) in Utah; as land was planted to sugarbeets, the nematodes became serious pests. Since the 1960's *Claviceps* (ergot) in southeastern India has become important with the introduction of male sterile lines of *Sorghum*. Likewise, *Alternaria triticina* Prasada and Prabhu, a species unknown until 15 years ago, now causes significant damage to dwarf Mexican wheats (L. R. Batra, unpublished data) in Panjab.

Many animal parasites are transported by movements of man and animals. Man regularly transports malaria and schistosomiasis, two of his most important parasitic deseases; gastrointestinal nematodes of ruminants are regularly imported in domestic stock and in animals for zoological parks. This movement presents potential dangers to livestock.

Many parasites occur in equilibrium with natural hosts but are injurious to secondary hosts. It is evident that *Parelaphostrongylus tenuis* (Dougherty), a nematode parasite of the brain of *Odocoileus virginianus* (Zimmermann) (white-tailed deer), is an important cause of the failure of *Cervus canadensis* (Erxleben) (North American elk or wapiti) and *Alces*

alces americana (Clinton) (moose) to re-establish in the northeastern United States (*3, 97*). *Fascioloides magna* (Bassi) Ward (large American fluke), a normal parasite of Cervidae, produces no clinical disease in deer or cattle, but in *Ovis aries* L. (domestic sheep) it usually causes fatal liver damage (*40*). In the western and southwestern United States, the arterial worm *Elaeophora schneideri* Wehr and Dikmans occurs in *Odocoileus hemionus* (Rafinesque) (mule deer) and in white-tailed deer without clinical disease; in domestic sheep it causes mucosal lesions and severe dermatitis, and in North American elk it commonly causes blindness, necrosis, and death from brain damage (*54*).

Angiostrongylus cantonensis (Chen) Dougherty, a nematode that develops in the brain and matures in the intestine of rats but with larval stages in snails and slugs, can infect man when larvae in the molluscs are ingested. The disease eosinophilic meningoencephalitis occurs in Pacific Basin islands including Hawaii and in southeast Asia. The etiology of the disease was elucidated after 1960 when MayBelle Chitwood of the Animal Parasitology Institute, Beltsville, identified a worm from the brain of man and predicted, from its classification, the life cycle and kind of intermediate hosts that would be involved (*1, 94*).

An imported species of parasite may replace less virulent species causing disease where previously host and parasite were in equilibrium. In Great Britain *Nematodirus battus* Crofton and Thomas causes clinical gastrointestinal nematodiasis in domestic sheep. After sheep from Great Britain were imported into Norway, *N. battus* was observed (*53*) gradually to replace the native *Nematodirus* spp within three years and to produce clinical disease not previously observed. We can predict that importation of *N. battus* into the United States would cause similar problems.

New diseases can appear when man comes in contact with parasites of other animals. In 1964 the first documented case of intestinal infection of man with a nematode of the genus *Capillaria* was reported (*20, 21*). Subsequently an epidemic occurred in coastal regions of the Philippine Islands, with more than 100 deaths over several years. This nematode has the rare characteristic of producing embryonated eggs *in utero*, a characteristic otherwise known only in a few species of *Capillaria* of bats and fish. Larval stages have been found in freshwater fish eaten raw by the people that were infected, and internal auto-infection is known to occur in monkeys and man (*26*). The only natural definitive host known as yet, however, is man.

Transport of fungi from one region to another may result in a serious epidemic requiring quarantine. There are four species of *Zizania* (wild rice), all closely related, three in North America and one in eastern Asia. The American species are economically important and also play a role in the balance of nature. The Texan *Z. texana* Hitchc., moreover, is an endangered species (E. E. Terrell, unpublished data). The Asiatic *Z. latifolia* (Griseb.) Turcz. ex Stapf has a parasitic fungus, *Ustilago esculenta* H. and P. Sydow, which causes the lower culms to swell; in China, these swollen culms are eaten as a vegetable. *Z. latifolia* has rarely been introduced into the United States, and there is no evidence that the fungus is established here. However additional introductions of the fungus might have serious consequences; its potential effect on the North American *Zizania* is unknown, but it would be wise to regard the potential effect as devastating and therefore to exclude *Z. latifolia* by quarantine or other means.

Shifts in agricultural import practice lead to the importation of insect pests new to our fauna. This concern underlies a significant activity of the USDA, yet we still know very little about the native pests in many parts of the world. Andean potato tuber borers, *Premnotrypes* spp (weevil subfamily Leptopiinae), were first described in 1914, and by 1956 Kuschel (*68*) recognized 11 species. However, the systematics of this group remains poorly understood; in addition, numerous other weevils (Cylindrorhininae, Listroderinae) in this region may also prove to be signficant enemies of potato. As cold-adapted organisms these beetles, if introduced into North America, could cause serious crop damage.

With the increased production of banana, various otiorhynchine and brachyderine weevils, notably *Philicoptus* spp, have in the last decade emerged as serious pests in the Philippines. Efforts at control are hindered by lack of knowledge of systematics and natural history. The species involved are polyphagous, flightless, localized endemic forms, characteristics shared by many of the pest species that have become established in North America (for example, *Otiorhynchus* spp). Therefore, aside from the immediate problem of dealing with these pests in the Philippines, there is a real danger of their establishment in other tropical and subtropical areas where, without natural controls, the beetles may be able to disperse unchecked.

With increasing international commerce we cannot surely prevent, but only delay, the introduction of foreign pests. However, it is imprudent freely to permit new entries of well-established pests, because such new immigrants may be of greater virulence—having evolved in their primary centers of diversity. With all pest introductions, there is delay in combatting outbreaks and spread because of the lack of adequate biosystematic and behavioral information.

Search for biological control agents. Once a native or exotic pest species is established, the search for effective predators or parasites becomes a priority concern; immigrants leave their natural enemies behind, and native pests may outrace theirs. Biosystematic information provides a data base from which to extract clues about where to look for parasite or predator species, be they natural enemies of the pest or enemies of related species. For introduced pests, the best place to search is in the primary center of diversity of the pest. If a native pest has no effective parasites, then the best place to search is in an area where its closest relatives are concentrated.

Alternanthera philoxeroides (Mart.) Griseb. (alligator weed), an important aquatic pest introduced from South America, has effectively been controlled in the southeastern United States by three host-specific insects (*23*). The search for these insects in South America was begun after consideration of the probable center of origin of *Alternanthera*. Similarly, *Chondrilla juncea* L. has spread worldwide from the probable Eurasian center of diversity of the genus, with a progressive decrease in the number of host-specific organisms in the new areas of establishment. The Australian population of introduced *C. juncea* is being controlled by an Italian race of a rust, *Puccinia chondrillina* Bubak and Syd. Many other examples of successful biological control of weeds, supported by biosystematic data, may be cited.

Biosystematic resources have had considerable impact on many biological control programs and they have often been the only reliable sources of essential data on distribution, natural enemies, and diagnostic features. For example Gordon's work (*47*) on the New World Epilachninae has provided a systematic background that enables a search for effective parasites of *Epilachna varivestis* Mulsant (Mexican bean beetle), a joint USDA-State of Maryland project. Efficient parasites have previously been located in India, but they are not adapted for overwintering in a cold climate and must be reinoculated each season. Search for parasites with desired characteristics will be concentrated in high altitude Andean regions where the numerous *Epilachna* spp are inactive for parts of the year.

Misidentifications have frequently caused the failure of otherwise well-executed biological control projects (*29*, see also Rosen, chapter 2, this volume). Although incorrect determinations have sometimes been the result of human error, they normally reflect inadequate systematic knowledge of the pest or natural enemy. Surprisingly, incorrect identification of the pests themselves often has brought about many problems. For example, the mealybug *Planococcus kenyae* (Le Pelley) caused extensive damage to coffee in Kenya and was the target of a major biological control program. Unfortunately the mealybug was misidentified on two successive occasions, first as *P. citri* (Risso) and later as *P. lilacinis* (Cockerell). After each incorrect determination parasites were collected and imported without success. Le Pelley (*70*) realized that the mealybug was different and described it as new; he then located an indigenous range of the pest and introduced effective parasites (*71*).

In many instances the biological control specialist has drawn attention to the fact that a pest previously believed to be one taxon actually is a complex. This information is usually based on field data, particularly on host preference by parasites. For example Bartlett (*5*) noted differences in the susceptibility of various "races" of the black scale, *Saissetia oleae* (Olivier); these differences were causing significant difficulties in the success of the biological control programs of the black scale. Since then, De Lotto (*30*) discovered that *S. oleae* is actually a complex of three species: *S. oleae, S. miranda* (Cockerell and Parrott), and *S. neglecta* De Lotto. Although diagnostic, morphological differences exist among these species, they are so slight that they were previously overlooked. Because the parasites were able to detect differences in hosts, the systematist was alerted to the fact that these seemingly minor variations could be used as diagnostic characters.

Reported cases of cryptic or sibling species are frequent (*96*) although we may be unable readily to distinguish these taxa on a morphological basis. They are real indeed, and failure to recognize them may result in a considerable waste of funds and effort in biological control programs.

Effectiveness of natural enemies may be augmented through intraspecific and interspecific hybridization, although this field remains largely unexplored with respect to entomophagous insects. Thus, it should be possible to produce an effective parasite on a particular organism by combining attributes of different parasites effective on related targets; to improve fecundity; or to modify environmental tolerance.

Search for pollinators. The introduced *Apis mellifera* L. (honey bee) is a good general pollinator but it is less effective than some solitary bees on certain crops. *Megachile pacifica* (Panzer), accidentally introduced into the United States around 1940 from Eurasia, the center of diversity for *Medicago sativa* L. (alfalfa), has become an important, commercially-managed pollinator for this crop in North America and elsewhere (*8*). Other promising Eurasian alfalfa pollinators of the genera *Melitta, Pithitis,* and *Anthidium* are being investigated. *Osmia cornifrons* (Rad.), a Japanese species managed for over thirty years in its center of origin to pollinate apples and pears, was introduced at Beltsville this year. It is more active than honey bees in cool weather and it need not be fed after its activity in spring (S. W. T. Batra, unpublished data). *Peponapis* and *Xenoglossa* bees, which specialize in *Cucurbita* pollination and are highly effective, were left behind when the squashes and gourds were transported from the western hemisphere; their introduction to pollinate this crop may be worthwhile.

OPPORTUNITIES TO IMPROVE PREDICTIVENESS

Opportunities to improve the predictive power of agricultural biosystematics depend upon an improved foundation in basic research. What problems must we pursue in relation to biotic diversity, genetic and ecotypic variability, and coevolution, and how can we turn them to our advantage?

Biotic diversity. Exploration, introduction, and domestication of food plants and animals is a continuous process; many food resources remain underexploited, and others await discovery. Exploitation for other desirable characteristics, especially for drugs and chemicals, is far less complete. Opportunities to increase our resource base for germplasm are rapidly diminishing as tropical forest ecosystems are destroyed, so our need to complete this inventory is urgent and imminent. We need broader collaboration with ethnobiologists working in remote areas and agricultural ecosystematists (see Duke, Chapter 4, this volume), to pinpoint poorly known biota that have particular potential for domestication (*22*). Taxonomic data help define geographic areas in most urgent need of exploration and help narrow the scope of exploration to manageable limits.

We must emphasize the need for continued and expanded support to obtain quality data from all taxonomic groups and geographic areas. The following are illustrative: (1) Of the 15,000 Fungi Imperfecti, only a few hundred are known from the sexual stages; the remaining are known from asexual or conidial forms. Consequently, hypotheses about relationships are either imperfect or lacking (see Luttrell, Chapter 11, this volume; and Kendrick, Chapter 7, this volume); (2) The plant-feeding beetle family Curculionidae contains some 55,000 currently recognized species worldwide, a number reflecting perhaps only 10 to 20 percent of the total fauna. Biological data are lacking for the vast majority. The state of our knowledge is such that we are still unable to make satisfactory predictions as to which species, species groups, or genera are likely to be of agricultural importance.

Genetic and ecotypic variability. A principal characteristic of the biparental species is its variability, both within populations and in a geographic context; highly mutable or ecotypically plastic uniparental species are also found. A general prediction we can make is that any newly introduced biotic or abiotic pressure will result in unexpected adaptation. Thus, selection for resistance to pesticides, altered susceptibility to desirable or undesirable parasites, altered ability to compete successfully, or similar problems will result from environmental disturbances such as cultivation of native plants or introduction of desirable or undesirable exotics.

Morphological relationships enable predictions that agriculturally unknown species, closely related to a well-known cultivated species, also have agricultural potential, or that a little-known species may be susceptible or resistant to particular pests. Opportunity to make such predictions is proportionate to available information. Recently, chemical and karyotype relationships have come to augment such knowledge, and advances in serology and protein electrophoresis further elucidate the relatedness of many crops (*115*). More biosystematic data are needed for many crops, even important well-known ones, such as barley, beans, corn, peanuts, potatoes, rice, sorghum, sugarcane, tomato and wheat. Even the taxonomic status of some *Triticum* spp, *Oryza* spp, and *Solanum* spp is incomplete or unknown .

Two principal factors complicate a better understanding of variability in plant-parasitic nematodes: (1) We have insufficient data on normal variation within species and the mechanisms involved; and (2) Undescribed taxa may be incorrectly identified as known forms (*106*).

Races are known to exist in many major plant parasites. There are at least four races of *Heterodera glycines* Ichinohe (soybean cyst nematode), six or more of *Globodera rostochiensis* (golden nematode of potato) (*32, 60, 61*), at least 20 of *H. avenae* Wollenweber (oat cyst nematode), about 12 of *Ditylenchus dipsaci* (Kuhn) Filipjevo (stem and bulb nematode), and one or more each in several species of *Meloidogyne* (root-knot nematode). Mechanisms of evolution of new races of plant nematodes are not clearly known; hybridization is involved with some as in stem and bulb nematodes, and gene frequencies which change in time in response to selection forces within field populations are thought to account for others, such as *H. glycines.* Regardless of the mechanisms involved, new races do arise under field conditions and often in a resistant crop, which is thus no longer resistant. This is a major concern of plant breeders and nematologists working together to keep resistant varieties available to growers. Since most races are presently indistinguishable except by lengthy host tests, we urgently need more advanced and detailed morphological studies supplemented by both scanning electron microscopy and biochemical data in order readily to identify a given race.

Four main races of soybean cyst nematodes, *Heterodera glycines,* are known. They are determined mainly by their host-variety response (*31, 95, 98, 107*), though morphology of races 1, 2, and 3 is also important (*44*). In response to these findings, Golden *et al.* (*45*) developed standard procedures for testing and characterizing races of this nematode.

Host-adapted races with characteristics distinct from those of other apparently conspecific populations are known for many animal parasite species (*18, 48*). For most naturally occurring races of animal parasites, conspecificity is difficult to determine. We have very little knowledge of the range of variability within even the common parasites of humans and their domestic animals.

Haley (*48*) suggested the use of laboratory-developed, host-adapted races to study the interaction of environment and genotype in nematodes, in susceptible and in previously resistant hosts. Using a laboratory-developed model consisting of a parent race from rats and a hamster-adapted race of *Nippostrongylus brasiliensis* (Travassos) Lane, morphological differences (*73*) similar to those found in *Haemonchus* of sheep, cattle, and other ruminants were found (*18*). Both races had longer spicules in hamsters

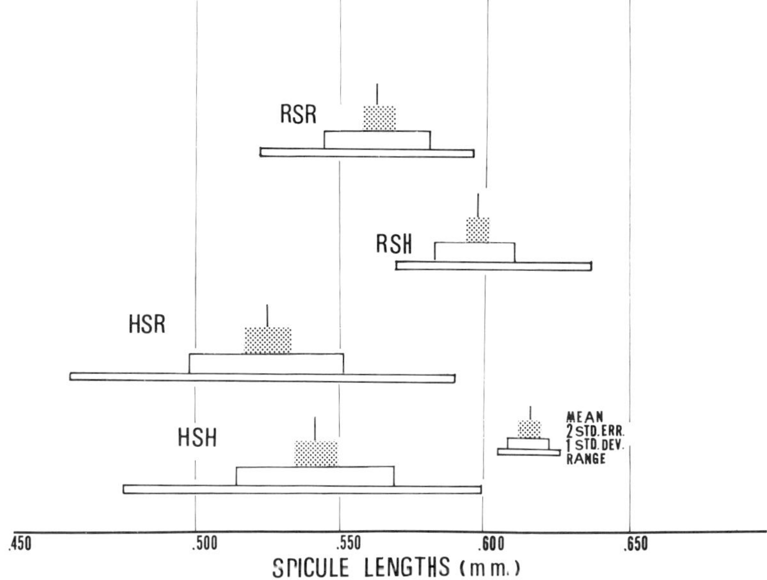

Figure 16.2. Spicule lengths of two strains of *Nippostrongylus brasiliensis* collected from the intestines on the sixth to the tenth days of the infections. The rat strain was from rats (RSR) and hamsters (RSH). The hamster-adapted strain was from rats (HSR) and hamsters (HSH) (after Lichtenfels, ref. *73*).

than in rats, but the hamster-adapted race had shorter spicules in both hosts (Fig. 16.2). If only populations in their own hosts had been compared, the differences in spicule length would have been partially masked by the effects of host species (i.e., genetic differences would have been masked by effects of the environment). The size differences between the races resulted from different growth patterns that developed after 150 nematode generations due to an earlier immune response in the hamster than in the rat.

We cannot hope to acquire adequate distributional data on which to base quarantine policy if we cannot recognize species to determine whether or not they occur in this country. Systematists currently disagree, for example, on whether *Taenia (Multiceps) serialis* (Gervais) Stiles and Stevenson and *Taenia (Multiceps) multiceps* Leske are separate species. The larval form of the first species, believed to have become extinct in the United States (*12*), causes "gid," a central nervous system disorder in cattle and sheep, with the adult form being a tapeworm in dogs. The second species has a larval form in rabbits with the adult in dogs, and it does occur, in the United States.

Biological characteristics of our most important introduced pests have been intensively studied. However, the taxonomic value of such studies has often been neglected with the unfortunate result that definitive taxonomic statements still cannot be made. Severe taxonomic problems may arise if a pest is introduced once and subsequently differentiates into numerous distinctive forms or is introduced on multiple occasions from biologically distinctive source populations. The alfalfa weevil was apparently introduced into North America from three source areas: eastern and western races of *Hypera postica* (Gyllenhal) are widespread, biologically differentiated, and partially intersterile (personal communication from R. F. W. Schroder); and a third form in the southwest [recognized by some workers as a separate species, *H. brunneipennis* (Boheman)] is distinguished biologically but lacks precise morphological definition. How should these races be treated taxonomically? In general, no amount of intensive study of American populations alone can resolve the taxonomic problem posed by such introductions—since introduction of biologically intermediate forms might destroy intersterility barriers. Answers must be sought, once again, in primary centers of diversity, using diverse procedures such as clinal analysis ((*121*), hybridization studies, and karyotypes and electrophoretic analyses.

If a "species" includes populations having notably different and disjunctive biological characteristics, it may be a complex containing two or more reproductively isolated biological species. If such a "species" includes pest races, particularly in a biological control context, it becomes imperative to understand just what the biological limits of the species are and what the most vulnerable period in its life cycle is. For example, the imported fire ant in the southeastern United States was long considered to be one species, but actually two species (*Solenopsis richteri* Forel and *S. invicta* Buren) had been imported (*15*); this is critical information for investigators seeking specific parasites for biological control.

Heritable and environmentally induced variations are common among the fungi. Many fungi are pleomorphic, having more than one sporiferous or otherwise morphologically distinct stage in life. Thus, many parasites of warm-blooded animals are yeastlike at about 37°C but are filamentous at room temperature. Similarly, many species (from diverse genera, families, and orders) that are mutualistic with insects are yeastlike under the influence of their macrosymbionts but otherwise are filamentous (*6, 7, 9*). Some common insect pests of stored grains or grain products (*Lasioderma, Oryzaephilus, Sitophilus*) cannot survive without their species-specific, endosymbiotic fungi and/or bacteria (*66*). *Dacus oleae* (Gmelin) (olive fly), *Hylemyia* (potato fly), and *Blatta orientalis* L. (oriental cockroach) are other important pests with endosymbionts (*14*). These pests, usually controlled by pesticides, can also be checked by low, experimental doses of certain antibiotics and sulfa drugs aimed at the microsymbionts. Unfortunately, most such microsymbionts, contained in specialized pockets called mycetomes, are still uncharacterized (*14*).

Apart from studies of *Neurospora, Schizophyllum, Aspergillus,* and the economically important *Saccharomyces cerevisiae* Meyen ex Hansen, opportunities to investigate speciation in fungi are most promising in research on variability of obligate parasites. In addition to hybridization and mutation, parasexuality and somatic recombination are very significant in bringing about variation (*16, 28, 38*). Such infraspecific taxa as varieties or *formae specialis,* races, and biotypes are known for many rusts, smuts, powdery mildews, downy mildews, and other agriculturally important fungi. Concommitant with the development and release of newly tailored crops, we must also vigorously investigate variability and speciation of these groups.

Detection of races and the ever-present genetic variability in plant pathogens has provided 1) a definitive proof of the hypothesis of coevolution of hosts and parasites or "gene-for-gene interdependence"; 2) an impetus and a reason to preserve susceptible genotypes, which must be available before one can recognize resistance in domesticated species or their wild relatives; and 3) an incentive to maintain a continuous, vigorous, permanent program of breeding for major crops. How long any new resistant variety takes to succumb to a virulent race depends on its genetic variability with respect to pathogenicity.

Coevolution. Coevolution in our text refers to the reciprocal adaptations expressed between food-supplying and food-consuming organisms. It refers to both supraspecific and infraspecific relationships, therefore representing facets and extensions of the phenomena of diversity and variability treated above. We here examine some of the problems posed by coevolution (*50*) and offer some suggestions as to how to turn them to our advantage, with particular reference to plant-parasitic nematodes and animal parasites.

Relationships among parasites have long been used to help clarify evolutionary relationships among their hosts (*81*). Thus, taxonomic relationships among lice or fleas may clarify ordinal or family relationships among their mammalian or avian hosts. As another example, the bruchid seed beetle assemblage associated with seeds of *Cassia* section *Cassia* is entirely different from that associated with other sections of *Cassia,* implying that the relationship of *Cassia* section *Cassia* to other sections of the genus is remote (*122*). Holm (*55*) placed the grass genera *Molinia* and *Spartina* respectively in the tribes Arundineae and Chlorideae from analyses of host ranges of *Puccinia* spp.

What factors govern a gradual coadaptation between food and prey, as opposed to a major host shift? Most members of the seed beetle genus *Caryedes* attack members of the fabaceous legume tribe Phaseoleae, but members of one species group attack a section of the caesalpiniaceous legume genus *Bauhinia* (*64*). Why? We assume that defensive mechanisms of these divergent plant groups are sufficiently similar to have permitted what on the surface would appear to be a major host shift.

Valuable data on the migratory history and origin of various races of man have been provided by their parasite fauna (*27*). Parasites are being used as biological tags better to understand migratory patterns and systematic relationships of birds and fish. Mayr (*77*) pointed out that "Sibling species often differ in the number of the kind of parasites they carry" and cited the following examples: 1) two sibling species of California *Octopus* were initially distinguished on the basis of mesozoan parasites (*90*); 2) two sibling termites *(Nasutitermes)* were initially distinguished by differences in staphylinid beetle faunas in their nests (*34*); 3) two sibling turbellarians were initially distinguished by different levels of parasitization by a peritrichan (*91, 92*); and 4) sibling species have been distinguished by parasite susceptibility in fish (*63*), wasps (*13*) and homopterous insects (*69, 76*).

Larval parasitic helminths are posing severe tests for current classification systems. As our food needs increase, we exploit new sources, often invertebrates, and notably seafood. Clams and crabs, only recently heavily exploited (for example, surf clams, *Spisula solidissima* Dillwyn), have become important. We are finding parasites in these new foods, especially larval forms (*75*). Similarly some microorganisms found in Chesapeake Bay bivalves must be evaluated as potential parasites of animals and man (*52*). Man (*74*) and swine (*4, 113*) are known to become infected by larval nematodes obtained from fish. We should expect to find larval forms of vertebrate parasites in invertebrates. Much of the necessary information to identify larval specimens has not been developed, and critical research is needed to ascertain their medical and veterinary significance. Classification of larval forms will be possible only with systems that reflect evolutionary patterns.

By comparing parasites of Old and New World primates, it can be demonstrated that Old World primates are more similar in their parasite faunas and will probably be more suitable for research on medical problems of man. On the other hand, parasites of New World primates are more similar to those of dogs, rodents, and insectivores (*19*) and thus are less suitable for research on medical problems.

Coevolution is not restricted to host-parasite relationships. A specialized pollinator is required for successful fruit set in certain crops because coevolution between the two resulted in specialized adaptations. Under agricultural conditions the efficiency of a natural pollinator (*8*) may be reduced, as commercially grown crops may be less attractive because of artificial selection, or as monoculture may reduce the supply of alternate food sources for the pollinator.

Clearly, our search for biological control agents should be guided by understanding coevolutionary relationships. If we know that related parasites have very different hosts, we know that potential for useful discovery is relatively small; but if the related parasites tend to attack related hosts, then our potential for successful exploration is relatively great.

It is of utmost importance to have detailed information on the geographic distribution and frequency of races of important pests on all relatives of agricultural plants and animals on a world-wide basis. Unfortunately, extensive, reliable data have not been accumulated for most such pests. For example, we have little such data for the important plant-parasitic nematodes and fungi, except for some isolated species within a specific country. However, considerable information is available for the golden nematode of potato, perhaps our most studied plant nematode; this species is thought to have originated in or near Peru along with the potato, was probably introduced into Europe with potatoes from South America in the early 1600s, and has since spread to about 50 countries world-wide (*43, 101*). The soybean cyst nematode, described from Japan in 1952 (*58*), is of obscure origin and relatively limited distribution but may have developed in the Orient with the culture of soybeans and related crops; several races are now known, but the total picture is far from complete.

In studying host-adapted nematode races, Haley (*49*) found that significant adaptation to the golden hamster, *Mesocricetus auratus* (Waterhouse), occurred rapidly (within two to eight generations of the nematode *Nippostrongylus brasiliensis*), and that maximum development in hamsters was reached after 16 to 24 nematode generations. A similar adaptation to a changed environment occurs when parasites are exposed to antiparasitic treatment of the host. Adaptation of *Eimeria tennella* (Railliet and Lucet) coccidia of chickens, to antiparasitic treatment has been shown to occur within four to eight serial exposures to drugs (*78, 79, 80*). How will the phenomenon of adaptation be expressed with biological control agents? A successful parasite usually does not kill its host. Will the biological control agents lose much of their effectiveness after a few parasite generations? We should study the phenomenon of adaptation to determine how it can be slowed in order to extend the useful life of resistant hosts and chemical and biological control agents. Should we, for example, try to minimize contact between treated and untreated parasite populations, or would adaptation be slower if gene flow among parasite populations is permitted? Basic work is needed on the relative contributions of selection pressure and isolation versus gene flow in the evolutionary process (*33, 35*). Answers to these basic questions will have important implications for the management of resistant host varieties, antiparasitic treatments, and biological control agents.

SUMMARY AND CONCLUSIONS

Biosystematic contributions toward acquiring, improving, and protecting food resources are fundamental. Their value is most evident when we can apply them to forecast future events or predict results of future operations. They have major importance in forecasting results of crop and pest introductions and in the search for biocontrol agents and pollinators. Primary centers of diversity provide the broadest base from which to acquire, improve, and protect agricultural biota, and these need detailed exploration before they are destroyed by human encroachment. Data on host range and ecological amplitudes along with data on taxonomic relationships form the basis for predicting pest potential. Data on inoculum load, population dynamics, climatic conditions, and the like combine with taxonomic data to enable forecasting of epidemics.

Opportunites to improve predictiveness of systematics, in relation to problems posed by diversity, variability, and coevolution, increase in proportion to the data base as supplied by basic research. Even among relatively well-known cultivated and pest species, much additional work needs to be done to counter weak spots, whether obvious or subtle, in our defences against existing or future depredations. For example, short-lived phenomena such as endemic pest outbreaks, wherever they occur, need critical study to discover susceptible genomes and thereby identify and select genes for resistance. As exemplified by the 1970 outbreak of southern corn blight in the United States, some overseas pest problems of today will become our problems of tomorrow.

Improved communication and collaboration, including computer linkages, are needed among all agricultural disciplines to coordinate existing data not widely available to systematists and to focus on particular, timely systematic problems. Prospects are bright for agricultural biosystematics to increase in fundamental importance and application.

ACKNOWLEDGMENTS

We thank the following Agricultural Research Service (ARS) scientists for suggestions: J. Duke, G. Gordh, R. D. Gordon, L. Knutson, E. E. Leppik, P. L. Lentz, and R. F. W. Schroder, and we likewise thank M. J. Shannon, of the Animal and Plant Health Inspection Service. Douglass Miller and S. W. T. Batra, ARS, respectively wrote paragraphs on scale insects and on pollinators and biological control of weeds.

LITERATURE CITED

1. Alicata, J. E. 1962. *Angiostrongylus cantonensis (Nematoda: Metastrongylidae) as a causative agent of eosinophilic meningitis of man in Hawaii and Tahiti*. Can. J. Zool. 40: 5–8.

2. Allen, M. W., and S. A. Sher. 1967. *Taxonomic problems concerning the phytopathogenic nematodes*. Ann. Rev. Phytopathol. 5: 247–264.

3. Anderson, R. C. 1971. *Lungworms*. Pages 81–126 *in* J. W. Davis and R. C. Anderson, eds. *Parasitic diseases of wild mammals*. Iowa State University Press, Ames.

4. Ashizawa, H., D. Nosaka, S. Tateyama, and M. Usui. 1973. *Studies on the swine gastric-anisakiasis. III. Pathohistological findings in the parasitic areas*. Bull. Fac. Agr., Miyazaki Univ. 20: 191–201.

5. Bartlett, B. R. 1960. *Biological races of the black scale, Saissetia oleae, and their specific parasites*. Ann. Entomol. Soc. Amer. 53: 383–385.

6. Batra, L. R. 1966. *Ambrosia fungi: extent of specificity to ambrosia beetles*. Science 153: 193–195.

7. Batra, L. R. 1967. *Ambrosia fungi: a taxonomic revision and nutritional studies of some species.* Mycologia 59: 976–1017.

8. Batra, L. R., S. W. T. Batra, and G. E. Bohart. 1973. *The mycoflora of domesticated and wild bees (Apoidea).* Mycopath. Mycol. Appl. 49: 13–44.

9. Batra, S. W. T., and L. R. Batra. 1967. *The fungus gardens of insects.* Sci. Amer. 217: 112–120.

10. Becklund, W. W. 1964. *Revised check list of internal and external parasites of domestic animals in the United States and possessions and in Canada.* Amer. J. Vet. Res. 35: 1380–1416.

11. Becklund, W. W. 1968. *Ticks of veterinary significance found on imports in the United States.* J. Parasitol. 54: 622–628.

12. Becklund, W. W. 1970. *Current knowledge of the gid bladder worm, Coenurus cerebralis (=Taenia multiceps), in North American domestic sheep, Ovis aries.* Proc. Helm. Soc. Wash. 37: 200–203.

13. Bohart, G. E. 1942. *Notes on some feeding and hibernation habits of California Polistes.* Pan-Pacific Entomol. 18: 30.

14. Buchner, P. 1965. *Endosymbiosis of animals with plant microorganisms.* Wiley, New York. 909 pp.

15. Buren, W. F. 1972. *Revisionary studies on the taxonomy of the imported fire ants.* J. Georgia Entomol. Soc. 7: 1–26.

16. Byrde, R. J. W., and C. V. Cutting, [eds.]. 1973. *Fungal pathogenicity and the plant's response.* Academic Press, New York. 500 pp.

17. Campbell, R. W., M. G. Miller, E. J. Duda, C. E. Biazak, and R. J. Sloan. 1976. *Man's activities and subsequent gypsy moth egg-mass diversity along the forest edge.* Environ. Entomol. 5: 273–276.

18. Chitwood, M. B. 1957. *Intraspecific variation in parasitic nematodes.* Syst. Zool. 6: 19–23.

19. Chitwood, M. B. 1970. *Comparative relationships of some parasites of man and old and new world subhuman primates.* Lab. Animal Care 20: 389–394.

20. Chitwood, M. B., C. Valesquez, and N. P. Salazar. 1966. *Physiological changes in a species of Capillaria (Trichuroidea) causing a fatal case of human intestinal capillariasis.* First Int. Cong. Parasitol., Proc., 2: 797–798.

21. Chitwood, M. B., C. Velesquez, and N. G. Salazar. 1968. *Capillaria philippinensis sp. n. (Nematoda: Trichinellida), from the intestine of man in the Philippines.* J. Parasitol. 54: 368–371.

22. Committee on genetic vulnerability of major crops, National Research Council. 1972. *Genetic vulnerability of major crops.* U.S. Natl. Acad. Sci., Washington. 307 pp.

23. Coulson, J. R. 1977. *Biological control of alligator weed, 1959–1972, A review and evaluation.* U. S. Dept. Agric. Tech. Bull. 1547, 98 pp.

24. Creech, J. L., and L. P. Reitz. 1971. *Plant germplasm now and for tomorrow.* Adv. Agron. 23: 1–49.

25. Cronquist, A. 1969. *On the relationship between taxonomy and evolution.* Taxon 18: 177–187.

26. Cross, J. H., T. Banzon, M. D. Clarke, V. Basaca-Servilla, R. H. Watten, and J. J. Dizon. 1972. *Studies on the experimental transmission of Capillaria philippinensis in monkeys.* Trans. Royal Soc. Trop. Med. & Hyg. 66: 819–827.

27. Darling, S. T. 1921. *The distribution of hookworms in the zoological regions.* Science 53: 323–324.

28. Day, P. R. 1960. *Variation in phytopathogenic fungi.* Ann. Rev. Microbiol. 14: 1–16.

29. DeBach, P. 1960. *The importance of taxonomy to biological control as illustrated by the cryptic history of Aphytis holoxanthus n.sp., a parasite of Chrysomphalus aonidum, and Aphytis coheni n.sp., a parasite of Aonidiella aurantii.* Ann. Entomol. Soc. Amer. 53: 701–705.

30. DeLotto, G. 1971. *A preliminary note on the black scales of North and Central America.* Bull. Entomol. Res. 61: 325–326.

31. Duclos, L. A., and J. M. Epps. 1970. *A new threat to soybeans from the cyst nematode.* Soybean Digest 30: 10-11.

32. Dunnett, J. M. 1957. *Variation in pathogenicity of the potato root eelworm (Heterodera rostochiensis Woll.) and its significance in potato breeding.* Euphytica 6: 77-89.

33. Ehrlich, P. R., and P. H. Raven. 1969. *Differentiation of populations.* Science 165: 1228-1232.

34. Emerson, A. E. 1935. *Termitophile distribution and quantitative characters as indicators of physiological speciation in British Guiana termites.* Ann. Entomol. Soc. Amer. 28: 369-395.

35. Endler, J. A. 1973. *Gene flow and population differentiation.* Science 179: 243-250.

36. Erwin, D. C., G. A. Zentmyer, J. Galindo, and J. S. Niederhauser. 1963. *Variation in the genus Phytophthora.* Ann. Rev. Phytopathol. 1: 375-396.

37. Evans, A. A. F., and R. N. Perry. 1976. *Survival strategies in nematodes.* Pages 382-424 *in* Neil A. Croll, ed. *The Organization of Nematodes.* Academic Press, New York.

38. Fincham, J. R. S., and P. R. Day. 1971. *Funal genetics.* Blackwell, Oxford. 402 pp.

39. Flor, H. H. 1971. *Current status of the gene-for-gene concept.* Ann. Rev. Phytopathol. 9: 275-296.

40. Foreyt, W. J., and A. C. Todd. 1976. *Development of the large American liver fluke, Fascioloides magna, in white-tailed deer, cattle, and sheep.* J. Parasitol. 62: 26-32.

41. Frankel, O. H., and E. Bennett, eds. 1970. *Genetic resources in plants—their exploration and conservation.* Davis Co., Philadelphia. 554 pp.

42. Gallegly, M. E. 1968. *Genetics of pathogenicity of Phytophthora infestans.* Ann. Rev. Phytopathol. 6: 375-396.

43. Golden, A. M., and D. M. Ellington. 1972. *Redescription of Heterodera rostochiensis (Nematoda: Heteroderidae) with a key and notes on closely related species.* Proc. Helm. Soc. Wash. 39: 64-78.

44. Golden, A. M., and J. M. Epps. 1965. *Morphological variations in the soybean cyst nematode.* Nematologica 11: 3844 (Abstr.)

45. Golden, A. M., J. M. Epps, R. D. Riggs, L. A. Duclos, J. A. Fox, and R. L. Bernard. 1970. *Terminology and identity of infraspecific forms of the soybean cyst nematode (Heterodera glycines).* Plant Dis. Reptr. 54: 544-546.

46. Goodey, J. B., M. T. Franklin, and D. J. Hooper. 1965. *T. Goodey's "The nematode parasites of plants catalogued under their hosts."* Edit. 3: Commonwealth Agricultural Bureaux. 214 pp.

47. Gordon, R. D. 1975. *A revision of the Epilachninae of the Western Hemisphere.* U.S. Dept. Agric. Tech. Bull. 1493. 409 pp.

48. Haley, A. J. 1962. *Role of host relationships in the systematics of helminth parasites.* J. Parasitol. 48: 671-678.

49. Haley, A. J. 1966. *Biology of the rat nematode, Nippostrongylus brasiliensis (Travassos, 1914). IV. Characteristics of N. brasiliensis after 1 to 40 serial passages in the Syrian hamster.* J. Parasitol. 52: 109-116.

50. Harlan, J. R. 1976. *Diseases as a factor in plant evolution.* Ann. Rev. Phytopathol. 14: 31-51.

51. Harrar, J. G. 1961. *Socioeconomic factors that limit needed food production and consumption.* Fed. Proc. 381-383.

52. Harshbarger, J. C., S. C. Chang, and S. V. Otto. 1977. *Chlamydiae (with phages), mycoplasmas, and rickettsiae in Chesapeake Bay bivalves.* Science 196: 666-668.

53. Helle, O. 1969. *The introduction of Nematodirus battus (Croften and Thomas, 1951) into a new environment.* Vet. Rec. 84: 157-160.

54. Hibler, C. P., and J. L. Adcock. 1971. *Elaeophorosis.* Pages 263-278, *in* J. D. Davis and R. C. Anderson, eds. *Parasitic diseases of wild mammals.* Iowa State Univ. Press, Ames.

55. Holm, L. 1969. *An uredinological approach to some problems in angiosperm taxonomy.* Nytt Mag. Botan. 16: 147-150.

56. Holton, C. S., J. A. Hoffman, and R. Duran. 1968. *Variation in Smut fungi*. Ann. Rev. Phytopathol. 6: 213-242.
57. Hooker, 1967. *The genetics and expression of resistance in plants to rusts of the genus Puccinia*. Ann. Rev. Phytopathol. 5: 163-182.
58. Ichinohe, M. 1952. *On the Soy Bean Nematode, Heterodera glycines n.sp., from Japan*. Mag. Appl. Zool. 17: 1-4.
59. Isaac, I. 1967. *Speciation in Verticillium*. Ann. Rev. Phytopathol. 5: 201-222.
60. Jones, F. G. W. 1975. *Rothamsted Report for 1975*. Part 1. 191-212.
61. Jones, F. G. W., and K. Pawelska. 1963. *The behaviour of populations of potato-root eelworm (Heterodera rostochiensis Woll.) towards some resistant tuberous and other Solanum species*. Ann. Appl. Biol. 51: 277-294.
62. Kalkstein, L. S. 1976. *Effects of climatic stress upon outbreaks of the southern pine beetle*. Environ. Entomol. 5: 653-658.
63. Keleher, J. J. 1952. *Growth and Triaenophorus parasitism in relation to taxonomy of Lake Winnipeg ciscoes (Leucichthys)*. J. Fish. Res. Board Canada 8: 469-478.
64. Kingsolver, J. M., and D. R. Whitehead. 1974. *Classification and comparative biology of the seed beetle genus Caryedes Hummel (Coleoptera: Bruchidae)*. Trans. Amer. Entomol. Soc. 100: 341-436.
65. Knipling, E. F. 1972. *Simulated population models to appraise the potential for suppressing sugarcane borer populations by strategic releases of parasite Lixophaga diatraeae*. Environ. Entomol. 1: 1-6.
66. Koch, A. 1967. *Insects and their endosymbionts*. Pages 1-106 *in* S. M. Henry, ed. *Symbiosis II*. Academic Press, New York.
67. Kuckuck, H. 1962. *Vavilov's Genzentrentheorie im heutigen Sicht*. 3rd Cong. Eur. Assoc. Res. Pl. Breed. Eucarpia, [Paris] 1962: 177-196.
68. Kuschel, G. 1956. *Revision de los Premnotrypini y adiciones a los Bagoini*. Bol. Mus. Nac. Hist. Nat. (Chile) 26: 187-235.
69. Lal, K. B. 1934. *Insect Parasites of Psyllidae*. Parasitology 26: 325-334.
70. Le Pelley, R. H. 1935. *The common coffee mealy-bug of Kenya*. Stylops 4: 185-188.
71. Le Pelley, R. H. 1968. *Pest of coffee*. Longmans, Green and Co., London. 590 pp.
72. Leppik, E. E. 1970. *Gene centers of plants as sources of disease resistance*. Ann. Rev. Phytopathol. 8: 323-344.
73. Lichtenfels, J. R. 1971. *Changes in the phenotype of the rat nematode, Nippostrongylus brasiliensis, after one and 150 nematode generations in hamsters*. J. Parasitol. 57: 517-525.
74. Lichtenfels, J. R., and F. P. Brancato. 1976. *Anisakid larva from the throat of an Alaskan Eskimo*. Amer. J. Trop. Med. & Hyg. 25: 691-693.
75. Lichtenfels, J. R., J. G. Kern, D. E. Zwerner, J. W. Bier, and P. A. Madden. 1976. *Anisakid nematode in shellfish of Atlantic continental shelf of North America*. Trans. Amer. Micros. Soc. 95: 265-266.
76. Maramorosch, K. 1958. *Studies of aster yellows virus transmission by the leafhopper species Macrosteles fascifrons Stal and M. laevis Ribaut*. 10th Int. Cong. Entomol., Proc., 3: 221-228.
77. Mayr, E. 1963. *Animal species and evolution*. Harvard Univ. Press, Cambridge, Massachusetts, 797 pp.
78. McLoughlin, D. K. 1970. *Efficacy of buquinolate against ten strains of Eimeria tenella and the development of a resistant strain*. Avian Dis. 14: 126-130.
79. McLoughlin, D. K., and M. B. Chute. 1971. *Efficacy of decoquinate against eleven strains of Eimeria tenella and development of a decoquinate-resistant strain*. Avian Dis. 15: 342-345.
80. McManus, E. C., W. C. Campbell, and A. C. Cuckler. 1968. Development of resistance to quinoline coccidiostats under field and laboratory conditions. J. Parasitol. 54: 1190-1193.
81. Meeuse, A. D. J. 1973. *Co-evolution of plant hosts and their parasites as a taxonomic tool*. Pages 289-316 *in* V. H. Heywood, ed. *Taxonomy and Ecology*. Academic Press, New York.

82. Metcalf, C. L., and W. P. Flint. 1951. *Destructive and useful insects*. McGraw-Hill, New York. 1070 pp.

83. Michel, J. F., M. B. Lancaster, and C. Hong. 1972. *Host induced effects on the vulval flap of Ostertagia ostertagi*. Int. J. Parasitol. 2: 305-317.

84. Miller, C. A., and T. R. Renault. 1976. *Incidence of parasitoids attacking endemic spruce budworm (Lepidoptera: Tortricidae) populations in New Brunswick*. Can. Entomol. 108: 1045-1052.

85. Miller, P. R. 1969. *Effect of environment on plant diseases*. Phytoprotection 50: 81-94.

86. Miller, P. R., and M. J. O'Brien. 1957. *Prediction of plant disease epidemics*. Ann. Rev. Microbiol. 11: 77-110.

87. Moseman, J. G. 1966. *Genetics of Powdery Mildews*. Ann. Rev. Phytopathol. 4: 269-290.

88. Ollerenshaw, C. B., and L. P. Smith. 1966. *An empirical approach to forecasting the incidence of nematodiriasis over England and Wales*. Vet. Rec. 79: 536-540.

89. Parmeter, J. R., Jr., W. C. Snyder, and R. E. Reichle. 1963. *Heterokaryosis and variability in plant-pathogenic fungi*. Ann. Rev. Phytopathol. 1: 51-76.

90. Pickford, G. E., and B. H. McConnaughey. 1949. *The Octopus bimaculatus problem: a study in sibling species*. Bull. Bingham Oceanographic Coll., Peabody Mus. Nat. Hist., Yale Univ. 12: 1-66.

91. Reynoldson, T. B. 1948. *British species of Polycelis (Platy-helminthes)*. Nature 162: 620-621.

92. Reynoldson, T. B. 1956. *The population dynamics of host specificity in Urceolaria mitra (Peritricha) epizoic on freshwater triclads*. J. Animal Ecol. 25: 127-143.

93. Roane, C. W. 1973. *Trends in breeding for disease resistance in crops*. Ann. Rev. Phytopathol. 11: 463-486.

94. Rosen, L., R. Chappell, G. L. Laqueur, G. D. Wallace, and P. P. Weinstein. 1962. *Eosinophilic meningoencephalitis caused by a metastrongylid lung-worm of rats*. J. Amer. Med. Assoc. 179: 620-624.

95. Ross, J. P. 1962. *Physiological strains of Heterodera glycines*. Plant Dis. Reptr. 46: 766-769.

96. Schlinger, E. I., and R. L. Doutt. 1964. *Systematics in relation to biological control*. Pages 247-280 *in* P. DeBach, and E. I. Schlinger, eds. *Biological control of insect pests and weeds*. Chapman and Hall, London.

97. Severinghaus, C. W., and R. W. Darrow. 1976. *Failure of elk to survive in the Adirondacks*. N.Y. Fish and Game J. 23: 98-99.

98. Smart, G. C., Jr. 1964. *Physiological strains and one additional host of the soybean cyst nematode, Heterodera glycines*. Plant Dis. Reptr. 48: 542-543.

99. Spaulding, P. 1958. *Diseases of foreign forest trees growing in the United States*. U. S. Dept. Agric. Handb. 139. 118 pp.

100. Spaulding, P. 1961. *Foreign diseases of forest trees of the world*. U. S. Dept. Agric. Handb. 197. 361 pp.

101. Spears, J. F. 1968. *The Golden Nematode Handbook*. U. S. Dept. Agric. Handb. 353. 81 pp.

102. Stakman, E. C., and J. J. Christensen. 1960. *The problems of breeding resistant varieties*. Pages 567-624 *in* J. G. Horsfall and A. E. Dimond, eds. *Plant Pathology*. Academic Press, New York.

103. Stecker, P. G. 1968. *The Merck index*. Ed. 8. Merck and Co., Rahway, N. J. 1713 pp.

104. Stevenson, F. J., and H. A. Jones. 1953. *Some sources of resistance in crop plants*. Pages 192-216 *in Plant diseases, the yearbook of agriculture, 1953*. U. S. Dept. Agric. Washington, D. C.

105. Stolp, H., M. P. Starr, and N. L. Bargent. 1965. *Problems in speciation of phytopathogenic pseudomonads and xanthomonads*. Ann. Rev. Phytopathol. 3: 231-269.

106. Stone, A. R. 1972. *Heterodera pallida n. sp. (Nematoda: Heteroderidae), a second species of potato cyst nematode*. Nematologica 18: 591-606.

107. Sugiyama S., K. Hiroma, T. Miyahara, and K. Kogubun. 1968. *Studies on the resis-*

tance of soybean varieties to soybean cyst nematode. II. Difference of physiological strains of the nematode from Kariwano and Kikyogahara. Japan. J. Breeding 18: 206-212.

108. Terrell, E. E. 1968. *A taxonomic revision of the genus Lolium.* U. S. Dept. Agric. Tech. Bull. 1392. 65 pp.

109. Thorne, G. 1935. *The sugar beet nematode and other indigenous nemic parasites of shadscale.* J. Agric. Res. 51: 509-516.

110. Trevino, G. S. 1975. *Foreign animal disease control programs in the United States.* J. Amer. Vet. Med. Assoc. 167: 459-462.

111. United States Department of Agriculture. 1960. *Index of plant diseases in the United States.* U. S. Government Printing Office, Washington, D. C. 531 pp.

112. United States Department of Agriculture. 1966-1976. *Index-Catalogue of Medical and Veterinary Zoology.* Supplements 16-20. U. S. Government Printing Office, Washington, D. C. [about 8000 pp.].

113. Usui, M., H. Ashizawa, D. Nosaka, and S. Tateyama. 1973. *Studies on the swine gastric-anisakiasis. I. Morphological findings on worms.* Bull. Fac. Agric., Miyazaki Univ. 20: 169-177.

114. Vanderplank, J. E. 1975. *Principles of plant infection.* Academic Press, New York. 216 pp.

115. Vaughan, J. G. 1977. *A multidisciplinary study of the taxonomy and origin of Brassica crops.* BioScience 27: 35-40.

116. Vavilov, N. I. 1928. *Geographische Genzentren unserer Kulturpflanzen.* Int. Kong. G. Vererb. Wiss. [1927] Z. Indukt. Abstramm. Vererblehre Suppl. 1: 342-369.

117. Vavilov, N. I. 1949-50. *The origin, variation, immunity and breeding of cultivated plants.* [Transl. from Russian by K. S. Chester]. Chronica Botanica 13: 1-364.

118. Walter, J. M. 1967. *Hereditary resistance to disease in tomato.* Ann. Rev. Phytopathol. 5: 131-162.

119. Watson, A. J. 1971. *Foreign bacterial and fungus diseases of food, forage, and fiber crops: an annotated list.* U. S. Dept. Agric. Handb. 418. 111 pp.

120. Watson, I. A. 1970. *Changes in virulence and population shifts in plant pathogens.* Ann. Rev. Phytopathol. 8: 209-230.

121. Whitehead, D. R. 1972. *Classification, phylogeny, and zoogeography of Schizogenius Putzeys (Coleoptera: Carabidae: Scaritini).* Quaest. Entomol. 8: 131-348.

122. Whitehead, D. R., and J. M. Kingsolver. 1975. *Megasennius, a new genus for Acanthoscelides muricatus (Sharp) (Coleoptera: Bruchidae), a seed predator of Cassia grandis L. (Caesalpiniaceae) in Central America.* Entomol. Soc. Wash., Proc., 77: 460-465.

123. Zadoks, J. C. 1972. *Methodology of epidemiological research.* Ann. Rev. Phytopathol. 10: 253-276.

124. Zeven, A. C., and P. M. Zhukovsky. 1975. *Dictionary of cultivated plants and their centres of diversity.* Centre for Agric. Pub. & Doc. Wageningen. 219 pp.

125. Zhukovsky, P. M. 1965. *Main gene centres of cultivated plants and their wild relatives within the territory of the U.S.S.R.* Euphytica 14: 177-188.

six

GENERAL OVERVIEW

17] Biosystematics, Taxonomy, and the Practical Man

(an after-banquet lecture)

by J. HESLOP-HARRISON*

ABSTRACT

Users of taxonomy who deal with living organisms in some practical way are not as yet extracting all the benefits they might from the newer insights into the sources, causes, and nature of biological variation now being gained from biosystematics. Biologists work within a classificatory and nomenclatural system based upon orthodox Linnean-style systematics, and it is not now conceivable that this should be replaced by any other for the general taxonomy of organisms. Yet it does not provide an effective vehicle for communicating complex biosystematic data, particularly at the infraspecific level. The requirement now is to make such data more readily accessible to potential users. At a cost, this could be done for the whole of the living kingdoms by exploiting the potential of electronic data processing methods. We are a long way from being able to do this; but it is already realistic to think of dealing with the data of agriculturally-oriented systematics in this way. It should now be the aim to set up the appropriate systems, keyed into the general taxonomic system, to handle biosystematic data pertinent to economically important groups. The returns in greater effectiveness and efficiency would quickly repay the initial expenditures involved.

INTRODUCTION

"The truths of science," wrote Lancelot Hogben in his *Science for the Citizen (3)*, "are the recipes for human action." In 1938 when these words were written, biosystematics had not been named as a discipline; the collection of essays to be published the following year under the title *The New Systematics* (*4*) and under the editorship of Julian Huxley was still an embryo; and biologists were still inclined to raise an eyebrow at the quaint idea that an experimental element might find its way into taxonomy. Now, nearly forty years later, we meet here at Beltsville to review the role of the new systematics in agriculture—to see how *its* truths are being, or can be, adopted as recipes for human action to help to meet the most urgent of contemporary needs, that of feeding and maintaining our own species.

*Royal Society Research Professor, Welsh Plant Breeding Station, University of Wales, Aberystwyth SY23 3EB, U. K.

The new systematics holds the center of the stage in this symposium; but I propose to begin this talk by saying something about the *old* systematics—classical, orthodox or Linnean-style taxonomy, or whatever we care to call it—and its functions, for we should not forget that it continues to serve a multitude of users. Those "users"—who as E. S. Lutrell (Chapter 11, this volume) has reminded us also include "makers"—form a heterogeneous group. In effect they constitute all of those many groups of people whose job it is to deal with organisms in some practical way, including our fellow biologists and laboratory workers as well as the producers of our food and the guardians of our health. Their *needs* are also heterogeneous, and perhaps many do not themselves know what they really are. But there is certainly one need common to all. Broadly speaking, the clients of taxonomy and taxonomic services begin with the same demand—for a *name*. Now Shakespeare was undoubtedly right in asserting that the properties of the genus *Rosa* are not affected by the taxonomic nomenclature we care to apply to its species; but that naming plays a vital part in taxonomic communication will need no emphasis to anyone attending this symposium. The institutes providing taxonomic services—like Kew, and like Beltsville—score their productivity by reference to the numbers of identifications carried out and names supplied; and the do-it-yourself user is inclined to rank the value of floras and faunas by the utility of their keys and the number of times they lead him to a tolerably "right" answer.

The demand for a name for an organism is, in all fairness, wholly rational and thoroughly well justified, given certain assumptions. The name, so we are often told, provides the key to the literature. Given the name, the practical man can hope to do many things. He can hope to find out where an organism stands in the scheme of things; he can hope to learn how to culture or kill it; and he is equipped with the means of conveying his own experience to others. W. B. Turrill, whose concepts of "alpha" and "omega" taxonomy are now ingrained in the literature, talked of taxonomy becoming the repository of a very considerable part of the totality of biological knowledge (*11*). In a literal sense this scarcely has much meaning; but I have always supposed that Turrill's intention was to convey the idea that knowledge about organisms was only of *value* if it were systematized, and that the job of providing the system and so giving the user that essential guidance needed in his attempts to navigate through the ocean of biological fact, surmise, and guess was firmly that of the taxonomist.
lying behind the faith in the name of an organism as providing access to the whole chapter of knowledge pertaining to it are thus seen to be related in the main to the concept of the species. This is almost wholly so for the vast bulk of higher organisms, plant and animal, and in a substantial

THE CHALLENGE OF BIOSYSTEMATICS AND TAXONOMY

The "system" that taxonomists have been constructing for upwards of two hundred years takes as its fundamental group the *species,* the unit to which the "name"—legally, now, the "binomial"—is attached. The assumptions measure still also for microrganisms. I do not propose to invade the territory surveyed by Arthur Cronquist in his opening address (Chapter 1, this volume) in any depth, but my own theme requires that I should make some comment on the matter of species.

Let me say immediately that I agree heartily with much of what Arthur Cronquist said, but there are some points where our views may diverge. For example, I, myself, do not now think that one need worry very much about the concept of the species. The greater part of the species "problem" is one of definition—not of the word, but of the ways it may appropriately be used in the different contexts in which its use is deemed desirable. The species is, in fact, already defined for us as a taxonomic category, a pigeon-hole in the hierarchical structure of classification determined by philosophical ideas current during the lifetime of Linnaeus, and stabilized subsequently by the accretionary labors of generations of plant and animal systematists. Experience tells us, in the main, that for a considerable range of practical purposes the device of a hierarchical system with the named species as its elementary category is a useful one. I would submit that it is, moreover, the only conceivable one now for a general taxonomy of organisms. This conclusion derives not only from the circumstance that so much intellectual capital has been sunk into it, but just as pertinently from the fact that no one has yet thought of a better system. There are *no* credible alternatives waiting to be dragged out, brushed off, and brought into use once the muddle-headed post-Linneans have been persuaded to quit the scene.

The species, then, is the taxonomic category bearing the binomial, and it has legal sanction as such through the two international codes of nomenclature. No congress or commission has sought to define it, or any other taxonomic category, *biologically;* and herein resides the strength and true value of this nomenclatural species concept. It predicates no special kind of variational unit, and, accordingly, in the absense of any rigid specification, it is available to be extended to all groups. Some years ago, at the time of the first international conference on chemotaxonomy (*2*), I commented that in the taxonomic context the question "What is a species?" is scarcely relevant, since it cannot be answered in any *useful* manner, and a question which cannot be answered usefully is not worth asking. The question that *is* highly relevant is: What kinds of variational unit, in the group under consideration, are most conveniently and usefully named as taxonomic species? Such a question obviously does permit an answer, even if in the particular context the answer should turn out to be "none," as has been found in various microorganism groups.

The user of taxonomy may have some grasp of these considerations; or on the other hand he may not. When he demands of his prospective informant, "What is this?", he stands a reasonable chance of getting an answer, such as: "This is *Lolium perenne.*" At first sight the encounter has been a successful one. Information has been requested, and information has been given. But neither party in the exchange has acknowledged the monstrous ellipsis. What the informant is *actually* conveying—or should be, if that person is worth his or her salt as a taxonomist—is, in précis, something like this: "In my opinion, this specimen falls within the variation range generally acknowledged as being encompassed within species X. Species X was established by an authority whose judgment I respect and follow. In the group in question species X is more or less commensurate as a variational unit with other species, and may be assumed to have similar biological properties. I am not telling you anything about those properties, nor am I committing myself to any statement about the existence, or otherwise, of infraspecific variation meriting acknowledgment in nomenclatural taxonomy. However, the name I have given you has been validly published, and in my opinion provides you with the best entry into the pertinent literature." Perhaps few users who are not themselves taxonomists will appreciate all of this, but the more sophisticated will sense a good deal of it, and it is one of the tasks of those who teach taxonomy to ensure that their students get to know enough of the working of the taxonomic system to understand what the name as an information package actually conveys. (As an aside one must surely add that a baleful effect of the decline or diminution of taxonomic teaching in university undergraduate courses is that the proportion of biology graduates who have such understanding is lower than it might be. I know of courses in applied areas of biology where no taxonomic theory at all is taught; yet graduates from such courses might well have the greatest need for some appreciation of taxonomic procedures in their professional lives subsequently.)

What, then, are the needs of the user beyond those served by a valid species identification for the organism he or she happens to be handling? The answer to this is critical for our appreciation of the role of biosystematics, for almost always the requirement is for data of a biosystematic character *not* encapsulated in orthodox taxonomic treatments, mostly concerning the genetical or genecological affinities of species or their internal variation, and mostly of a nature not readily assimilated into nomenclatural taxonomy at all.

I do not wish to be misinterpreted here, so before moving to the question of how biosystematic data are to be handled and communicated, I will say something about what biosystematics has done and can still do for orthodox taxonomy.

A minor irritation commonly suffered by workers in museums and herbaria is the repeated exhortation somehow to use the new insights into the sources and causes of biological variation to "improve" the classification of this or that group. Most of those responsible for dealing with the broader vistas of their craft—as apart from those whose hobby it is to delve into the niceties of tiny groups—know how chimerical such an idea can often be. Take once again the attitude toward species. As Arthur Cronquist reminded us (Chapter 1, this volume), now that the enthusiasms of the 1930s and 1940s have faded, it would be difficult to find an advocate of a species definition based upon experimental tests. Julian Huxley made the point in his prefatory chapter in *The New Systematics* of 1940 (*4*). "It is certainly right," Huxley wrote commenting on Dobzhansky's views, "to attempt a dynamic, instead of a static, definition by thinking of . . . species as stages in a process of evolutionary diversification; but it is impossible to insist on infertility as the sole criterion of this stage."

What biosystematics has done for taxonomists of a classical bent is to give them a better feeling for the *nature* of species, and so a better understanding of what classical taxonomy as it has existed for more than 200 years has been all about, revealing more clearly the reasons for its successes and failures. From 1940 onward, Ernst Mayr has developed for us the concept of the *biological species,* achieving something he probably did not expect, for there can be no doubt that in doing so he has mainly managed to provide an insight into what kind of variational unit the species of ordinary taxonomy tends to be in sexual groups. "Species," Mayr (*7*) wrote in 1942, "are groups of actually or potentially interbreeding populations which are reproductively isolated from other such groups." This "definition" is not, and others developed from it in the postwar years are not specifications for species, but descriptions of what the "good" species of orthodox, Linnean-style taxonomy turn out to be, and these were first defined on the basis of comparative morphology. Mayr, indeed, managed to focus and up-date ideas which in truth had their origins many years earlier. His description differs from that of Lindley (*6*) of almost a century before—"A species is an assemblage of individuals . . . capable of reproduction by seed without change, breeding freely together and producing perfect seed from which progeny can be reared. . . ."—only in acknowledging that the *potential* for interbreeding is to be taken into account, with which Lindley surely would have concurred. Lindley, in that post-Linnean, pre-Darwinian period of the mid-19th century, was seeking for the biological meaning of the species of his nomenclatural taxonomy; and a century and a half earlier still John Ray (*8*) had shown similar intent in seeking to lay hold of the essential nature of species as they were known to him by referring to their propensity for "breeding true."

The point I seek to make here is simply that so far as the procedures of taxonomy at and around the level of species are concerned, the newer knowledge of the nature of biological variation has not had any really profound effect, at least in sexual groups. It has, indeed, provided the sanction for much of what taxonomists have been doing all along. There will no doubt be some who will wince at these assertions. Surely there has been a taxonomic revolution? Surely we all know that the activities of museum and herbarium taxonomists are mostly no more than pathetic 19th century carry-overs? But the facts speak otherwise. Species and genus definitions—the parts of the taxonomist's activities that impinge most on the user because of their impact on nomenclature—have been refined and in some groups made more rational through the better understanding of the nature of the variation pattern—through biosystematics, if you will—but there has been no revolution. And for much of the work of general taxonomy, for the specialist struggling to bring order to the hordes of Diptera awaiting description and naming, or to the teeming flora of tropical forests, what biosystematics has brought is a useful package of case law which can be scanned for analogies, knowing that for the great bulk of the world's biota there is no hope of calling directly on experimental methods.

At the same time, there is no doubt that experimental biosystematics has brought to many taxonomists of a classical bent a fuller appreciation of the very great *limitations* of their discipline. We have heard many references to sibling and cryptic species in the course of this symposium. These fall entirely outside of the competence of normal taxonomic procedures and are best left alone by museum and herbarium workers. One might press the case further, for I believe there is only the slenderest of arguments for extending Linnean-type taxonomy based upon the meticulous description of "species" into groups with asexual, clonal modes of reproduction, where the concept of the biological species is essentially meaningless. Correspondingly, there are often very grave difficulties in the way of dealing with infraspecific variation by the procedures of nomenclatural taxonomy. It is not only that the techniques to detect and define such variation often lie outside the competences of the museum and herbarium, but rather more fundamentally that infraspecific variation is frequently multi-dimensional and not of a character that would even *allow* treatment by the establishment, description, and naming of subspecific taxa.

So what must be done to systematize and make available the data of biosystematics? Earlier I made the point that we have no credible alternative to our present system for a *general* taxonomy of organisms, and I do not depart now from the conviction that what we have is not merely adequate but eminently successful in its function of providing a broad overview of the organisms of the world. But at the levels we are now discussing, I believe there is no alternative but to put the formal taxonomic approach aside and turn toward other ways of handling and communicating the data.

A quarter of a century ago this matter came under searching discussion in various meetings and conferences, and the case was rather strongly supported for the separate treatment of the data of what was then commonly called experimental taxonomy by the creation of "adjunct taxonomies," with special category systems designed to accommodate variational units defined by criteria other than those of comparative morphology. Such systems had already been proposed for plants by Turesson (*10*) in his well-known ecotype-ecospecies-coenospecies hierarchy; and on the animal side ideas of how to deal with infraspecific variation both conceptually and practically by means other than those of nomenclatural taxonomy had been advanced by Rensch (*9*), Kleinschmidt (*5*), and others. Gilmour and I (*1*), twenty-three years ago now, tried to work out a flexible category system based on the neutral term "deme," and we sweated a lot of blood—and extracted a lot of fun—in the task of inventing hypothetical situations and seeing how they might be treated in such a quasi-taxonomy. This and other exercises of the period did, I believe, fulfill a useful function; and I still think there are circumstances when biosystematists would do well to heed at least the arguments of that time, particularly when they find themselves tempted to tinker with the orthodox taxonomy of the groups in which they happen to be working.

But the whole situation has really undergone radical change since those days, and now we can see that a real revolution has taken place: It is the technological revolution in data handling that has been discussed at this symposium. We have now available the competence to store, retrieve, and reassemble in any pattern we may wish, any set of information about living organisms, and the process is relatively painless. The potential is so great that many of us have not yet come to grasp its implications, and certainly there is no widespread acknowledgment yet of the fact that with tools of the capacity of those available now many procedures and practices—even, may I say it, the hallowed habit of hard-copy publication—are simply no longer necessary. For the most part they represent a disastrously wasteful expenditure of human effort.

T. G. Gautier (Chapter 12, this volume) showed us the potential power of electronic data processing in the handling of the data of orthodox taxonomy, and such applications as there have been so far have indicated that the value of herbarium and museum collections in research and in practical application will be enormously enhanced when it becomes possible to make the huge stores of data they hold more directly accessible by computerization. But as T. G. Gautier and other speakers have stressed, the possibilities extend far further than this. There already exist systems of great flexibility, such as TAXIR and SELGEM, and some have already been applied in the setting up of crop data-banks. The potential is there for the development of systems that will select and convey special packages of data including all relevant parameters—genetical, ecological, cytological, agronomic, or economic—tailored for the individual user. Such systems could deal with aspects of variation that pass beyond the capacity of orthodox nomenclatural taxonomy, such as intersecting clines, varying interfertility relationships, complex polymorphisms, host-parasite specializations and the like; and could, for example, permute and recombine data so as to reveal causal links which could enormously extend our understanding of the biology of the organisms concerned.

CONCLUSIONS

So this appears to me to be the future for biosystematics in agriculture: research to extend the totality of knowledge of organisms important to man and their interrelationships; and the accumulation of the information not in a random manner in book- and journal-filled mausoleums, but accessibly in data files, available to be called upon by users in such depth and in such combinations as each may need for his or her particular problem. I foresee this revolution as progressing rather rapidly, but also not to be without its difficulties. To point to one task that will attend it: There will be a certain job of education to do among biosystematists and their clients, to apprise them of what can and should be done. But to explore this would take us too far.

What of the costs of exploiting the potentials of biosystematics for agriculture in the ways we are now envisaging? Three years ago at the Royal Botanic Gardens, Kew, we held a major international conference on electronic data processing for taxonomic collections. The take-home message of that conference was that there was exceptional scope for the application of data-handling methods in herbaria and museums, but that the serious problems were economic, related largely to the scale of the operation. At a cost of $2 per entry, putting the Kew herbarium on file would mean an expenditure approaching $10 million. I do not see, therefore, a rapid shift to mechanized processing for the existing data banks of orthodox taxonomy, desirable though this would be. But the case is different in the field of agriculturally-oriented biosystematics. The economical thing here is to move straight to mechanized systems, and reap the benefits right from the start.

Just how compatible is the data system that we are now envisaging going to be with taxonomy as we have known it for so long? The answer is surely wholly so, provided that each discipline operates within its own envelope. At the risk of boring you with repetition, I must say once again that orthodox taxonomy continues to have its own vital part to play and its own services to render. It must key into the new systems, and they in their turn must establish a proper relationship with it. I certainly do not advocate watertight compartments, nor even commensalism: symbiosis is what should be aimed for. Symposia like this one, attended by so wide a spectrum of scientists, involved in the study of biological variation from so many angles, and in unearthing the truths that will indeed, as Hogben (*3*) suggested, become the recipes for human action, provide one way to achieve the objective.

LITERATURE CITED

1. Gilmour, J. S. L., and J. Heslop-Harrison. 1954. *The deme terminology and the units of micro-evolutionary change.* Genetica 27: 147–161.

2. Heslop-Harrison, J. 1963. *Species concepts: theoretical and practical aspects.* Pages 17–40 *in* T. Swain, ed. *Chemical plant taxonomy.* Academic Press, London, New York.

3. Hogben, L. 1938. *Science for the citizen.* Allen & Unwin, London. 372 pp.

4. Huxley, J., ed. 1940. *The new systematics.* Clarendon Press, Oxford. 583 pp.

5. Kleinschmidt, O. 1930. *The formenkreis theory and the progress of the organic world* (translated from the German: *Die Formenkreislehre und das Weltwerden des Lebens.* Gebauer-Schwetschke. 1926. 188 pp.) H. F. and G. Witherby, London.

6. Lindley, J. 1846. *The vegetable kingdom; or, the structure, classification, and uses of plants.* Bradbury and Evans, London. 908 pp.

7. Mayr, E. 1942. *Systematics and the origin of species from the viewpoint of a zoologist.* Columbia Univ. Press, New York. 334 pp.

8. Ray, J. 1686/1704. *Historia plantarum; species hactenus editas aliasque insuper multas noviter and descriptas complectens* (titles of volumes vary). 3 vols. Smith and Walford, London.

9. Rensch, B. 1933. *Zoologische Systematik und Artbildungs Problem.* Verh. Deutsch. Zool. Ges. (1933): 19–83.

10. Turesson, G. 1922. *The genotypical response of the plant species to the habitat.* Hereditas 3: 211–350.

11. Turrill, W. B. 1938. *The expansion of taxonomy with special reference to the Spermatophyta.* Biol. Rev. 13: 342–373.

18] Summary and Appraisal of the Symposium

by EUGENE MUNROE*

ABSTRACT

This symposium has dealt with biosystematics as applied to agriculture, both taken in the broad sense, though with some special reference to the Beltsville research program. The subject was not new, yet the review was timely. Striking progress has been made in a number of areas, but integration is needed, both of different approaches to the discipline and of the discipline with its applications. Despite the large number of species and groups, the complexity of the interactions among them, and the broad range of applications of biosystematics to agriculture and management of the environment, such integration now appears possible, chiefly because of advances in electronic data processing. However, the cost, sophistication, and centralization of large computerized systems limits their usefulness, as does their vulnerability to breakdown and their susceptibility to abuse. Bureaucracy and political interference, though probably inseparable from major public support in a responsible democratic system, are already threatening some aspects of scientific progress. Scientists, together with well-motivated politicians and administrators, must find ways to contain these threats. They will be wise to keep simpler and less costly arrangements for research and communication in continued operation, as well as those that depend more completely on high technology. The ultimate safeguard for science lies, however, in the integrity and character of the individual scientist. Indeed, intellectual integrity and the courage to defend it are fundamentally what society expects of science. The problems of biosystematics are large compared to our resources for solving them. We need a study to estimate the size of the discrepancy. We also need flexible and decentralized priorities for research, improved methods, systems and technology, and a larger work force. With dedication we can meet the challenge.

INTRODUCTION

I will begin by stating for the permanent record what a fine job the organizers did in planning and organizing this symposium. Not only were

*Biosystematics Research Institute, Agriculture Canada, Ottawa, Ontario K1A 0C6.

the logistics handled pleasantly and smoothly but the selection and arrangement of contributions made for a lively and varied program. The alternation of general with specific topics was particularly useful in maintaining interest and in relating practice to theory. I think I speak for all when I say that the occasion was agreeable, informative, and thought-provoking.

The stated objects of the symposium were to review the importance of taxonomy and systematics in agriculture, to expose all participants to a variety of new developments in systematics and its applications, and to help establish priorities for change in response to a changing world. The expressed hopes were that information would be provided on a range of disciplines and on the links between them, and that a balance would be achieved between the basic and the applied, the predictive and the practical, with due attention to current social and economic needs.

So comprehensive a task could have been approached in many ways. I confess that I found the range of presentations somewhat different from what I had expected. In this pragmatic decade and this mission-oriented center it was pleasant to hear more about organisms, taxonomy, and biosystematics, and less about agriculture, management, and political constraints than in many other contemporary conferences. The latter subjects were not ignored, but they were subordinated to the technological aspect of the discussion.

Even a major symposium on a defined field is necessarily only a sampling. In evaluating this one we should be conscious of its limitations, its biases, and its omissions. First, we are biological systematists talking mainly to and for our fellow systematists about systematics. We have a continuing problem of linking ideas with those of other scientists and of non-scientists. Though we have made a start in this direction, it is doubtful that we have reached the halfway house. Second, we speakers are a restricted and non-random sample of the biosystematic world. We have been selected to represent fields pertinent to the scientific program at Beltsville, and further according to availability, presentability, and the other factors that plague program committees. Important segments of subject matter, of theory, and of philosophy are inevitably left out. I will be referring to some of these, but each reader will think of others for himself. Finally, as in every symposium, time was too short. Every paper could have been expanded to a book, every discussion period to a symposium of its own. I hope some of these potential books will be written. The discussions will have to be continued in the minds of participants and readers of this volume, now that the symposium is over.

I have been assigned the twofold task of summarizing the program and appraising it. Rather than carry out these duties in a detailed and literal way, I will respond obliquely by giving some general impressions of the broad subject matter and of where I think it takes us. The individual contributions speak for themselves and are accompanied by abstracts. I will therefore cite specific points only as illustrations. They will be selected on a scale of pertinence to my remarks, not of importance or merit. Accordingly I hope that any apparent value judgments will be received with indulgence.

THE SYMPOSIUM IN GENERAL

First we should define what we are talking about. The organizers did not tell us what they meant by either "biosystematics" or "agriculture," and opinions seemed to differ at least about the former.

Some seemed to think of biosystematics as being an approach to systematics based on the characteristics of the living organism: experimental, behavioral, or genetic systematics would fall within this concept. Others used it for what was formerly called "the new systematics," that is, the whole array of indirect, theoretical, experimental, and modern methods as opposed to "conventional" systematics. Some years ago (*9*) I made up my own definition. I took systematics to be the general science of classification, and biosystematics to be that part of it which fell within the sphere of biology. As this idea was shared by some of the contributors to this symposium, and as this definition includes the others and their variants, I think it will be convenient to continue its use here. Anyone speaking of a more restricted concept can refer to "chemical biosystematics" and so forth.

For agriculture it would be appropriate to use the U. S. Department of Agriculture's definition, except that I am not quite sure what it is. It is obviously a wide one, because it includes forestry. I suppose it to be something like "the management of land-based living things for the production of food and fiber and for other useful purposes," but this is only my approximation, and the official concept of a farm probably includes things as wet as a cress bed or a duckpond. The exact limits don't matter as we will have to stray beyond them to talk about such things as land use and protection of the natural environment.

Turning now to the symposium itself, I find that I had a recurring sense of *déjas vu* throughout it. This came partly from a considerable number of specific links between the subject matter and my personal experience or interests. Though my career of choice is that of an alpha museum taxonomist, I have spent some time as an administrator and science advisor. During the Second World War I served as a medical entomologist and then, and later, had some casual contact with Protozoa. I also worked for three years as a nematode ecologist. Like Arthur Cronquist, I was a student when Th. Dobzhansky (*2*), Richard Goldschmidt (*3*), and C. D. Darlington (*1*) were initiating a new era in systematic and evolutionary studies. I was imbued with the dynamic species concept and with the relationship of cytogenetics to classification. My doctoral thesis was on biogeography, written under the tutelage of Wm. T. M. Forbes, a believer in continental drift and in the Pleistocene decimation of the Amazonian biota. I have subsequently written on biogeography (*7*), on the status of biosystematics (*6, 8*), and on the relationship of both to ecology (*5, 9*). These papers included some thoughts about trends in experimental and computerized biosystematics. In a less likely development I emerged briefly as an armchair expert on biological control (*10*), and had to consider among other things the great importance of the specific habits of insect parasites. I now work at a basic level in an applied agricultural biosystematics institute. At that institute we have had the honor of an extended visit from Willi Hennig, several visits from Lars Brundin, and a short visit from Robert Sokal. All have left their mark. My colleagues include chemical, physiological, and genetic systematists, whom I consult from time to time, and a number of mycologists, of whom Bryce Kendrick formerly was one. It is, then, not surprising that the proceedings from time to time struck a familiar note.

Apart from these individual recollections, however, the meeting seemed to echo the past in a much more general way. Professor J. Heslop-Harrison (Chapter 17, this volume) reminded us that the form and scope of this symposium are remarkably similar to those of *The New Systematics,* edited by the late Sir Julian Huxley more than 35 years ago (*4*). Then, as now, systematists were exploring the implications of new and interdisciplinary approaches. Then, as now, there was a problem of applying the new methods in practice and of striking a balance with conventional taxonomy. Then, as now, the value of phylogenetic classification was in question. Then, as now, rate genes were being proposed as a basis for dyschronic or saltatory evolution. Then, as now, there was a difficulty in reconciling the dynamic species concept with the facts of apomixis and of bridging the gap to evidence based mainly on dead specimens.

The same ground has been reworked wholly or in part in a number of subsequent papers and collections. As might be expected, we have increased our knowledge and the techniques for acquiring it very greatly, yet we still seem to be working on the same major problems, even if we have also added several new ones. In fact biosystematics appears to be a growing and proliferating set of puzzles rather than one that is being contained and tidied up.

I will consider the specific topics of the symposium under four general headings: the taxonomy of particular groups, theoretical questions, technological advances, and organizational and practical problems.

TAXONOMY OF PARTICULAR GROUPS

The groups emphasized were protozoans, nematodes, and fungi. Arthropods were considered mainly as illustrations of theoretical or general problems. Non-protozoan microorganisms, photosynthesizing plants, flatworms, annelids, mollusks, and chordates were not discussed or hardly so, possibly because their taxonomy is not actively studied at Beltsville. The contributors dealt partly with species problems, partly with higher classification. Norman P. Levine's lighthearted but penetrating paper spanned both, and showed how far protozoan systematics has progressed since I last knew anything about it. There is a surprising variation in the level at which work is done in different taxa. In part, no doubt, this reflects the different characteristics of the material, and in part different levels of interest and support. By and large, this group of participants showed strong awareness of the practical implications of their work, probably because of the direct service responsibilities that many of them carry.

Two general points should be made here. The first is the forbidding magnitude of the task that confronts special taxonomy. Twenty-five years ago, Sabrosky (*11*) estimated that the number of species of insects then named was between 625,000 and 1,500,000. He considered this to be 80 to 90 percent of the described animal species. Thirteen years ago (*8*) I put the possible number of species of living organisms at 10,000,000. Now, extrapolating from the better known groups and biotas, I would double that guess. Of how many species may have existed in the past I have no idea. Compact descriptive monographs of even 10,000,000 species would require about 50,000 volumes of 300 pages each, neglecting illustrations.

The second point is the importance, and the difficulty, of bringing modern concepts and methods out of the realm of theory and into the actual process of classifying species and their variations. I will be returning to both these points in later paragraphs.

THEORETICAL QUESTIONS

The theoretical subjects we discussed in this symposium fall into several categories: the species, its nature, definition, and structure; principles and methods of classification, especially phyletic classification; biogeography as related to ecology and agriculture; and the predictive value of biosystematics for agricultural ends.

The species. At this level, it was disappointing for me to hear yet another debate on the semantic question of what should be called a species. Arthur Cronquist quoted the old "joke" that a species is what a competent taxonomist calls a species. For me this is no joke, but a statement of fact. True, we might nowadays say "systematist" or even "biosystematist" rather than "taxonomist," but we do proceed in these matters by consensus. The Linnean system of nomenclature is at best a simple framework on which the complex web of biosystematic data can be hung. The frame is more or less uniform, but the web differs in fineness and weave in its different parts, and also in the scrutiny and use to which different parts are subjected. Not only are local distortions permissible, but there is no test by which we could detect and prevent them. Objectively, the characters and their patterns of distribution differ in different groups. Subjectively, we have always had splitters and lumpers in every group. Indeed, either splitting or lumping may be preferable, depending on the needs of the problem. Many concepts will fit within the same nomenclature. What is important is that we define our concepts and adhere to the definitions. I do not worry much about freedom in this direction leading to chaos. If we as individuals stray too far from the usage of our contemporaries or from the dictates of convenience, our work will soon be ignored.

More significant than this question of terms is the progress being made in our knowledge of the genetic and geographical structure of species. After the initial impetus given by Dobzhansky's synthesis, the science of population genetics seemed for a long time not quite to get off the ground. The gap from the unrealistically simple Hardy-Weinberg formula and Sewall Wright equations to the complexities of populations in nature appeared unbridgeable. Many interesting phenomena were discovered—introgressive hybridization, balanced polymorphism, fugitive species, density-dependent changes in gene frequency, to name a few. Striking individual cases were investigated, such as the ecogenetics of *Heliconius* and a steadily accreting body of information on *Drosophila.* But the transition from this stage to widespread genetic analysis of the variation of species in nature, such as we had hoped to see in our generation, was not made.

Now it appears from studies such as those reported by G. L. Bush and G. B. Kitto (Chapter 6, this volume) and by L. H. Throckmorton (Chapter 13, this volume) that there is both an explanation and some hope of an answer. On the one hand the genetic diversity of natural populations is much greater than we had ever supposed. On the other hand powerful new methods make it possible for us to acquire data and to analyze complex situations much more effectively than we could in the past. Though the lively discussion that followed these two papers showed that not all questions of method or interpretation are settled, in this case the debate was over concepts and procedures, not words, and we may hope it will prove intellectually productive.

The conclusions of these papers seemed to be directed mainly to the sequence and dates of phylogenetic separations. Obviously, however, the same approaches can be applied to ecological and geographical genetics and their short-term dynamics. We will hope to see this connection made and also that between the cryptic genetics revealed by molecules and the patent genetics of polymorphism, clines, reproductive barriers, and species characters that is of prime interest to practicing systematists.

Principles and methods of classification. I found the discussion of this subject rather one-sided, not because of inadequacy of presentation, and certainly not because of any lack of vigor during the discussion periods, but because certain important points of view were unrepresented or underrepresented. In particular, pheneticists and other non-phyletic classifiers were not heard from. The phyletic or cladistic method was described in the Hennigian version by D. H. Kavanaugh (Chapter 8, this volume), and the ensuing comments tended to deal with the fine points of this methodology and with the differences between it and its close philosophical relatives.

It was Hennig's stated intention to make systematics an exact science and thereby to achieve for it the recognition it deserves. Among the steps he took were: development of an epistemological foundation; adoption of a new terminology, mostly for existing concepts; introduction of a considerable degree of rigor into the procedures for inferring phylogeny; and formulation of some ideas that either were genuinely new or had not been widely appreciated, for example, the different significance of synapomorphy and symplesiomorphy and the use of sister groups as the foundation of higher classification.

The value of Hennigian epistemology depends on one's philosophical preferences: it differs, for instance, from the likewise erudite and highly developed epistemological foundation established by Sokal and Sneath (*14, 15*) for their phenetic methods, though Sneath (*13*) has emphasized that the scientific structures erected on these different foundations are not necessarily out of harmony.

Hennig's terminology seems to have had an ambivalent effect. It slowed the initial acceptance of his work, but once the "in words" were mastered, they cast a certain glamor, and created a mystique. In plain German or plain English the new ideas would have stood out more clearly from the old, and they might have slipped more smoothly into the mainstream of biosystematic thought.

I hope I have not given the impression of disparaging Hennig's work, which was timely and catalytic, even if his brand of monophyly sometimes leads to bizarre results. Hennig rendered a valuable service in promoting precision and clarity in phylogenetic thought. Newer work based on his method has made further progress in that direction, by identifying assumptions and hypotheses more clearly, thus exposing them to subsequent critical reexamination (*12*). Other studies, which I will not identify, show that adherence to formal Hennigian methods will not counteract an injudicious choice of characters for analysis. Human logic, like that of the computer, follows the rule, "garbage in, garbage out."

I suppose I personally belong to the eclectic or, as D. H. Kavanaugh termed it, the "trust me" school of phyleticists. I would rather say "try me," because I think I can rely on my colleagues' zeal in ferreting out any stupidities I commit. Whether this is a worse motto than "trust Hennig," I leave to others to decide.

As an eclecticist I prefer a many-sided approach, a multispective if not an omnispective one. I therefore welcomed the quite different route—in level, in characters and in strategy—taken by the genetic-molecular phyleticists. The two papers in this field, already mentioned in connection with the genetic structure of species, were closely related in fundamentals, even though some differences in particulars were evident. It is exciting that genetic variability and the number of allele differences between individuals or populations can be estimated, that the sequence of molecular and genetic changes in a given lineage can be determined, and that these findings can be related to taxonomy, and plausibly to phyletic and even to geological time.

G. L. Bush (Chapter 6, this volume) correctly observed that it is important to get these new methods out into the world of practical taxonomy. There are, however, difficulties to be overcome. Taxonomists must be induced to take notice, in itself no mean feat. Then time and facilities must be found. Even the relatively simple procedure of gel electrophoresis is expensive, full of pitfalls, and extremely time-consuming. At our institute we have had to confine our investigation by this and related means to only one genus, and even then progress is slow. I do not know how we could secure funds or arrange priorities to get into amino-acid sequence analysis, immunology, or DNA hybridization. It may be only coincidence that G. L. Bush and L. H. Throckmorton (Chapter 13, this volume) also reported on only a single genus each.

Another more elementary but very important problem is that molecular methods so far depend on living material, often in substantial quantity, whereas museums and herbaria for the most part contain dead material, the quantity of which is often severely restricted. The development of comparably sensitive techniques that could utilize materials recoverable from museum specimens, for instance structural proteins or polysaccharides, would be a major breakthrough. In many cases, too, it seems that the conclusions of deductive, chemically-based genetic interpretation would be both surer and clearer if linked to direct genetic and cytological findings. To be fair it must be said that some of the studies mentioned were cytological in approach.

Biogeography, ecology, and agriculture. I turn with pleasure to the field of biogeography. Not only were the presentations excellent, but they lie very close to my own past interests. I must, in fact, discipline myself not to discuss them in detail, lest over-enthusiasm prevent me from ever finishing. B. B. Simpson (Chapter 9, this volume) reviewed broadly and with admirable clarity the modern theory of the dynamics of biotas, and went on to describe how numbers are being put into some of the equations and how they can be applied in conservation and resource management. J. A. Duke (Chapter 4, this volume) showed that the old ideas of the potential areas of species and the bioclimatic analogs of habitats can be developed into a powerful tool for the acclimation and deployment of useful plants and the containment of weeds. Not content with theorizing about the value of ecosystematic information, he has gone on to secure and tabulate hard data on an impressive scale. I thought that to the uninformed the tables might give a false impression of immutability, for, as J. A. Duke obviously knows, biological species have a capacity to transcend previously established limits that is not shared by their chemical or mineral counterparts. I thought also that Drs. Simpson and Duke have, from different directions, come very close to reaching common ground in studying the effects of mixing and modifying natural and managed communities or organisms. If a meeting of minds actually takes place, it may well prove to be highly synergistic.

Predictive value of biosystematics. The prospect of quantifying ecosystematic models leads directly to the predictive value of systematics for agriculture, dealt with from various aspects by L. R. Batra and his co-authors, and touched upon in other papers. The most striking feature is the variety of levels at which systematics has this predictive impact. From the molecule and the gene, through the population, the cryptic or sibling species, and the species "distinguishable by ordinary means," thence along one path to the higher taxonomic categories and along another to communities and biotas and their dynamics, all have predictive potential for various facets of agricultural planning and practice. How different this is from the naive view that the main function of systematics in agriculture is to identify a small number of useful and harmful species!

An unsatisfying aspect of this discussion was that the quantitative element was, in fact, small. L. H. Throckmorton's genetic uncertainty was at leased based on calculation of the discrepancy between actual and possible genotypes. J. A. Duke's ecological amplitudes, too, are quantitative and based on measurement. The equations of island geography, though quantitative in form, are highly stochastic; they will be more useful when their spread has been better determined. If Vavilov's principle has been quantified we were not told of it. It was not indicated whether the U.S. Department of Agriculture's program to detect 124 foreign crop pests (Foote, Chapter 15, this volume) is based on a statistically evaluated sampling design, and, if so, how effective the net is calculated to be, or how far a pest will build up before it is likely to be detected. I hasten to say that these examples are mentioned because they were the first to come to hand. Most of the discussion of predictiveness was qualitative and by examples. Quantification will be necessary before biosystematics can reach its full potential for prediction in agricultural planning.

General Comments. So much for the theoretical discussion. I think we heard too little about some things: mating systems, ethology as related to identification and to population structure, coevolution, pheromones, chemoreception, and the senses in general, to name a few. I felt that some of the most important lines of investigation are still proceeding along their own tunnels, and that if we could break down the walls between them we would be mining the biosystematic lode more effectively. In particular I deplore evolutionary and ecological elitism, too intent on elegant and fashionable theory to communicate effectively with scientific commoners, and its converse, systematic stodginess, too preoccupied with detail and routine to contribute to or apply the exciting new developments in theoretical biology. Some of the world's better theorists suffer from the former fault, some of its better taxonomists from the latter. Fortunately these characteristics are not evident in the symposium participants themselves or in the host institution.

Despite these strictures, I believe that the range and quality of the theoretical discussion were on the whole most gratifying, and that the case for a broad biosystematic program, below, at and above the species level, was clearly made.

TECHNOLOGICAL ADVANCES

Technological advances in biosystematics are, of course, increasingly numerous. The most important ones discussed in the present symposium were in molecular methods and in the application of electronic data processing (EDP). I have already given some attention to the former, and I will therefore concentrate in this section on EDP. Three main areas of use were discussed: in bibliographies and literature searches, in databanks, and in identification of specimens.

Electronic retrieval of bibliographic information is no longer a novelty. For recent literature of the better known species there are now fairly good data bases, and searches are apt to produce an embarrassingly large number of titles. The titles can be sorted as well as retrieved electronically, though not so far as sensitively as by human inspection. Identification of titles does not yet relieve one of the work of obtaining the documents. Automatic addressing of reprint requests and automatic photocopying of identified documents are no doubt already technically feasible, but I doubt if the cost has yet reached the economic level. Some progress has also been made toward computerized taxonomic catalogs. These have the advantage of being relatively easily updated, as well as the merits of quick and convenient search, tabulation, analysis, and printout.

The problems and promise of databanks were very well outlined by T. G. Gautier (Chapter 12, this volume). I do not think I could improve on his abstract.

C. R. Gunn and L. R. LaSota (Chapter 14, this volume) described three uses of the computer in identification, namely: in the construction of taxonomic keys; in the "conversational" mode in applying keys or tables to the determination of specimens; and, so far mainly potentially, to the automatic detection of diagnostic characters from such visual arrays as scanning electron microscope images. In a paper already referred to (*8*) I suggested that chromatograms and similar two-dimensional presentations might be suitable for automatic scanning if they could be satisfactorily standardized and calibrated. The same would apply to the electrophoretic separations discussed during this symposium. I wonder also why holography, with its three-dimensional capability and very high information content, has not been used more extensively for taxonomic records and as computer input.

We have already made impressive strides in applying EDP to biosystematics. However, the metaphor is perhaps a good one, because I feel that the projects discussed, with the possible exception of automated recognition of diagnostic characters, are somewhat pedestrian. They aim to do work that is already being done, but to do it faster, less laboriously, and possibly more efficiently. But, then, biosystematics is characterized by problems involving astronomical amounts of data and unbelievably complex relationships. And, it is exactly such problems that the memory and analytic power of computers are best fitted to deal with.

It follows, therefore, that the most economical and rewarding use of EDP in biosystematics ought to be on problems that are orders of magnitude larger and more sophisticated than those we have tackled by conventional methods. Integrating major theoretical fields, putting numbers in the great equations, extracting unsuspected regularities from biosystematically ordered data: it is in such activities that EDP can break new ground and justify the massive investment of work and money that it demands. Naturally databanks and the retrieval of published information are essential components of such higher order systems, but we should be looking beyond the building blocks to the grand design. Quantitative understanding of real ecosystems and their changes would be of inestimable value in the planning of practical agriculture and in evaluating its wider interactions. Agriculture has a real stake in developing the facilities needed for such understanding.

Like other high technology, EDP unfortunately has its dark side. This also was touched on by T. G. Gautier. If a substantial part of the available biosystematic data is pooled in a centralized databank, and if research and description depends on the availablity of these data, who will control the right of access, and the machines and programs needed for readout and analysis? The possibilities of abuse are obvious, and I doubt that final solutions have yet been found.

In summary, many biosystematic problems are so complicated and data-intensive, either in themselves or in their applications, that EDP appears to offer the only hope of solving them. However, this will be costly, will change priorities and work-styles, and will threaten the traditional rights of biosystematists. I see no way of avoiding these difficulties. We will have to meet them as they come, and try to overcome the dangers and disadvantages of progress while reaping its benefits.

ORGANIZATIONAL AND PRACTICAL PROBLEMS

I said at the outset that it was pleasant to hear so little about practical and political considerations, but, as with many pleasant things, I am not sure it was altogether healthy. Perhaps we gave too little attention to agriculture, to money and resources, to bureaucracy, to politics, and to the moral position of the biosystematist. As a mild corrective, I propose to devote a few paragraphs to these subjects. I will consider them under four headings: biosystematics and agriculture; resources, feasibility, and priorities; problems of organization; and the responsibilities of the biosystematist.

Biosystematics and agriculture. Agricultural research is an applied science and as such it serves a clientele. It has in fact a primary clientele consisting of farmers, foresters, and the agricultural community, and a secondary one consisting of everyone else affected by agriculture, that is, the whole public. What should these clienteles expect from biosystematics? To answer we must briefly recall the capabilities of the discipline and see how they can meet the demands and needs of agriculture and of the public.

This symposium showed that biosystematics is useful at different levels and in different ways. *Species identification* gives the primary scientific name, and with it entrance to the literature and the whole apparatus of information retrieval. Knowledge of *infraspecific variation* and its causes is used in predicting the behavior of particular strains, in developing and improving useful varieties, and in a wide range of pest control measures that depend on host resistance, pest susceptibility, or combinations and interactions of the two. *Higher classification* is useful directly in facilitating species identification and in maximizing predictiveness by grouping like species. It is useful indirectly through its contribution to evolutionary, biogeographic and ecological thought. In a sense agriculture is evolution, biogeography and ecology speeded up, altered, and to some degree managed, by man. Understanding of long-term processes in nature can therefore give us insight into the likely consequences of the increasingly profound influence we exert on the living environment.

If we turn now to the demands and needs of our clientele, it is clear first that the demands and the needs are not necessarily the same, and second that those of the farmer and agriculture are not necessarily the same as those of the public at large. Farmers are practical people with urgent problems. They want to see productivity increased, risks reduced, pests wiped out. They would rather see it this year than next year or ten years hence. Their demands are apt to be for visible progress with perceived problems on a short time scale. This is the message they pass on to the county agent, to the industrial representative, and to the politician.

Though these are the perceived needs of the grower, which government and commercial agencies neglect at their risk, they are, of course, only part of the real needs. Long-term policies, and the development of really effective control measures, require a different time scale and a far greater depth of research. Not only that, even immediate work, to be reliable and informative, needs a banking of research and technology that extends far beyond the immediate and apparent problems. The useful variety has to be developed from a rich and flexible gene pool, the pest species has to be distinguished from a swarm of related species, which in turn have to be viewed in a broad perspective of the group in question. Farmers, though sometimes demanding, are not stupid. They do not support an organization of the size and sophistication of the Agricultural Research Service to tell them answers that are under their noses, or that they could learn by contract research for a relatively small expenditure. Certainly they expect service work from government—as much as they can get—but they also expect deep understanding, advance warning, new concepts and methods, such as only a large national organization has the resources to provide.

What, then, of the second clientele, the public? In part its interests are the same as those of agriculture. It is in the public interest to have a thriving food and fiber industry. But other aspects of agricultural activity may conflict with some aspects of the public interest and even the larger interests of the farmer himself. It is in nobody's best interest to have polluted water, poisoned land, and depeleted ozone. It is contrary to the interest of many to have depauperated wildlife, diminishing fisheries, or brown and ugly landscapes. Where interests are in conflict, passions tend to run high. What is the corrective agent? It is factual knowledge. It is easy to argue with your neighbor, but hard to argue against facts. Credible scientific data will defuse many a political or policy time bomb.

Regarding environmental questions, and those of long-term, self-sustaining, biological, organic, or integrated management methods, the ability to monitor the status and trends of species in ecological communities is of paramount importance. I am afraid that in most cases this cannot yet be done with any degree of reliability, first because the taxonomic knowledge in many key groups is inadequate, second because reliable sampling systems have not been worked out, and third because manpower and facilities to do the sampling have not been funded. All these factors argue for a greatly increased investment in biosystematics to support environmental studies, both to protect the public against unwise land management practices and to protect the farmer and forester against ill-found criticism and claims.

As it happens, the research and reference facilities needed to carry out this monitoring mission are very closely related to those already devoted—in the United States on an impressive scale—to the identification of agriculturally important species and their close relatives. The same groups, often the same individual experts, are involved in both problems. R. H. Foote (Chapter 15, this volume) described to us the close relationship that exists between the National Museum of Natural History of the Smithsonian Institution and the Systematic Entomology Laboratory of the Agricultural Research Service. A similar close relationship to the emerging environmental mission would strengthen all three efforts individually and collectively. This situation, of course, is not unique to the United States.

In simple words, the agricultural mission is best served by a biosystematic organization, whether centralized or dispersed, that combines ready service capability with the breadth and depth of research backing that will assure accuracy and insight, and preferably with cross-links and shared resources connecting it with other missions with related requirements.

Resources, feasibility, and priorities. This symposium was so brimming over with good ideas, with positive suggestions, that we hardly ever saw a dollar sign, or heard a cautionary note about schedules, physical facilities, or manpower. To the extent that our object was to stimulate one another to renewed efforts, perhaps this was good; but we all know that in the real world these constraints exist and that in these times they are becoming increasingly severe.

I do not intend to give a lecture on budgeting, or to go into the fine points of grantsmanship or manpower and resource forecasting. However, I did find it deeply disturbing that scientists could talk at length about the benefits of new methods and new ideas and give hardly an inkling of their cost, in money, in manpower, in materials, even in relative terms. Yet agricultural biosystematics is a money-oriented applied science, no less than is civil engineering. If we have services to sell, useful concepts to offer, we should be prepared to talk in terms of costs; otherwise, we cannot expect the public to "buy" our ideas and our offerings of additional services.

Let me not sound too discouraging. We may seem to be in wintry times for funding, but they are mild compared to the "ice age" of the thirties. The new technologies we are talking about will require extensive new funding. We will not be able to cut back significantly on resources for conventional taxonomy, because the input to the new structures will depend for some years on conventional work, probably on an increased scale. But if modern biosystematics has the power and significance that we believe it has, the funding will be forthcoming. It will be forthcoming provided that the products justify it. Costing and planning must be rational; operations must be honest and efficient. The Mohole and other foundered projects bear witness to that. But with common sense and a good product, eventual support is assured.

A word now about manpower. How can we talk about biosystematics without talking about biosystematists? Yet there was in this symposium nothing on training, nothing on supply, and nothing on estimated demand. Perhaps everyone found it too depressing to mention. Perhaps they felt a surplus of top-quality graduates could be taken for granted. In many cases such sentiments have preceded times of acute shortage. I do not make any predictions, I simply say that the subject should be considered. A recent study of entomological manpower in Canada has been the subject of an open meeting. I will not try to anticipate the published results, except to say that systematists are in shorter supply than are many other specialists. How far these Canadian findings will be relevant to other countries, I cannot say.

Resources never multiply as fast as ideas do. That is why we have to set priorities. "Priorities" literally mean things done first, but we can also interpret the concept of priorities as doing some things and not others or by doing some things on a larger scale than others. All are viable approaches to choice, but selection of the wrong approach may cause needless damage to a program.

Whatever the basis, choices must be made in biosystematics, as in all science. We must choose what resources will be devoted to the study of different groups, different regions and different systematic levels. We must choose what of the many possible new methods we will support, what connections we will encourage to other biological fields. In agriculture part of our choice will be dictated by utility and demand. We must give priority to agriculturally important organisms and we must provide an adequate identification service. We also must consider human problems. We cannot retain biosystematists and take on new research projects as easily as we can change dies on a power tool or modify spray schedules on a field. Several years of expertise go into the mastery of even a relatively limited field of biosystematic study. We must plan ahead and provide the flexibility to deal with at least the taxonomic groups most likely to be of practical importance. Finally, we must give sufficient priority to quality. We must have enough basic research in active progress to make sure that our biosystematics is based on understanding and not on mechanical regurgitation of casually digested observations. Only intelligent and experienced systematics will yield reliable information useful to the clientele.

Problems of organization. It is not long since research was mainly an individual enterprise. Perhaps it remained so a little longer in taxonomy than in the majority of scientific disciplines. That day is now passing, even in biosystematics. Though there is still room for the individual student, both the approaches and the material needs of modern work demand organization and cooperation to an increasing degree. Large collections, large libraries, and costly apparatus are required for most studies, demanding at the very least an institutional connection. Multidisciplinary problems require the formation of teams of workers with complementary backgrounds. Pooling of information in databanks and analyses on the scale made possible by EDP will tend toward still larger and more expensive aggregations.

In our society the only source from which such large enterprises can be funded is the public treasury. A greater and greater share of the financial support for every type of research institution comes from government sources. Government funding also implies government control. Government control brings with it the danger of abuses: passive abuses such as bureaucracy in administration and expedience in policy-setting, and active abuses such as corruption, nepotism, and political interference. But who can say that bureaucratic administration and political control are wrong?

They are the foundation of democratic government. If we are not be be responsible to the public through their elected and appointed representatives for public funds, to whom shall we be responsible?

I do not really know the answer. I am sure that science will have to live with bureaucracy and live with politics for a long time to come. The association will probably become closer rather than more distant. The challenge is to work out rules, structures, arrangements, that will make the association an effective and fruitful one rather than one that is sterile and mutually frustrating. Scientists must learn to understand the problems of politicians; administrators, and particularly politicians, must learn, or remember, that the problems of science really are rather special and that the methods suitable for a military establishment or a taxation apparatus are not wholly appropriate for running a scientific laboratory. Science arose in warfare with established religion; now it seems to be engaged in an equally desperate struggle with entrenched officialdom. Intellectual honesty prevailed in the former conflict; perhaps it will prove equally effective armor in the present one.

Nonetheless, the centralized state-control of science, be it biosystematics or any other discipline, presents grave intellectual risks. Scientists need to counter these risks by keeping other facilities and other vehicles open, even if they are smaller and less opulent. If the central databank and computer facilities became inoperative, inaccessible, or prohibitively expensive, it would be very desirable still to have a printed literature, inexpensive journals, dispersed collections, and informal networks for consultation and calculation. A great deal can still be done by personal enterprise, particularly the enterprise of resourceful and cooperative individuals. Such enterprise is still the best insurance against the dead hand of bureaucracy, or even the dead hand of a complacent scientific establishment.

Before leaving the subject of organization, I will mention one more topic—scientific parochialism. Governments have geographic jurisdictions, and they are generally reluctant to spend money outside of them. Organisms, on the other hand, are no respecters of mere political boundaries. It is true that much valuable work can be done on local biotas. Jurisdiction-minded administrators and politicians would like to believe that an adequate biosystematic picture can be built up from a mosaic of such local studies. Unfortunately this is far from the truth. Systematists certainly have a responsibility to investigate their local biotas. But to interpret these requires a dimension of cross-comparison that can hardly be explored by studies carried out at home. Some of us must travel and see differences for ourselves. Some of us must carry out global biosystematics. Some centers must maintain collections of international scope. At any particular time it is a moral obligation of the most scientifically developed and affluent countries to take the lead in this. For a while it appeared as though this would cause a permanent and unfair concentration of the main facilities in a few locations, but now it seems that the global distribution of wealth and expertise is more flexible than many had supposed. To my mind the important thing is what is done, not where it is done. I believe the healthiest arrangement is that under which different institutions in different countries each develop global perspectives in certain problems, according to their inclinations, interests, and resources. It may be, however, that government-supported scientists will need to fight a continuing battle to defend this point of view.

Responsibilities of the biosystematist. I said that intellectual honesty is the best armor of science. In the last analysis the future of science will depend on the character of individual scientists. What are the moral responsibilities of the scientist, or specifically of the biosystematist? I can only give you my opinion.

The biosystematist is on the one hand an intellectual and an expert, on the other hand a person, a citizen and often a public servant. He or she has responsibilities in all these capacities and at times they may be in real or apparent conflict. I believe that the biosystematist's fundamental professional responsibility is *intellectual integrity*. This is what makes the trained person a scientist; this is what has earned public respect for science and what has made possible its operational success.

Intellectual integrity, though fundamental and essential, is in itself hardly sufficient. As a person, the scientist has a responsibility to be *humane*. The true scientist will not undertake wantonly destructive, dangerous, or cruel activities simply to gratify curiosity. He or she may also have specific responsibilities as a citizen, employee, etc., for instance in the guarding of state secrets. The moral issues here may become more complex. Response to them will depend on the scientists's own and the prevailing ideas of right and wrong.

Often conflicts between secular and scientific responsibility will prove more apparent than real. Scientists are employed as such in expectation of the very quality of intellectual honesty that I mentioned. If they allow themselves to be subverted, they serve neither their masters nor themselves.

Formerly scientists were reputed to live in "ivory towers." Now they must be *aware*. They must be in touch with their times, in order to identify threats, and more positively in order to recognize practical needs and opportunities for service. *Service* is a responsibility every scientist must expect to shoulder. The scientist must serve the public, to justify its respect and support. The scientist must serve his or her discipline, certainly by contributing to it intellectually, but also to the extent of his or her talents and abilities, by serving and supporting its organizations. Disciplinary associations and societies prove to be an important safeguard for scientists in the modern world. If we do not organize our own affairs, others may do it for us, and not necessarily to our benefit. Scientists have a responsibility to *communicate,* not only within their discipline but also to the public. Confidential employment may impose restrictions on certain individuals, but usually there will be some open channels even in this circumstance. In general, however, the public has a right and a need to know what science and the scientist are doing.

In sum, science is supported because it is credible and useful. Its credibility makes it the arbiter of subjective conflicts; its usefulness makes it the basis of technological advance. Both qualities depend on its objectivity, its verifiability, and its power to unify concepts. We must strive for a science that sees deeply and commands confidence, a science that serves human knowledge and human welfare.

CONCLUSIONS

Biosystematics, even in its direct agricultural applications, is so vast a subject that it is unlikely that present resources of people, materials, and systems can deal with it adequately and uniformly in any reasonable period of time. The difficulty is multiplied by the indirect effects and interactions of agriculture and the desirability of using agricultural biosystematic facilities for the study of related non-agricultural problems.

Should we despair? I think not. We can respond to the challenge in several ways. First, we should measure it. So far as I know, no study has been made to estimate the amount of information needed to complete the main descriptive and cataloging tasks of biosystematics to defined levels of acceptability, or to determine the resource requirements of typical analytic projects and contributions to multidisciplinary and applied activities outside the biosystematic field. Such an estimate, though perhaps crude and tentative, would be invaluable if not essential for intelligent advance planning.

Second, we must certainly set priorities. In this utilitarian, nationalistic, and political age there is a danger that we will set them too narrowly and too rigidly and that we will relate them too closely to short-term goals. In my opinion, priorities are more successfully set piecemeal, by institutions, teams or individuals responding to perceived needs and opportunities, than centrally by some authority or council. For one thing, imposed priorities generate resistance, whereas discovered ones inspire enthusiasm. For another, it is hard to plan centrally with enough knowledge, in enough detail, and without bias, or to administer the plans as flexibly and efficiently as the planners intended. Also, though the establishment and its conventional wisdom are often right, they have a record of serious mistakes in judging innovations and future needs. On the other hand, the scientific community reacts positively to challenges and incentives, concentrating on important and well-funded areas while still leaving room for individual enterprise and insight even in an increasingly organized society.

Third, we must improve our methods, and especially capitalize on new technology. Every increase in the efficiency of gathering, recording, processing and disseminating data improves our capacity to cope with biosystematic problems, and increases the ratio of thought to drudgery in our efforts to solve them. We must of course not let technology become an end in itself, or let it impose barriers of price or control to the orderly progress of research. But, properly utilized, the rapid advances in instrumentation and data processing give hope of at least narrowing the gap between the magnitude of biosystematic problems and our ability to solve them.

Fourth, we must find ways of putting more people to work on biosystematics. We have come through a period of declining opportunities in professional science; many graduates trained as biosystematists have had to find other kinds of employment. Some of these experts can be brought back into biosystematic work if the need is realized and if positions or contracts are made available. That will depend on public and scientific appreciation of the importance of biosystematics. Much, though not all, the responsibility for developing this appreciation lies with biosystematists themselves. Like others with a good product to sell, they must make it and its virtues known to potential clients. This symposium has been one step in the right direction. However, the charge also falls to informed users to press for satisfaction of their needs. Agriculturists, resource managers, environmentalists, and the nature-oriented public must insist on the development of increased biosystematic facilities, services, and information.

Full-time employment is not the only way of increasing the biosystematic work force. Plenty of good biosystematic research is done by scientists with other primary orientations, either as a sideline or as a component of more general investigations. Good work is done by amateurs in a number of groups, and in some countries, including Britain and Japan, a very detailed knowledge of the fauna has been developed largely through amateur activity. Some would argue that these ancillary efforts may dilute the quality of research. I do not regard this as a serious danger. The self-criticism of the scientific community supplies its own corrective, and the enthusiasm of the part-time or spare-time biosystematist may well be a necessary antidote to the "dry rot" that sometimes affects official and academic institutions.

But what are we really trying to do? We are trying to arrive at the best representation we can of the living world in terms of the similarities and differences of its components, to arrange the components logically in terms of regularities in the similarities and differences, and to use this ordered arrangement in order to retrieve information, to derive scientific principles and to plan practical action. The task is hard enough and important enough to demand all the energy, all the intelligence, all the common sense, all the integrity, and all the dedication that we can put into it. Let us prove that we can master it.

ACKNOWLEDGMENTS

I thank the Agricultural Research Service, United States Department of Agriculture, for funding my attendance at this symposium. I appreciate the many kindnesses extended to me by the organizers of the symposium, and also the courtesy of those participants who sent me advance texts or drafts of their presentations.

LITERATURE CITED

1. Darlington, C. D. 1939. *The evolution of genetic systems*. Cambridge University Press. 149 pp.
2. Dobzhansky, Th. 1937. *Genetics and the origin of species*. Columbia University Press, New York. 352 pp.
3. Goldschmidt, R. 1938. *Physiological genetics*. McGraw-Hill, New York and London. 338 pp.
4. Huxley, J. S., ed. 1940. *The new systematics*. Clarendon Press, Oxford. 583 pp.
5. Munroe, E. 1958. *Systematics as a tool in ecology*. Tenth Int. Congr. Entomol., Proc., 2: 663-667.
6. Munroe, E. 1960. *An assessment of the contribution of experimental taxonomy to the classification of insects*. Rev. Can. de Biol. 19: 293-319.
7. Munroe, E. 1963. *Perspectives in biogeography*. Can. Entomol. 95: 209-308.
8. Munroe, E. 1964. *Problems and trends in systematics*. Can. Entomol. 96: 368-377.
9. Munroe, E. 1964. *Biosystematics and dynamic ecology*. Univ. of Toronto, Roy. Ontario Museum, Life Sci. Contrib. 59. 17 pp.
10. Munroe, E. 1972. *Status and potential of biological control in Canada*. Pages 213-255 *in* P. S. Corbet and R. M. Prentice, eds. *Biological control programmes against insects and weeds in Canada 1959-1968*. Commonwealth Inst. Biol. Control, Tech. Commun. 4.
11. Sabrosky, C. W. 1952. *How many insects are there?* Pages 1-7 *in* A. Stefferud, ed. *Insects*. U.S. Dept. Agric. Yearbook of Agric. 1952.
12. Smith, I. M. 1976. *A study of the systematics of the water mite family Pionidae (Prostigmata: Parasitengana)*. Mem. Entomol. Soc. Can. 98. 249 pp.
13. Sneath, P. H. A. 1964. *Numerical taxonomy: introduction*. Pages 43-45 *in* V. H. Heywood and J. McNeill, eds. *Phenetic and phylogenetic classification*. Systematics Assoc., Publ. 6.
14. Sokal, R. 1962. *Typology and empiricism in taxonomy*. J. Theor. Biol. 3: 230-267.
15. Sokal, R., and P. H. A. Sneath. 1963. *Principles of numerical taxonomy*. Freeman, San Francisco and London. 359 pp.

Index of Authors*

*Senior authors of literature cited are listed; full citations appear on page numbers given in italics.

Index of Subjects